APOHA

APOHA

BUDDHIST NOMINALISM AND HUMAN COGNITION

EDITED BY
Mark Siderits
Tom Tillemans
Arindam Chakrabarti

Columbia University Press New York

Columbia University Press
Publishers Since 1893
New York Chichester, West Sussex

Library of Congress Cataloging-in-Publication Data

Apoha : Buddhist nominalism and human cognition / edited by Mark Siderits, Tom Tillemans, and Arindam Chakrabarti.

p. cm.

Includes bibliographical references and index.

ISBN 978-0-231-15360-7 (cloth : alk. paper) — ISBN 978-0-231-15361-4 (pbk. : alk. paper) — ISBN 978-0-231-52738-5 (ebook)

1. Buddhist logic. 2. Nominalism. 3. Negation (Logic) I. Siderits, Mark, 1946– II. Tillemans, Tom J. F. III. Chakrabarti, Arindam.

BC25.A65 2011

181′.043—dc22

2010049385

Columbia University Press books are printed on permanent and durable acid-free paper.
This book is printed on paper with recycled content.
Printed in the United States of America

Contents

Preface vii

Introduction 1
• ARINDAM CHAKRABARTI AND MARK SIDERITS

1. How to Talk About Ineffable Things: Dignāga and Dharmakīrti on Apoha 50
• TOM TILLEMANS

2. Dignāga's Apoha Theory: Its Presuppositions and Main Theoretical Implications 64
• OLE PIND

3. Key Features of Dharmakīrti's Apoha Theory 84
• JOHN D. DUNNE

4. Dharmakīrti's Discussion of Circularity 109
• PASCALE HUGON

5. Apoha Theory as an Approach to Understanding Human Cognition 125
• SHŌRYŪ KATSURA

6. The Apoha Theory as Referred to in the *Nyāyamañjarī* 134
• MASAAKI HATTORI

7. Constructing the Content of Awareness Events 149
• PARIMAL G. PATIL

8. The Apoha Theory of Meaning: A Critical Account 170
• PRABAL KUMAR SEN

9. Apoha as a Naturalized Account of Concept Formation 207
• GEORGES DREYFUS
10. Apoha, Feature-Placing, and Sensory Content 228
• JONARDON GANERI
11. Funes and Categorization in an Abstraction-Free World 247
• AMITA CHATTERJEE
12. Apoha Semantics: Some Simpleminded Questions and Doubts 258
• BOB HALE
13. Classical Semantics and Apoha Semantics 273
• BRENDAN S. GILLON
14. Śrughna by Dusk 283
• MARK SIDERITS

Bibliography 305
List of Contributors 321
Index 325

"Without Brackets: A Minimally Annotated Translation of Ratnakīrti's *Demonstration of Exclusion*," translated by Parimal G. Patil, is available on *Apoha*'s page at http://cup.columbia.edu/.

Preface

This is a book about the Buddhist nominalist theory of apoha. The basic idea of the apoha theory is that a general term like "cow" refers to all those things that are not non-cows. This is one of those philosophical ideas that might seem too smart by half. Students of Indian and Buddhist philosophy have often wondered whether the theory could be any more than a facile logical trick, one that ultimately fails to solve the problem it addresses: explaining our ability to use general terms without supposing that universals and other equally odd abstract entities exist. For four days in May 2006, fourteen scholars gathered at Crêt Bérard, a retreat in the hills above Lausanne, Switzerland, to try to answer this question. While the basic idea of the apoha theory is simple, its history and development in India and Tibet are quite complex. Among the scholars attending the conference were experts in various facets of that history. But the gathering also included scholars working in such diverse areas as philosophy of language, linguistics, and cognitive science. The aim was to try to arrive at a better understanding of what the apoha theorists were actually saying and then see if their view turns out to be a promising approach to the study of human cognition.

The conference at Crêt Bérard spurred much excitement and a great deal of subsequent discussion. The papers collected here represent the fruits of that discussion. While all the contributors save one were at the conference, none of the papers in this volume was presented there. Some of these papers are descendants of conference presentations, refined in the light of much discussion and debate. But other papers were only written after the conference, reflecting new insights that grew out of that discussion and debate. This is not your usual volume of conference proceedings; this is collaborative research on a problem that until now has received only sporadic attention from individual scholars. Our hope is that by combining philosophical and textual-historical approaches we have allowed the ideas in this Buddhist material to find their rightful place in a contemporary discussion.

There are many who contributed to the success of the project. The conference would not have been possible without the generous support of the Elisabeth de Boer Fund of the University of Lausanne. Thanks are also due to Thomas Doctor and Heidrun Köppl, who worked tirelessly to ensure that everything ran smoothly, not only during the conference but at arrival and departure as well. We also wish to acknowledge the efforts of the staff of Crêt Bérard, who made all the conference participants feel welcome in their charming surroundings. Thanks are due as well to the editors of *Acta Asiatica* for granting permission to reprint (a slightly amended version of) an article by Masaaki Hattori that first appeared in the pages of their publication. Some of the editorial work involved in assembling this volume was made possible thanks to HK research support provided by Seoul National University and support from the National Institute of Advanced Study, Bangalore. Wendy Lochner, our editor at Columbia University Press, was an early supporter of this project. She deserves our thanks not just for her help in bringing this volume to fruition but also for her efforts to advance comparative philosophical research in general. We also wish to express our gratitude to Christine Mortlock, Michael Haskell, and Robert Demke for their help in the production process.

"Without Brackets: A Minimally Annotated Translation of Ratnakīrti's *Demonstration of Exclusion*," translated by Parimal G. Patil, is available on *Apoha*'s page at http:// cup.columbia.edu/.

APOHA

Introduction

Arindam Chakrabarti and Mark Siderits

This is a book about the apoha theory of Buddhist nominalism.[1] The apoha theory is first and foremost an approach to the problem of universals—the problem of the one over many. That problem is one of explaining how it is possible, when we see a pot, to think of it as *a* pot and call it by the name "pot," a name that applies to many other particular pots. What is the one thing, being-a-pot, that this particular shares with many other particulars? Is there really such a thing in the world, over and above the individual pots, or is it just a mental construction of some sort? To hold the first alternative is to be a realist about universals, to hold the second is to be a nominalist. The apoha theory is a distinctive Buddhist approach to being a nominalist.

To fully appreciate the apoha theory one must understand the problem of universals. The usual practice, in introducing the problem of universals, is to start with Plato and Aristotle. In this introduction, we start with classical Indian philosophy instead. The problem of universals has played important roles in both the Western and the Indian traditions, and we suspect that the terrain may have been more fully explored on the Indian side. The first part of this introduction is a brief survey of classical Indian approaches to the problem. Part 2 then takes up some modern and contemporary Western approaches to the related problem of what concepts are, and how we acquire and manifest our mastery of them. Part 3 uses the discussions of the first and second parts to construct a taxonomy of possible approaches to the problem of universals and locate the individual papers on the apoha theory that make up the rest of the book in that taxonomy.

PART 1: ARGUMENTS FOR AND AGAINST UNIVERSALS IN CLASSICAL INDIAN PHILOSOPHY

That the word for an ontological type or a category, in Sanskrit, is *padārtha*, literally, "meaning of a word," deserves serious attention and reflection. Somewhat like Frege and his commentator Dummett, but nearly two thousand years before them, Indian philosophers of language saw clearly that a theory of meaning ultimately boils down to metaphysics. Apoha, historically, is Dignāga's answer to the question, "What does a word mean?" By the time it is elaborated by Dharmakīrti in his autocommentary on the chapter on Inference-for-oneself, this theory of meaning becomes partly a complex metaphysics of relationless, propertyless bare particulars, and partly a psychology of imagined generalities concocted out of habitual practice-guiding exclusions. These bare particulars serve as the ultimate referents and constituents of a lived world where actions are undertaken on the basis of word-signified demarcations-from-the-other. To get the thrust of the underlying nominalist attack against mind-independent universals, we must first survey the semantic roots of Indian realisms concerning universals.

From the time that Patañjali wrote his "Great Commentary" to Pāṇini's grammar, semantic analysis of case roles and verbs embroiled the Sanskrit grammarians in ontology. In Pāṇini's grammar and its early commentaries (between fourth and second centuries B.C.E.), three crucial technical terms for a universal—*sāmānya*, *jāti*, and *ākṛti*—were already explicitly in use. The device of adding a *-tva* or *-tā* (roughly equivalent to the English "-ness") to any nominal root, "X," yields, as meaning, the property of being (an) X, shared by all Xs. From "substance" (*dravya*) one can thus mechanically abstract "substanceness" (*dravya-tva*), from "real" (*sat*), "reality" (*sattā*), from "humans," "humanity." With this device in place, it was natural to make the distinction between an individual substance and the property that makes it what it is, its abstract essence. But to parse our talk of concrete cows rather than of the bovine essence, the Grammarians drew the distinction between talking about one particular cow and talking about *any cow* or *a cow* in general (*Vyākaraṇa-Mahābhāṣya* on Pānini's Astdh, sūtra 1.2.58 and 1.2.64). The distinction between the general and the particular also came up for discussion in the context of the logic of pluralization. What allows one to say "trees" or "men" instead of using the word for a tree or man as many times as the number of trees or men one referred to? Because the direct meaning of a common noun is the shared universal property of the referents, one could eliminate all but one remaining (*ekaśeṣa*) occurrence of that word, when speaking generally of all instances or any instance. One could also issue universalizable moral imperatives such as "*a cow* ought not to be killed," which, Patañjali jokes, is not obeyed by simply sparing the life of an individual cow.

Jāti (a word which, in modern Indian vernaculars, has come to mean a class, caste, or even a nation, and which is the Sanskrit counterpart of the Latin *genera*) is used by Pānini for a shared property of all the particulars of one natural kind, which serves also to distinguish any one of them from things of other kinds. The particulars are called *vyakti*—a word that etymologically suggests a distinct concrete tokening of common and uncommon properties. The problem with this generalist theory of meaning—defended by Vājapyāyana (perhaps third century B.C.E.)—was that when, in a descriptive or prescriptive sentence, the action denoted by the verb has to hook up with what the noun means, what is meant by the noun has to be a particular. For, after all, no one can bind cowness with a rope, cut the tree essence, or have lunch with humanity.

Thus, in Indian semantics the dispute between those who insisted that a word primarily means a universal and their rivals who held that it must be particular substances that are the first meanings of words is at least twenty-two centuries old. The word often used for a universal by Patañjali was *ākṛti* (literally, "shape"), which is more reminiscent of form than of a property. In answer to the basic question "what is a word?" Patañjali considers the option, "is it that which remains nondistinct among distinct individuals, untorn when individuals are torn down?" and answers, "no, that is not the word, that is only the universal form (*ākṛti*)."

The need to switch to imperishable universals as meanings was felt both by the Grammarians and the Mīmāṃsā school of Vedic hermeneutics, for whom the authority of authorless sentences of the Vedas rested on their being eternal. The relation between words and objects was said to be "entrenched" and permanent. If perishable particular horses, cows, humans, and plants were the meanings of words, how could they be the eternally connected meanings of these beginningless Vedic words? The word *gauḥ* (cow) is therefore best taken to be eternally connected to the timeless bovine essence.

The first clear recognition of the need to postulate universals might also have come as much from reflecting on the generality or repeatability of the audible words themselves as from the theory of meaning. That there could be many pronunciations or distinguishable phonations of the same word was seen to be an unquestionable example of the one-in-many. That naturally went hand in hand with the idea of the real word type existing timelessly, independent of its temporal, perishable token utterances. Later, in the philosophy of Bhartṛhari (sixth century C.E., sometimes called a "linguistic nondualist"), word universals and meaning universals and our natural tendency to superimpose the former on the latter were elaborately discussed. Out of these discussions there eventually emerged the Jaina notion of a vertical universal (*urdhvatā-sāmānya*), as against the more common property universal, which was termed "horizontal universal" (*tiryak-sāmānya*). As in Descartes's example of the piece of wax, a single substance that assumes different forms at different times (first hard and white, then soft and colorless), the vertical universal is a case of one over many where the one is a single substance, while the many are different forms. In the case of the horizontal universal, by contrast, the one is a form while the many are the distinct substances (or other particulars) in which that form occurs. Here we see clear recognition of the fact that the problem of universals is fundamentally a problem of explaining sameness in difference.

The Hot Topics for Debate

Between the fifth and fifteenth centuries C.E., the debate between mainstream Nyāya-Vaiśeṣika and Mīmāṃsā realists and dissident Buddhist nominalists raged around the existence of eternal essences. The major points of disputation were:

1. Must we explain the use of a common noun or the experience of community across a plurality of particulars by postulating a single real property inherent in each of those particulars? (Vaiśeṣika and Mīmāṃsā said yes with some caveats, Buddhists said no.)
2. Is the property totally distinct from the individuals that exemplify it? (Vaiśeṣika said yes, and Bhāṭṭa Mīmāṃsā said yes and no.)
3. Does a universal exist only in all its own instances or are universals omnipresent? (A trick question of the Buddhist nominalist, answered cautiously by the Vaiśeṣika.)
4. Do universals have any role in causation? (Vaiśeṣika said that they can cause our awareness of them, while for Buddhists anything that is eternal must be causally inert, hence nonexistent. For Udayana [eleventh-

century Nyāya-Vaiśeṣika], nomic relations of necessary concomitance are ontologically founded upon the universals inhering in causes and effects.)

5. Can the work that is done by universals be done by relations of resemblance between particulars? (Vaiśeṣika said no, Jainism and Maddhva Vedānta said yes.)

Classical Nyāya-Vaiśeṣika Realism About Universals

Universals come to occupy a crucial role as the fourth type of real in the scheme of six basic categories of reals or "things-meant-by-words" (*padārthas*; notice again the semantic orientation) listed in the *Vaiśeṣikasūtras* of Kaṇāda. In that canonical scheme, after three types of unrepeatables—substances, particular qualities, and motions—come common properties. Although substances, qualities, and motions are entities of different types, they share one common property: they are all real. What is this realness that is common to all substances, qualities, and motions? Realness is a generic essence present in many substances, many qualities, and many motions. It is a universal, the highest one. Then there are less general features as well, the substancehood shared by all substances, the qualityhood common to all qualities, and the motionhood inherent in all motions. These second-tier universals are called "common-uncommon" since they function as defining properties belonging to all the members of the class to be defined and lacked by all others.

The *Vaiśeṣikasūtra*'s word for universal is *sāmānya* (the phonetic resemblance with "sameness" may not be entirely accidental), meaning "what is common." The word for an individuator or particularity is *viśeṣa*, which means uncommon feature or specialty, the difference maker. Flowerness might be a common property, shared by roses, jasmines, and sunflowers. But the same property would be a difference maker when you compare a rose with fruits, seeds, stones, and animals, since none of these except the rose has flowerness. Hence Kanāda's aphorism: "Universal and particularity depend upon understanding" (VS 1.2.3). Commentators hasten to point out that this formulation does not mean that universals are subjective or invented by our ways of understanding the world. All it means is that we find out by the verdict of our understanding whether some property is a pure universal or also a demarcator, as shown above.

Four broad arguments are proposed by these staunch realists for proving the existence of universals.

1. The evidence of sense perception is the strongest of all. Unless it leads to logical inconsistency, we must admit some common recurrent entity in each of those many things that sense perception shows to be of the same kind. This class character, the basis for our sense of sameness, is a universal for it is the same one found in many.
2. The argument from the meaning of general words, which runs as follows. A learnable common noun such as "bird" can denote an unlimited number of particulars of enormous variety. How the same word with the same meaning can correctly apply to so many diverse particulars calls for explanation. The explanation must lie in a distinction between reference (*śakya*) and sense (*śakyatāvacchedaka*). Thanks to the existence of an objective universal, for example birdness, which serves as the shared sense, the same word can distributively refer to all birds or any bird. This does not boil down to one or the other of the two early extreme views that the bare particular or the pure universal is the primary meaning of a word. It is the balanced view that the meaning of a word is a particular possessing a general property, something serving as the common mode of presentation of its unlimited number of referents.
3. Then we have the argument from lawlike causal connections. Fire is a substance, but when it causes burning, its causal efficacy is not determined by its simply being a substance, for then any substance would burn. To explain what makes fire—and not any other substance—the cause of burning, we need to postulate fireness as the property that limits the causality of fire toward this effect. With the advent of extremely technical New Nyāya (around the thirteenth century) the need to have limitors (*avacchedaka*) of causehood and effecthood became the standard ground for ontological commitment to universals.
4. Admission of universals also helped Nyāya solve the problem of justifying the inductive leap from observation of a few cases to a universal generalization covering all cases of a concomitance (e.g., where there is smoke there is fire). The common property observed in a few instances (the smokiness I perceive in the kitchen and at the bonfire) can, as it were, put us in direct perceptual touch with all the other instances where it also inheres, not in their individual details but in a generic way. Here the universal itself is supposed to play the role of the operative connection (linking bridge) between the sense organ and the apparently unobserved instances of that universal.

With all these supporting arguments for the universal's existence, the precise definition offered by Nyāya-Vaiśeṣika came down to this: "A universal is that which, being eternal, is inherent in many." Not any quality inhering in a substance is a universal. A wish inheres in a soul, but it is a short-lived episode, not eternal, hence not a universal. Colors are not universals in this system because they are unrepeatable tropes clinging to the particular surfaces. All colors share the universal colorhood. But two apples of exactly the same shade of red have two distinct red colors in them, just as each of them would have a distinct falling motion if they both fell. A universal must subsist wholly in each of its instances by the special relation of inherence. A universal must be wholly inherent in each of its instances. The word "inherent" must be taken seriously. A single string may be running through many flowers, but it—more precisely different segments of it—are only in contact with the flowers. The whole string cannot inhere in any one of them. This "single thread" analogy, therefore, is not entirely a happy one. We shall see later how Buddhist logicians refused to accept this idea of a single property running through many distinct objects. After asserting that "each entity is self-confined; they do not mix themselves with others, each of them being intrinsically unconnected to any other," Dharmakīrti gives a cartoonlike analogy: "Even if they are cognized together, by a cognition projecting itself as generality embracing, they would not be like the idols of spirits linked by a single string attached to their necks, there being nothing single [neither a property nor a relation] among these discrete particulars."[2]

What then is inherence? According to the orthodox realists, it is a kind of being-in, the converse of which is an intimate "having." Humanity inheres in me, just in case I have humanity. Now, *having* can be of many kinds. Things *have* qualities and motions. Wholes *have* parts. I *have* a pen in my hand. Rich people *have* big houses. The logical structure of each of these relations, of characterization, constitution, contact, and ownership, however, is utterly different. All four are more or less aptly reportable by the use of the preposition "in" or "of": the taste is in the apple; the room is or consists in the walls, roof, and floor; the pen is in between the fingers; and the house is of the rich merchant. Yet one initial grouping could be made to clarify their distinct structures. The taste and the room cannot exist without the apple or the room parts. The taste cannot float about on its own, minus the apple.[3] The room cannot stand independently of the walls. But that very pen can easily exist untouched by the hand, and that house can change hands. So the first two relations hold between pairs that are "incapable of standing apart from one another" (*a-yutasiddha*), whereas the other two relations hold between pairs that are "capable of standing apart from one another" (*yutasiddha*). However tightly my ring is stuck to my finger, it is not inherent in it as inseparably as fingerness is inherent in my fingers. It is no physical glue but a metaphysical inseparability that joins the goatness to the goat, ties up the running and the black color of the goat to the goat, and binds the goat to its body parts. The kind of being-inseparably-in that connects the universal to its instances has to be distinguished from the way a berry lies in a bowl. For the sake of economy—the principle of not multiplying entities beyond necessity—the mainstream Nyāya-Vaiśeṣika metaphysicians posit only a single such relation as enough to link up innumerable pairs of universals and particulars, qualities and substances, and wholes and parts. For systemic reasons, this relation is supposed to be eternal as well. And this is inherence (*samavāya*). Even other universal-friendly realists, such as the Bhāṭṭa Mīmāṃsakas, give the Vaiśeṣika much grief over this peculiar theory of exemplification. The Bhāṭṭas themselves take the relation between a universal and its own exemplifier to be identity-in-difference. The Buddhist logician finds both inherence and identity-in-difference equally unpalatable.

Though you cannot experience Vaiśeṣika universals by themselves, they are ontologically independent of their instances. Even when all cows are destroyed in the world, cowness will still be around, for otherwise the possibility of a fresh cow coming to be remains inexplicable.

Real Universals (*jāti*) and Titular Properties (*upādhi*): On Being a Cook

Though all universals are common features, not all common features corresponding to multiply applicable descriptions are, strictly speaking, universals. Even hardcore realists about universals feel the need for population control in the world of universals. Being a brahmin (a member of the priestly intellectual class) is taken to be a natural kind by Nyāya-Vaiśeṣika in the face of vehement opposition by anticaste Buddhists and Jainas. But being a cook or being a tailor would not be considered a natural or real universal, even though it is a common feature of cooks or tailors. Nyāya-Vaiśeṣika philosophers suggest six tests that an alleged (semantically suggested) property must pass to count as a genuine universal. These tests or hurdles are called "universal blockers."

1. If a property has only a single exemplifier then it is not a universal. "Being the Statue of Liberty" is not a universal and neither is timehood because there is no more than one Statue of Liberty or one time.

2. If two properties have exactly the same extension, for example, the property of being a homo sapiens and the property of humanity, they cannot be two distinct universals.
3. The domains of two universals can be either completely disjoint or one of them completely included in the other. They cannot be partially intersecting and partially excluding each other. Thus, being material and having a limited size cannot both be universals in Vaiśeṣika ontology, because while many things have both properties, open space is supposed to be material yet not limited in size, while the internal sense organ is supposed to be limited in size but immaterial.[4]
4. A regress-generating property is not a universal. Universalhood is not a universal, although all universals seem to have that property in common, for one could then multiply universals endlessly. Universals do not have further universals in them.
5. When the very nature of a characteristic is merely to distinguish its bearer, for example, an earth atom, from another particular of that kind, such an ultimate individuator should not be brought under a general category of individuatorhood, for that militates against its necessarily unique nature. Failing this test, the alleged generality "individuatorness" (*viśeṣatva*) does not qualify as a universal within Vaiśeṣika atomism.
6. The feature must bear inherence and no other relation to its bearer. Inherencehood is not a universal because, were it one, it would have to be related by inherence to inherence, which would be absurd. An absence cannot be a universal. Nor could the negativity common to all absences be a universal. Even though every rabbit is hornless, neither the absence of horn itself nor the absenceness of the absence resides in rabbits or absences by inherence.

Besides these, compound properties such as being a sturdy black cow or being either a cow or a buffalo are ruled out because universals are supposed to be simple.

What happens to the properties that get disqualified by a universal blocker? They are thrown into the mixed heap of titular, surplus, or imposed properties (*upādhi*). They might still be of much theoretical and practical use. Not only nonnatural generalities like being a New Yorker or being a carburetor, but even is-ness, knowability, and positive presence (shared by items of all six categories—substance, quality, motion, universal, inherence, and final individuator—but not found in absences) are merely titular properties. Knowability and existence (is-ness) are (intensionally) distinct properties, in spite of being coextensive, because they are not universals.

How Are Universals Known?

We need philosophical reasoning to grasp such deep universals as substancehood because many concrete instances of substancehood, such as time, atoms, and other people's souls, are not objects of perception. If the instances are perceptible, the universals must be directly perceptible as well. We see flowerness in a flower, just as we see its hue and smell its fragrance. According to Nyāya epistemology, to see Black Beauty as a horse we must first see its horseness (which is a perceived universal, though it is not perceived to be a universal).

But many strong arguments could be given against the perceptibility of universals. Let us examine several. First, if properties were perceived, we would perceive them even at the time of encountering the first exemplifier, but we do not. Hence, properties are abstracted, not seen. Both premises of this argument, of course, could be questioned. For the empirical knowledge of a common property to dawn gradually, a recognition must take place in the second, third, and subsequent sightings of the instances. To be faithful to the form of that recognition, "I have seen this sort of animal before," is to admit that even in the first instance that sortal property was seen.

Here is another antiperception argument. If properties were objects of perception, they would be causes of perception, but they are not. Therefore, they are not perceived. Again, both premises are rejected by Nyāya realists. Potness need not itself reflect light back into the retina for it to be causally relevant to the visual perception of potness. As long as the pot in which it inheres is in contact with the seeing eyes, it has a causally operative connection with the appropriate sense organ. If, of course, we define perception as prelinguistic and nonconceptual (as some Buddhists do), and universals are taken to be word-generated concepts, then to use that definition as an argument for the imperceptibility of universals would be crudely question begging.

Someone with Fregean sensibilities might propose another quick argument against the perceptibility of universals. Universals are not objects but functions. Therefore, they are not objects of perception. But there is a clear shift in the meaning of "object" between the premise and conclusion of this argument. In the West there is a basic resistance to admitting the sense perception of universals because universals are supposed to belong to the intelligible realm. In *The Problems of Philosophy*, Russell claimed that we have direct acquaintance with universals, but that acquaintance was not meant to be sensory. Only David Armstrong, whose view about universals comes very close to Nyāya-Vaiśeṣika realism, seems to have warmed up to the idea of perceiving universals.

Attacks from the Buddhist Nominalist

The Vaiśeṣikas' first argument for the existence of universals depends upon the generalization, "In every case, the sense of commonness or similarity felt by word users must be spawned by an objective universal." Surely this generalization is riddled with counterexamples. We have just seen how people feel a sense of similarity across the many cooks; yet Nyāya-Vaiśeṣika realists refuse to admit cookness as a universal. There is no good reason to posit these weird entities and every reason to eliminate them. So claimed the Yogācāra-Sautrāntika Buddhists. "It does not come there [from another place], it was not there already, nor is it produced afresh, and it has no parts, and even when it is elsewhere it does not leave the previous locus. Amazing indeed is this volley of follies!" (PV 1.152).[5] With this oft-quoted remark, Dharmakīrti summarizes his battery of objections against the Nyāya-Vaiśeṣika theory of universals. How can a universal remain the same while existing in distinct things and places? Does it scatter itself into parts or does it live in its entirety in each instance? When the locus moves, does it move? If cowness is everywhere, how can it be absent in a horse? If it is only where its instances are now, then how does it travel to a new place when a new cow is born there? It does not pervade the place where an individual is located, for then the place itself would be its instance, yet how can it manage to inhere in the individual that occupies that place? If the particular instance is needed as a manifestor of the ubiquitous universal, why can't we perceive the cow—its manifestor—independently of noticing the universal cowness? A lamp reveals the preexistent pot in a room, but you don't need to see the pot first before you notice the lamp!

Most of these difficulties, the realists retorted, suffer from a category mistake. They assume that a universal is just another kind of superparticular. But a universal is not a spatiotemporal thing, and that is why multiple location without divisibility is not a problem for it. In spite of such robust responses, Buddhist antirealism about universals became more and more trenchant in the second millennium, until we have such caustic remarks directed at the Vaiśeṣika realists as those of Paṇḍit Aśoka: "One can clearly see five fingers in one's own hand. One who commits himself to a sixth general entity fingerhood, side by side with the five fingers, might as well postulate horns on top of his head" (SD 101–2).

APOHA NOMINALISM IN A NUTSHELL: THE BUDDHIST EXCLUSIONIST ACCOUNT OF CONCEPT FORMATION

Buddhist logicians have an error theory about universals and permanent substances that they reduce to mental or physical particulars or simply eliminate. There are nothing but momentary quality particulars in the world. But the human mind, afflicted by perpetuation wishes and language-generated, deeply ingrained myths, has a tendency to cluster some of them together first in the fictional form of enduring substantial things (i.e., the mind constructs what the Jainas call vertical universals) and then further classify these "things" into types. This illusion of commonality, of course, has some pragmatic value because, except in thoughtless contemplative experience, our working cognitions of the world mostly take the form of predictive judgments or explanatory inferences on the basis of these apparently general features and their mutual connections. When a particular cow (which, in its turn, is a fictional cow shape superimposed on certain packets of quality tokens) is seen to be other than all other animals, the original indeterminate (concept-free) perceptual content somehow causally triggers this difference-obliterating tendency. The particular cow image is made to "fit" this linguistic and imaginative exclusion from the complementary class of horses, rabbits, pillars, and such things. The specificity of the particular cow—its numerical detailed differences from other cows—is ignored; instead, this mere exclusion from non-cows is foisted onto the perceptual content as a predicate. This exclusion masquerades as the universal cowness. To take Dharmakīrti's example, the universal "antipyreticness" is a useful figment of the imagination. In the external world, there is no single shared intrinsic property of different medicinal plants that all work as fever reducers except that they are other than those things that fail to relieve fever. Antipyreticness is an erroneous reification of this mere exclusion (apoha). This, in a nutshell, is the apoha nominalism of the Yogācāra Buddhist logicians, which is the topic of this collection of papers.

Milder Nominalisms: Resemblance Theories

In the midst of this great battle between realists and nominalists, the Jaina syncretists stepped in with their typical reconciliatory message: that every object of knowledge has an alternatively more-than-one (*anekānta*) nature, particularity and generality being just two of these. We cannot doubt that things do objectively resemble one another. These resemblances are real relations. But both the things and their mutual resemblances are particulars. Nothing has the burden of being strictly repeatable.

The Jainas reject the Buddhist version of nominalism, more or less on the same grounds that Kumārila Bhaṭṭa, the great Mīmāṃsaka, rejected it. Positive predicates, Kumārila had objected, cannot all be given a negative meaning. Since these exclusions are nonentities invented by erroneous imagination, to say that all our words mean them is to turn all words into empty terms. Indeed, since all exclusions are equally hollow in content, distinguishing one from another would be like trying to distinguish two nonexistents, one fictional fat man in the doorway from another bogus bald man in the doorway (to give Kumārila a Quinean example). Only those denials make sense which have something positive to deny. Since all descriptions capture only negations, this theory, ironically, strips our negations of all meaning, since there is nothing left to deny.

Although they use the Buddhist criticisms to do away with Vaiśeṣika realism about self- standing universals, the Jainas bring extremely pertinent charges against Buddhist nominalism. According to Dharmakīrti's "error theory," the projection of an external object-in-general meant by the word is possible because of a superimposition of the internally constructed exclusionary image form on external, nameless, uncategorized, positive particulars. But for a superimposition or false identification of the inner with the allegedly outer to happen, both the locus and the content of the error must be grasped by the same (error-committing) episode of awareness. Unless the rope is actually encountered (not as a rope) and the snake is recalled by the same piece of cognition, mistaking a rope for a snake is not possible. But what kind of perception will grasp both? Not a nonconceptual pure sensation, for it does not make any claims, hence makes no mistakes; and not the concept-laden ascertaining perception, for it never actually has access to any external or internal *particular*. So the apoha story is untellable under the assumptions of Buddhist epistemology. In its place, the Jainas propose a resemblance-based theory of perspective-dependent generality.

Prabhācandra anticipates the Russellian objection that at least all these resemblance relations would ultimately need a shared similarity universal. His answer is that, just as a Vaiśeṣika final individuator (*viśeṣa*) does not need another distinguisher, one resemblance does not need a higher-level resemblance or universal to explain why all those resemblances are similar. While accounting for the similarity between ground-level particulars, the similarities also account for their own similarity to each other.[6]

Contrasts with Western Metaphysics of Forms and Properties

It should be clear by now that there is no core theory of universals shared by all Indian philosophers. But we can discern five broad features that distinguish Indian theories of universals from their Western counterparts.

1. Even the strongest realist position, that of Nyāya-Vaiśeṣika, falls short of the ante rem realism of Plato's theory of ideas. Unlike Plato, Indian realists about universals were equally realists about the perceptible particulars of the external world. Earthly particulars were never thought to be less real copies of thinkable universals, even by those who believed in universals.
2. Even if we concede that Nyāya's universals were close to Aristotle's universal properties, which are immanent in concrete particulars, Aristotle could never agree that universals are themselves directly perceived by the same senses that grasp the corresponding particulars, which is the standard Nyāya position.
3. The peculiar form that nominalism took in the Indian Buddhist theory of word meanings as exclusions does not have any parallel in the West. We find an interestingly different counterpart of the Jaina and Maddhva theories of resemblance in Nelson Goodman, but apoha nominalism remains a unique contribution of Indian Buddhism.
4. Most Western realist accounts of universals take colors, smells, tangible textures, and such qualities, as well as relations such as "being larger than," as paradigm examples of universal properties. In Indian realist thought, these would count not as universals but as particulars. The distinction between such particular qualities (*guṇa*) and universal properties (*jāti*) has been sacrosanct. It is only very recently that the idea of quality particulars (or "tropes") has gained ground in Western analytic metaphysics. Neither are relations treated as genuine universals by any classical Indian realist.
5. Finally, the controversial and complex theory of inherence as a single concrete connector joining not only universals and their instances but also particular qualities to substances and, most puzzlingly, wholes to their parts, is totally foreign to Western realists.

PART 2: CONTEMPORARY WESTERN THEORIES OF CONCEPTS ON THE MARKET

Since the Sanskrit technical term *vikalpa*—central to apoha theory—is sometimes translated as "concept" (or "conceptual construction"), and with at least one (highly contentious) interpretation of apoha theory construing it as a version of conceptualism, it would be good to preface our investigation of the apoha theory with a quick survey of the available theories of concepts (and concept formation, and concept possession) in modern and contemporary Western philosophy. Unfortunately, like most frequently used words in a disputatious philosophical tradition, the word "concept" is so full of ambiguities that while all these theories are called "theories of concepts," it is far from clear that they are trying to explicate the same concept of a concept. Some of these theories are concerned with what it is that we can do when we possess a concept and how we manifest such possession; others are concerned with what kind of mental representation a concept is; and yet others are interested mainly in the psychosocial story of how we acquire concepts. But can these theories even have different concepts of a *concept*, if they do not have, at some level, the same concept of a concept? One suspects there is a paradox lurking here. Still, our prephilosophical idea of concepts can be unpacked with the help of the following characterizations. Something that is meant by a predicate (". . . is wise," ". . . is square," ". . . is a metal"), by a pluralizable common noun ("dog," or for that matter "concept" itself), and most directly by an abstract noun ("substanceness," "existence," "justice"), that which renders general propositions possible, that by which our thoughts are constituted, minimally, is called "a concept." Let us now look at an ahistorically ordered list of available theories of concepts so understood.

1. Classical definitionism: Beginning with Socrates, and fully matured in Aristotle's and Aquinas's doctrine of essences, the dominant Western account of concepts takes them as defining essences captured by necessary and sufficient conditions for counting as something. Being wet and being earth together constitute the concept of mud, because all and only mud is wet earth. The definition of a triangle captures the precise concept of a triangle. This is closely connected to the notions of a kind and a category, which are the types in which entities can be classified, divided, and defined according to definitions. Murphy (2004, 15) identifies three main claims of the classical theory: (a) that concepts are mentally represented as definitions, (b) that every object either does or does not fall under a concept,

unless it is an intrinsically vague or fuzzy concept like that of "big," and (c) that any item that satisfies the definition is as good an instance of a concept (an object falling under that concept) as any other, that all dogs, for example, are equally dogs and there is no typical/atypical distinction within members of the class of dogs. "The definition *is* the concept according to the classical theory."

This theory is widely seen to have been discredited by Wittgenstein's famous attack against it in *Philosophical Investigations,* showing our failure to find any properties had by all and only games. Definitions of all but the arbitrarily stipulated concepts of mathematics and logic are nearly impossible to agree upon. Once we start noticing borderline cases, most empirical and everyday concepts turn out to be vague, and the law of excluded middle (claim [b] above) seems to fail to apply to something even as basic as the concept of "alive." In addition to such "conceptual" warnings against the Aristotelian (or Vaiśeṣika) search for exact definitions, there is strong empirical evidence, from how people actually think, that concepts cannot be definitions. Different competing definitions of the same concepts seem to be entertained by different groups of people, who might for example debate whether a tomato is a fruit or a vegetable, whether flavor is a matter of taste or smell or both, whether tapestry is an art or a craft. There is no telling whether a low three-legged seat with a very small back is a chair or a stool, whether a fetus is a distinct person or not. There is no definition, yet there is a concept. So concepts are not definitions. Further, if concept *F* were the set of defining properties that all and only *F*s have in common, then analytic judgments explicating the concept of the subject by the predicate would be sharply distinguishable from synthetic ("ampliative") judgments where the concept of the predicate is added anew to the concept of the subject. But, as Quine has shown, the analytic-synthetic distinction cannot be sharply drawn. Hence concepts are not definitions.

For some, the final blow to the classical theory came when the empirical work of E. Rosch, C. B. Mervis, and others established the falsity of the third claim (c): some dogs, some chairs, some tomatoes are more typical than others and are more closely related to the common concepts of those items. A Chihuahua is not as typical a dog as a German shepherd, a white tomato the size of a cherry is not what one thinks of when one calls up the concept of a tomato, a penguin is not as typical a bird as a sparrow. These variations within the definitionally secured domain of a concept show what is called a "typicality effect." Typicality effects take the concept theorist away from the search for objective essences (recognition-transcendent satisfaction conditions) toward what people actually have in mind when they use a general word or a predicate, from what the word means to what individual users tend to mean by it.

2. The British Empiricist idea/image theory: Diametrically opposed to the classical common essence theory of concepts is the extreme empiricist theory of Hume that concepts are ideas, which are faint copies of sensory impressions. My concept of a dog is a smudged memory image of many impressions of individual dogs that I have received through my senses. Concepts, under this theory, are creatures of the individual mind. The difficulty with this view is one that Berkeley pointed out before Hume: since an image is a mental particular, however lacking in detail an idea-image may be, it cannot be a general idea. Thus, if one took the idea theory of concepts seriously, then by virtue of the reduction of concepts to particular mental contents, conceptualism about universals would be reduced to nominalism. If there is nothing but featureless particulars anywhere in the world, ideas in people's minds would also be featureless particulars.

The idea theory of Hume, therefore, does not give us any explanation of abstract general concepts, insisting instead that there aren't any. It also misconstrues the categorical distinction between particular sensations and general concepts as a distinction of degree of vivacity, as if a particular visual image of a cat, when its contour gets smudged and its colors paler, turns into the concept of a cat-in-general.[7] This is as dubious as the suggestion that an adjective or verb is a half-forgotten proper name. There are moreover Wittgensteinian reasons against such a theory of concepts (or meanings of words): it makes meanings and constituents of thought completely private. You and I do claim to have, sometimes, the same concept of a tree, but we can never have the same idea of a tree, let alone an identical mental particular called up by the word "tree." That makes the Humean version of the idea theory of concepts unattractive.

3. Wittgenstein's family resemblance theory: After rejecting the classical theory of definable essences corresponding to every meaningful general word, Wittgenstein's positive agenda was to motivate us to observe and describe the different ways in which a common noun or an adjective is used, without trying to insist, a priori, on some single thread of meaning running through all those uses. To master a word, and hence the concept-in-use attached to it, is to gradually develop command over a network of disjunctively woven criteria, an arbitrarily chosen subset of which can be used to identify the objects falling under it. The reason hopscotch is called a game may have nothing in common with the reason computer chess is called a game, though there could be intermediate examples of games such as cricket which have features in common with both hopscotch and chess.

These reasons do not form any unitary core of necessary conditions but neither are they subjective or mental such that each user has a private idea why they call some activity a game.

The problem with this powerful theory is that while its negative force, rejection of the classical Platonic and empiricist theories of concepts, is clear, its positive account of what it is to have the same concept as others, or how any of us picks up these loosely woven sets of criteria from diverse contexts of use, is far from clear. While Wittgenstein crucially distinguishes between following a meaning rule correctly and merely seeming to, he does not, for example, tell us how a misapplication of a concept is detected. How do we know whether someone is using the concept *game* or the concept *entertainment*, if the overlaps between the two show closer internal connection than the widely divergent cases of a game, which are held together by the concept *game*?

4. Kantian theory: Wittgenstein was not the first Western thinker to associate concepts with rule following. Already in Kant's *Critique of Pure Reason* concepts were defined as rules for synthesizing experiences. But what does Kant mean by a rule? First, it is a schematic recipe for pattern recognition, for running through and holding together certain bunches of sensory inputs as if they represent an object outside the sensation. Second, it is like a major premise of a syllogism, ready to subsume an instance or a subcategory under a more general predicate, to draw a "judgment" as a conclusion. Longuenesse (1998, 50) helps us here: "The two meanings of 'rule,' as rule for sensible synthesis [the concept as schema] and as discursive rule . . . or major premise of a possible syllogism . . . are indeed linked. Because one has generated a schema, one can obtain a discursive rule by reflection and apply this rule to appearances." Sensibility as a faculty is mere susceptibility to external stimuli. Mere sensing is not enough to give us an object outside the experience, let alone to form a general judgment about the object. Only understanding can organize the received sensory data into representations of this or that object of this or that category or sort. Understanding is the faculty that employs pure concepts and, with the help of imagination, generates empirical concepts. So, to have a concept is to know how to combine or synthesize an array of sensory stimuli such that the unified awareness could be intentionally directed toward an object, out there, from which the awareness distinguishes itself. Such is the concept of a substance, of any old tree or any water pot, which organizes sensations, memories, expectations, and permanent possibilities of perceptions such that the seeing or touching counts as an experience *of* a substance, *of* a tree, or *of* a pot.

About the actual psychological process of concept formation, Kant has an agnostic attitude: "The concept 'dog' signifies a rule according to which my imagination can delineate a figure of a four-footed animal in a general manner, without limitation to any single determinate figure such as experience, or any possible image that I can represent *in concreto*, actually presents. This schematism of our understanding, in its application to appearances and their mere form, is an art concealed in the depths of the human soul, whose real modes of activity nature is hardly likely ever to allow us to discover, and to have open to our gaze" (B 181).

Despite its obscurity, Kant's "rule of synthesis" account contains a valuable lesson that is uncannily similar to Dharmakīrti's account of the passage from direct perception of particulars to action-prompting ascertainment of objects. Though it is the senses that put us in contact with the external world, it is concepts that enable us to make claims about common objective targets of perception and practical activity. Through their generality, concepts make reidentification possible, and so become object makers.

5. Fregean function theories: Using his distinction between sense and reference, Gottlob Frege held that a concept is the reference of a predicate expression. More recently, Christopher Peacocke (1992b) has claimed that concepts are the senses of such expressions. Both theories share the view that a concept is like a predicate expression in that it serves as a function that takes us from one sort of object to another sort of object. According to Frege's semantic theory, the expression "__is red," for instance, plays the role of mapping objects onto truth-values. So when the blank indicated by "__" is filled by an expression that refers to a stoplight, it yields the truth-value True, while inserting an expression that refers to grass yields the truth-value False. The reference of such an expression is the function itself. For Frege this is as much a part of the world as are stoplights and grass. Otherwise, he believes, we could not explain how different speakers of the same language, or speakers of distinct languages, could all be said to grasp the same concept when they come to know that stoplights are red.

Peacocke's view is that a concept is not the function itself but its mode of presentation, the way in which it is grasped by those who are said to possess the concept. Like Frege, Peacocke thinks of a concept as something the possession of which enables the subject to represent an object in determinate ways, ways that may be true or false. Peacocke's view is also like Frege's in making concepts objective constituents of the extramental world. Both are clearly heirs of the Kantian insight that possession of a concept involves something like mastery of a rule, something that confers the ability to group together disparate mental presentations and to make inferences. The chief difficulty of both, from the perspective of our present concerns, is that in making concepts inhabitants of a mysterious "third realm" that is neither physical nor mental, they leave it unclear how concepts could play a causal role in human cognition. No Indian philosopher, whether a realist like the Naiyāyikas or a Buddhist nominalist, would accept such entities. While they might represent an elegant solution to any number of problems in formal semantics, they would strike such thinkers as ontological overkill. Frege's version also suffers from the paradoxical consequence that the expression "the concept *red*" does not refer to a concept. This is so because the reference of a definite description must be an object, something "complete" or saturated in nature, while concepts, as functions, are by nature "incomplete" or unsaturated. While Frege himself embraced this paradox, many see it as indicating a serious deficiency in the view.

6. Similarity-based theories: Bolstering Wittgenstein's family resemblance theory with empirical research, Eleanor Rosch and others came up with the influential thesis that instances fall under a concept, not in a yes/no fashion (as classical definition theory predicted), but in a probabilistic way, depending upon how many of the typical features are available in an instance. To be a dog, an animal need not have all the necessary conditions of being a dog—there being no such set of defining conditions at all—but only a sufficient number of conditions which are associated with the complex mental representation of a typical dog. Since the concept *robin* has many more of the structural elements of the superordinate concept *bird* than the concept *chicken* has, and the concept *penguin* has even less than *chicken*, robins are more readily recognized to be birds than chickens or penguins. Categorization, say, of household goods as furniture, thus turns out to be a feature-matching process shading off from the more typical instances of chairs and tables all the way to hat stands and ottomans.

The resulting prototype theory claims that a concept is constituted by a network of similarities with the best instances of a type. A concept, instead of being a fixed definition of a universal essence, may be a set of weighted feature representations most users of a word have in mind—a set which could be added to or subtracted from continuously. Picking up the concept *egg* and the concept *eggplant* from white, oval-shaped and dark purple "best examples," respectively, a child may fit into a community of users of the words "egg" and "eggplant" by becoming ready to also include purple eggs and white eggplants, as color moves away from the core set of criterial attributes to the periphery, through other more salient bases of resemblance.

Of all the resemblance-based theories of concepts, prototype theory is the most empirically well corroborated and statistically computed.

A recent variant on this basic approach is Jesse Prinz's (2002) proxytype theory, which purports to provide a theoretical framework for explaining how the ability to use concepts might be realized in the brain. Taking as a model D. Marr's computationalist approach to the neurophysiology of visual processing, Prinz sees concept formation as a matter of the formation of networks of stored images that exhibit dynamic interactivity and context sensitivity and yet are ultimately traceable to repeated occurrences of sensory stimulation. The result might be called neo-Lockean, but unlike Locke's theory there is built into the theory a way of accounting for the flexibility with which we deploy such concepts as *dog*. Not only can it explain the typicality effects that prototype theory is designed to handle, but it can also account for variations across contexts of concept application. For on Prinz's theory a context determines which subset of the set of features in the network stands proxy for the concept as a whole.

Prototype theory, proxytype theory, and the closely related exemplar theory all do a far better job of accounting for the actual practice of concept application than classical definition theory and its offshoots. But such approaches are not without their own shortcomings. One is the "pet-fish" problem. A dog or a cat is a typical pet; a middle-sized trout or a tuna is a typical fish. And no one can deny that the concept *pet fish* is composed of these two concepts. Yet, the typical pet fish, a guppy or a goldfish, does not remotely resemble a dog or a cat, a trout or a tuna, and has features not computable through resemblance from the intersection of the weighted group of features of the constituent prototypes. An adequate theory of concepts must account for their compositionality, the fact that we form complex concepts by somehow putting together their simpler constituents. Proxytype theory might claim the ability to handle this problem, but it is not clear how this will work. And in any event, all such approaches have one major flaw from the perspective of the Buddhist nominalist. They all take as given the ability to respond to similarity of stimulation. For the Buddhist nominalist, a key task for any theory of concepts is to explain how, in a world of pure particulars, certain stimulations come to seem similar to subjects. A theory of concepts cannot then take such an ability as a given.

7. Theory theory of concepts: The fact that people possess concepts that come with no feature- or exemplar-based stereotype and the fact that radically dissimilar stereotypes are associated by different users of the same concept throw into doubt the very basis of similarity-based theories of concepts. Mastering and deploying a concept may not be a matter of

feature matching or probabilistic frequency measurement in a space of perceptible resemblance to paradigm instances. Having and applying a concept might be more like theoretical problem solving. Concepts would then be minitheories for the categories they range over, theories that yield beliefs about causal connections and deep intrinsic properties of things falling under them. Gopnik and Meltzoff (1996) identify three characteristics of such a "theory": (1) Structurally, a theory is a system of abstract entities and laws, postulating perception-transcendent causal relations based on counterfactual reasoning. (2) Functionally, a theory enables the user to make predictions, give explanations of observable behavior, and give reasons for further taxonomies. (3) Dynamically, a theory tests itself with imagined counterevidence and initially defends itself by taming all recalcitrance, but is ultimately open to revision and theory change. Our concept of a natural kind like *whale* or a psychological or clinical phenomenon such as *depression* or *cancer* might be such a theory. A theory theory of concepts is, in a sense, an attempt to retrieve the realist insights behind the classical essentialist concept of a concept, noting that concept users have more faith in hidden essences underlying causal mechanisms than in observable superficial features or family resemblances.

One major problem with reducing concepts to theories is that theories themselves consist of, and hence presuppose the notion of, concepts. A theory theory of concepts would thus lead to either an infinite regress or circularity, if it identifies concepts with theories of one sort or other. Theory theorists might meet this charge by embracing concept holism: interdependent theories are determined by the entire system of beliefs held true by an individual. My concept *cancer* would be determined, not by its constituent elements or criteria, but by all the other theories and the inferential role of this cancer theory in producing my entire system of beliefs directly or indirectly involving cancer. This would entail the repugnant consequence that I cannot share my concept of cancer with anyone else who holds different theories and beliefs. This sort of holism would mean that any disagreement between two people on any topic entails that they are using different concepts. A second difficulty is that, like prototype theory, the theory theory has difficulty explaining compositionality (the pet-fish problem): combinations of concepts seem simple and straightforward compared to combinations of theories, something for which there is no good account.[8]

8. Fodor's informational atomism: Back to Hume? Over several decades Jerry Fodor has developed a provocatively "retro/remix" sort of theory of concepts building on what he now confesses to be Humean (psychological) billiard balls, setting his face against such major twentieth-century philos-

ophers of thought as Wittgenstein, Quine, Ryle, Davidson, Dummett, Putnam, and Peacocke, whom he lumps together as "conceptual pragmatists." According to Fodor's theory, our language of thought is made up of word-like mental particulars called "concepts," which are structureless, simple cognitive blobs directly referring to properties and entities in the world. Concepts, like Hume's "ideas," are mental representations that are not determined by our ability to do anything. Our possession of them need not be explained in terms of our knowing how to engage in verbal and nonverbal behavior. Most of these concepts are, according to Fodor, innate. And it is here that the archempiricist (copy of the sense impression) *idea* idea is cobbled together with the archrationalist Cartesian innatism and the semantic hook up between concepts and the world starts to look like an unexplained brute fact. *Dog* (an information-rich but unanalyzable concept) just ranges over real world dogs in virtue of being the mental particular that it is. "Cognitive processes are constituted by causal interactions among mental representations, that is, among semantically evaluable mental particulars" (Fodor 2006, 135). The information atoms that are Fodor's concepts, in turn, pick out external objects and properties and real relations. Fodor's concepts are avowedly mental particulars. This makes his view vulnerable to the concerns about communicability lying behind the private language argument.

9. The ability theory: We have seen that any variety of imagistic theory of concepts, as well as the holistic theory theory, leads to privacy and unsharability of concepts across individuals. Wittgenstein, while proving that no private system of rules can count as a language, also showed that mental representations are useless as explanations of what makes a word mean what it means. Mental-representation accounts of concept possession take concepts to be word analogues in a language of thought. (Dummett [1993b, 97] calls this "the code conception of language.") But if understanding is translating a string of words into a string of concepts, the same problems that arose with regard to the meanings of words would arise with regard to the meanings of concepts. Besides, saying, in an associationist way, that we understand the Sanskrit word *gauḥ* when the concept *cow* comes to mind in hearing those Sanskrit syllables is no help. "There is really no sense to speaking of a concept's coming to someone's mind. All we can think of is some image coming into mind which we take as in some way representing the concept, and this gets us no further forward, since we still have to ask in what his associating that concept with that image consists" (Dummett: 1993b, 98). Rejecting all Lockean representational theories of concepts, Dummett has proposed that concepts are best looked upon as recognitional capacities that, when exercised, constitute the understanding of sentences manifested in communicative practice. To have the concept of a mountain

lion is to be able to recognize one when one sees it, or at least to be able to justify accepting and rejecting statements using the word "mountain lion" as true or false. Since such abilities could be socially shared and publicly compared with norms, to identify concepts with such recognitional abilities would enable us to account for interpersonal concept sharing.

Of course, critics reject this ability theory on at least three powerful grounds. First, a concept (recall Frege's idea of the meaning of a predicate) is supposed to be applicable to or true of objects. But my ability to recognize marigolds is not true of those flowers. Second, abilities fail the compositionality test. Those with the ability to recognize pets and recognize fish do not necessarily recognize pet fish. Third, Fodor lambastes the recognitional-capacity account of concept possession as a kind of "concept pragmatism" that leads ultimately to "a kind of flirting with idealism" (2004, 31). In fact, a better charge against the ability view would be that it ignores the "inner story" and leaves no room for the distinction between acting socially as if one gets a concept and really getting (i.e., episodically having the right mental representation which is) a concept. While Dummett's embracing of meaning-theoretic antirealism tends to confirm the idealism alert, it is extremely hard to swallow the warning against idealism from Fodor, who is proposing to go back to Hume's private ideas theory of concepts (which led straight to phenomenalism). Against any broadly dispositional practical capacity or implicit knowledge theory of concept possession, Fodor says bluntly "concepts are mental particulars" (1998, 3). It should follow that word meanings are mental particulars. So the meaning of my words can never be the meaning of your words, because my mental particulars are not your mental particulars.

One thing we learn from a Dummettian ability theory's emphasis on the ability to manifest concept possession is this: Just as a theory of meaning should be a theory of understanding, similarly a theory of concepts should be a theory of concept possession and concept use. There is no point characterizing meanings or concepts in such a fashion (e.g., as involving some synaptic connections in the brain) that one cannot give any usable criterion for distinguishing someone who knows the meaning or possesses the concept from someone who does not. It remains to be seen whether Buddhist apoha theorists can provide this.

PART 3

The reader could be forgiven for sensing a disconnect between the contents of parts 1 and 2. For the most part the modern theories of concepts discussed in part 2 skirt the metaphysical issues that so exercised classical Indian philosophers. It might be thought that the downfall of classical definitionism explains this: that Indian realism requires real essences, so that in the wake of Wittgenstein's family-resemblance account, the field is left open to nominalism. But a sophisticated realism like that of Nyāya can agree that many of our concepts exhibit indeterminacy of application, typicality effects and the like, and still insist that some privileged set of concepts must carve nature at its joints (since otherwise there is no explaining the efficacy of concept-guided practice). And to say that nature has joints is to say that certain particulars by nature share something in common.

Perhaps it is thought that developments in modern logic—specifically the replacement of syllogistic logic by predicate calculus—have shown that belief in real universals rested on an improper understanding of the logic of our language. But even if we agree with Frege that predicate expressions are best thought of as functions or rules that take us from objects to truth-values, there is still the question how such rules come to be grasped.[9] The answer Quine (1953, 68) gives—by "inductive generalization"—marks him as what Armstrong (1978, 16) calls an "ostrich nominalist," someone who refuses to countenance universals but sees no need to give a reductive analysis of our seeming commitment to them. For one could learn to use "red" by induction from ostended instances only if the instances pointed to by the teacher were taken by both teacher and pupil to resemble one another. And if the felt resemblance has no basis in reality, it is a major miracle that the lesson can be shared among all the speakers of a language. But if the basis of felt resemblance among the instances is some sort of real resemblance, then it requires explaining why this does not yield commitment to universals. For, argues the realist, any two things resemble one another in some respect or other, so the instances used in teaching would have to resemble one another in some particular regard—for example, in redness.

So the metaphysical concerns that fueled the classical Indian debate and shaped the apoha theory have not been resolved in recent Western philosophy, they have just been set to one side. Still, the issues involved in the recent discussions are not without importance to our understanding and appreciation of the apoha theory. An apoha-theoretic nominalist owes us something the realist is under no obligation to provide. The realist can claim that we learn to use "red" from the ostended instances because when we perceive those particulars we also perceive the redness that inheres in each. A nominalist who eschews all talk of real resemblances (e.g., by claiming that the perceiving of resemblance is due to conceptualization) must instead posit concepts as devices that mediate between input from particulars and pattern-exploiting output (e.g., "learning by induction"). Then there will be questions to answer.

1. Are (at least some) concepts innate (Descartes, Fodor), or is concept possession something acquired only through experience? If the former, what explains the apparent congruence between innate concepts and the world? If the latter, how does this learning take place in the absence of real resemblances?
2. Are concepts mental particulars (Locke, Hume, Fodor), or are they to be understood instead as rules or schemata (Kant, Frege, similarity-based theories, theory theory)? If the former, how are we able to express our thoughts in a public language, and how does a particular have application to a many? If the latter, how are such rules mastered and how do we tell when such mastery has been achieved?
3. Are concepts the sorts of things that can enter into causal interaction with other things (Hume, Kant, prototype theory, etc.), or not (Frege, Peacocke). If the latter, then what explains the difference that concept possession makes to our worldly success? If the former, then if it is also true that only particulars are causally efficacious, how does a concept have application to a many?
4. How is the compositionality of concepts to be explained? How is it that from our grasp of such concepts as *pet*, *fish*, *blue*, and *lotus*, we are able to understand the expressions "pet fish" and "blue lotus"?

Even the most fully developed formulations of apoha theory fail to give clear-cut answers to all these questions. But since we wish to know not only what the apoha theorists said, but also whether what they said is philosophically significant, we shall want to work out, where possible, how they might answer them. The essays in this volume explore the apoha theory not merely as an historical artifact, but as a novel approach to an important philosophical problem. If this approach has real philosophical merit, it should yield viable answers to questions we might put to it concerning the concepts (*vikalpa*) that play a central role in the theory. At the same time, we will want to assess the theory's adequacy as a solution to the metaphysical puzzles that fueled the Indian debate over universals. Of these, perhaps the following are the most pressing:

5. Does the resort to exclusions in place of real universals succeed in explaining how particulars appear to naturally fall into kinds or classes

without in the end bringing in real resemblances? If so, how? If not, is there some account of why real resemblances need not be backed up by real universals?

6. Does the apoha theory really claim that the meanings of words are all negative in nature? If so, then how does it account for the fact that we take the meanings of words like "pot" and "yellow" to be positive? And how is it possible to arrive at anything positive entirely on the basis of negation? Does negation not require the existence of something positive to be negated? If the something positive is the particular, then won't the exclusion of what is distinct from that particular just give us back the particular itself and not something general? If the something positive is another concept, then insofar as concepts are really negations that are only taken as positive entities because of ignorance, how does this avoid circularity or infinite regress?

The papers in this collection attempt to answer these and related questions in two distinct ways: by examination of the historical record, looking at the works of individual apoha theorists and their critics, and by standing far enough back from the text-historical details to extract a somewhat idealized "apoha theory" and subjecting it to philosophical scrutiny. (Needless to say, each does some of both; but a given paper will inevitably put greater emphasis on one than on the other.) The first paper, by Tom Tillemans, draws a distinction between two types of apoha theory that is used extensively in other historically oriented papers. This is the distinction between "top-down" and "bottom-up" approaches. Tillemans points out that the Buddhist philosophical tradition within which the apoha theory developed (Yogācāra-Sautrāntika, the school of "Buddhist logic") draws a sharp distinction between the real particulars that make up the world and the general concepts we employ in thinking and talking about them. Given this radical scheme-content dichotomy, the question arises how speech and thought successfully mesh with the world. A top-down approach seeks to answer this question by starting with the resources of logic and language and showing how these can be used to pick out pure particulars when the latter lack all hint of generality or shared natures. A bottom-up strategy, by contrast, tries to bridge the gap by showing how the causally efficacious pure particulars could generate felt resemblances and thus give rise to general concepts. The top-down approach Tillemans identifies with Dignāga, the founder of the Yogācāra-Sautrāntika movement. The bottom-up approach he associates with Dharmakīrti, the "commentator" who significantly reshaped Dignāga's system.

While other authors see the top-down and bottom-up approaches as complementary, Tillemans does not. He claims that while Dignāga thought the nominalist could solve the problem of one over many by the ingenious use of two negations, Dharmakīrti essentially abandoned this approach in favor of a purely causal story that only pays lip service to the founder's views about exclusion. On this view, Dignāga's answer to (5) is that the realist's universal can be replaced by a conceptually constructed exclusion of the other, so that nominalism can avoid commitment to both universals and real resemblances. But Tillemans is skeptical that this solution will work independently of the widely held predicate nominalism, according to which there are no common reals, only common terms. So he denies that the resort to exclusions succeeds by itself in allowing the nominalist to circumvent universals or real resemblances.

Tillemans is less skeptical about the bottom-up approach he finds in Dharmakīrti. On this approach, the particular with which one is in perceptual contact causes a mental image that one then mistakes for a class character. It is, in other words, just a brute fact that two distinct particulars can cause the same judgment (e.g., "this is blue"). This brute fact does include a subjective component—it is due to "beginningless ignorance" that we make the same judgment about what are actually quite different entities. This supposedly absolves Dharmakīrti from the charge of smuggling universals into his account. Still Tillemans concedes that this will make Dharmakīrti look like an ostrich nominalist to some. His reply is that Dharmakīrti might better be thought of as a "happy nominalist," someone who feels no need to explain the utility of our judgments of sameness.

In distinguishing between the approaches of Dignāga and Dharmakīrti, Tillemans is principally interested in how these play out in their answers to (5) and only alludes in passing to some of our other questions, such as (4), the question about compositionality. (He takes Dignāga's approach to fail this test, but is silent on whether or how Dharmakīrti's causal approach fares any better.) But resemblances between his interpretation of Dharmakīrti and Hume would suggest that some answer to a question like (2), the question how a public language is possible if meanings are mental particulars, is called for. Other papers that follow Tillemans in his views about Dharmakīrti will have more to say about this.

Since the apoha theory seems to have begun with Dignāga, it is appropriate that the first detailed historical study in this collection be on the formulation of the theory presented in the fifth chapter of his *Pramāṇasamuccaya*, the chapter entitled "The Examination of Apoha." Ole Pind's contribution situates Dignāga's views squarely within the context of a debate with the Nyāya-Vaiśeṣika school over the nature of knowledge derived from verbal testimony. In classical Indian epistemology, a major topic of dispute is the number and nature of the means of knowledge (*pramāṇas*). Nyāya holds that when the testimony of a trustworthy source produces a true cognition in the hearer, this constitutes a distinct kind of cognitive instrument that is not to be assimilated to inference or any other sort of reliable epistemic process. Dignāga disagrees, holding that verbally acquired knowledge is just the product of a particular kind of inference. This disagreement stems in large part from the fact that while Nyāya is realist about universals, Dignāga and his school deny that there are universals, on the grounds that being eternal they would be causally inert. So where Nyāya could claim that a kind term (e.g., "cow") denotes a particular inhered in by a universal (e.g., a cow inhered in by cowness), Dignāga must devise some other account of how general terms refer and how sentences can bring about cognition of facts. His basic answer to the question how words refer is that just as inference can only put us in touch with facts indirectly, so words cannot denote particulars, only objects in general, and that this can only be achieved through exclusion.

This position is grounded in the view to which Tillemans alluded in claiming that Dignāga's school employs a strict scheme-content dichotomy: the view that perception takes as object the unique (and hence unconceptualizable) particular, while inference (which now includes any cognition that is verbally expressible) has as object a conceptual construction that is routinely but wrongly taken to be the same mind-independent real that is cognized in perception. Nyāya, by contrast, maintains that it is one and the same object (e.g., a fire) that is cognized both by perception and by inference. It is Dignāga's strict separation of these two objects of knowledge, plus his rejection of real universals, that results in his having to give some account (according to Tillemans, a top-down account) of how scheme and content mesh. As Pind makes clear, Dignāga approaches this task by focusing on how the meaning of a word can be learned.

Dignāga argues that in the absence of universals, the only positive entity that remains to serve as the object of linguistic cognition, the particular, cannot be the referent, since the particulars in the extension of a kind term are potentially infinite in number. But Dignāga also observes that such an account of word meaning leaves unexplained an interesting phenomenon: that there is uncertainty as we descend the taxonomic tree from determinable to determination, but not when we ascend the same tree. Suppose we know of some particular that it is a tree. This leaves room for doubt as to whether it is an elm, an oak, or a jackfruit tree; but from the knowledge that it is a tree, there is certainty that it is solid, a substance, and an existent. Dignāga is here employing the Nyāya taxonomy of universals, which starts with *existenceness* and descends through a variety of levels of determination to such particulars as a mango tree, an occurrence of scarlet, and the downward motion of a cow's tail. Now, any particular that is a mango tree is also a tree, a solid, a substance, and an existent. Hence, if the meaning of "solid" were the particulars that it denotes, then when we know of a particular that it is a solid, this should convey the information not only that it is a substance and an existent (ascending the taxonomic tree) but also that it is a tree and a mango tree (descending the taxonomic tree). In fact, it conveys the former but not the latter information. Hence, the meaning of a general term cannot be the particulars that are its extension.

Pind takes the point here to be that the grasp of word meaning must be, like what happens in inference, the grasp of something abstract, of a type and not of tokens. Awareness of the connection between word and object is not, properly speaking, inductive. But if there are no universals, what can this abstract object be? The suggestion is that since it cannot be anything positive, it must be a mere absence or lack. And here is a second place where it is clear Dignāga must have had Nyāya's taxonomic tree in mind. The idea is that since there can be no positive nature that is shared by all the mango trees, the abstract object that the word denotes must be what is picked out by the expression "that which is other than what is excluded by 'mango tree.'" The thought is simply that the determinable "tree" constitutes a field of determinations in which the nature of each determination ("elm," "oak," "mango," etc.) is fixed by contrast with the others in the field. The result is a mental construction (all mere absences being such) that can nonetheless be (mis)taken for something positive and quite real.

This approach may answer some questions, but it raises others. Pind takes Dignāga to have been chiefly concerned to answer question (6), the question whether word meanings are indeed wholly negative in nature (he answers in the affirmative), and thereby also to answer (5), how it is that particulars appear to fall into classes (through the generation of mentally constructed general natures). Dignāga saw confirming evidence for these answers in the fact that this approach also helps answer question (4) about compositionality. But although Dignāga uses the notion of a concept in his account, it is not clear just what he takes a concept to be. He appears to answer (2) by denying that concepts are themselves real particulars, but then one wonders how he would answer (3), the question about causal efficacy. And then there are the further questions stemming from his official answer to (6), questions having to do with the fact that since Dignāga's purely negative account of word meaning seems logically equivalent to the positive account of the realist, this approach to avoiding universals seems to lead to either circularity or infinite regress. It is just this charge that his critics took up and from which many of the later apoha theorists sought an escape.

John Dunne's contribution to this volume examines Dharmakīrti's revised formulation of the apoha theory and assesses its success as a nominalism. Where Dignāga sought to account for our awareness of a world that is structured in the form of a taxonomic tree, Dharmakīrti tries to explain how a world of pure particulars could cause us to form the false but nonetheless useful judgment that things naturally form kinds. So Dharmakīrti's enterprise is appropriately labeled "bottom-up" by contrast with Dignāga's "top-down" approach. The question is whether these are complementary: can Dharmakīrti be seen as filling in important blanks in Dignāga's account, or is he rather deploying a rival theory under the cover of a strategic use of Dignāga's terminology? As Dunne explains it, Dharmakīrti's chief concern is to describe a process whereby a mental image that is copied from a perceptual cognition could come to be taken as resembling other perceptions in such a way as to give rise to the sense that their objects form a kind. Dharmakīrti's first answer is that this depends on the objects each being taken to perform some function in which we take an interest. Since the objects are in fact unique particulars, it is actually false that they share the common nature of performing that function. Each fulfills it in its own specific way; it is only our interest in that function that makes us overlook these mutual differences and judge the objects to be alike in this respect. All they actually share is their common difference from those things that do not perform that function. And this difference being a mere negation, there is no temptation to take it for a real universal or resemblance. So it begins to look like Dharmakīrti is after all trying to explain how, in the context of a given set of human interests, general natures could be constructed in an entirely negative way, thereby providing a psychological model of Dignāga's semantic theory.

Dunne considers the objection that this account appears to presuppose that distinct particulars can all perform a function that we take to be the same. If the many particulars that we call "fire" do not in fact share the common power to cause heat, why should it prove useful for us to perceive them as all alike in this respect? Dharmakīrti's answer is that it is just the ultimate nature of each particular to cause an effect that we will judge to be the same. Dunne construes this appeal to the ultimate nature of things as a kind of concession that beyond a certain point the apoha approach must give way to the happy nominalist's way with the realist's one over many argument. This would lend credence to the claim that Dharmakīrti's apoha theory is importantly different from Dignāga's. But this might equally be taken as no more than Dharmakīrti's way of pointing out that since the ultimate nature of the particular is inexpressible, the question of what ultimately explains our judging one particular to resemble another is ill formed. For if it is true that all concepts involve exclusion, that no exclusion is ultimately real, and that all explanation involves conceptualization, then any attempt to explain our so judging that appeals to how things are apart from all conceptual construction is necessarily illegitimate. All this means is that our assessment of the apoha theory can only be based on its adequacy to what Dignāga and Dharmakīrti would call the conventional truth—how things are in a world that is in some sense already conceptually constructed.

On Dunne's account, Dharmakīrti gives the beginnings of answers to (2), the question whether concepts are mental particulars or rules, and (3), how concepts causally interact with other things. Dunne describes Dharmakīrtian concepts as "Janus-faced" insofar as they are at one and the same time mental particulars (copied images) and entities that apply to multiple instances (abstract types). If such a strategy can be made to work, it would effectively solve the dilemma posed by (3), the dilemma that abstractions lack causal efficacy while particulars lack generality. But can it be made to work? The idea seems to be that the mental image can be taken as resembling a multiplicity of perceptual images by virtue of its being indistinct or obscure. One might, though, wonder whether there could be such a thing as a visual image of a tiger with an indeterminate number of stripes. Dunne also suggests that the image's serving as a concept has to do with our taking it as resembling other images, and this would bring Dharmakīrti closer to thinking of concepts as rules or schemata. To the extent that exclusion can be seen as a mental operation, this way of thinking of concepts conforms well with Dignāga's negative semantics. But then it is unclear what role the mental image is to play in Dharmakīrti's theory of concepts. If mastery of a concept is the mastery of a kind of rule-governed behavior (namely, behavior that overlooks the differences among a class of particulars), then the particular mental image that is the other face of the Dharmakīrtian concept starts to look like something of an idle cog.

Dunne also represents Dharmakīrti as addressing (1), the question whether concepts are innate or acquired through experience. The answer comes in the theory of the "imprint" (*vāsanā*), which Dharmakīrti uses to explain how we come to see certain things as resembling. Now to call this mechanism an imprint is to suggest that the disposition is one we acquired through prior experience. But Dharmakīrti is aware that not all concept acquisition can come through language-mediated learning, since prelinguistic infants and nonhuman animals are able to respond differentially to certain classes of stimuli. And so Dharmakīrti speaks of certain imprints as stemming from "beginningless ignorance." Dunne takes Dharmakīrti to have just two such imprints in his theory, namely, the general disposition to find similarity and the disposition to mistake representations for objects (i.e., the error of implicitly accepting a naïve realist view of perception). But it would seem Dharmakīrti needs far more specific resemblance-perceiving dispositions to make his account work: the infant must be predisposed to perceive distinct nipples as "the same," for instance. And the claim that such dispositions result from "beginningless ignorance" need not be construed as calling them innate, at least not in the sense that contrasts with calling them acquired. Buddhist tradition holds that the series of rebirths prior to the present life has no beginning. On this view, a given disposition might be present in each life in the series (and hence "innate" in one sense) and yet acquired through experiences had in a prior life (hence in some sense "learned"). In the Buddhist context it is not clear that question (1) has a clear sense.

Question (6) poses a dilemma for those who take seriously the claim that the meaning of a word is something wholly negative in nature, an exclusion. Negation requires a negandum, and presumably this must be something that is positive in nature. But if it is the real particular, then the two negations involved in the apoha formula "not non-cow" return us to the particular with which we began and not a class character. The alternative is to take "non-cow" to denote something positive that is shared among all the things not called cows. But the class of non-cows is so heterogeneous that it is hard to imagine their sharing anything except the property of not being cows, so either the account is circular or it leads to an infinite regress. This was the challenge Dharmakīrti's opponents raised for his formulation of the apoha theory. It, and Dharmakīrti's response, are the subject of Pascale Hugon's contribution to this volume.

As Hugon describes it, Dharmakīrti's response seems puzzling. He chooses the second horn of the dilemma, but asserts that the realist is faced with the same difficulty. For the alleged interdependence of the two classes *cow* and *non-cow* means that one lacks independent access to the meaning of "cow" just as much as to "non-cow." The realist has a ready response to this challenge though: they can claim that the presence of the perceptible cowness in the cow and its absence from the non-cow give us a way of telling whether something belongs to the class of cows. But this then licenses Dharmakīrti to respond that the judgment of similarity performs the same function in the apoha theory. And since such a judgment relies not on real resemblances but instead on the interests of the subject, the apoha theory avoids commitment to entities that Dharmakīrti claims are ontologically problematic. So the tu quoque response is actually part of a complex and subtle dialectical strategy designed to bring out the extent to which apoha semantics mirrors that of the realist. In this respect, it might be observed, Dharmakīrti sounds very much like Dignāga, who quite cheerfully acknowledges that the "exclusion of the other" behaves just like a universal, all that is missing from the apoha theory being the realist's ontological baggage. The difference with Dignāga lies in Dharmakīrti's notion of the judgment of similarity. Hugon is herself silent on the question whether this is consistent with Dignāga's account and whether it is rationally defensible.

The subject of Shōryū Katsura's contribution is the account of the three meanings of the expression "exclusion of the other" (*anyāpoha*), which was first developed by the commentator Śākyabuddhi and further elaborated by Dharmottara. This account became increasingly important as the tradition begun by Dignāga sought to clarify and extend his and Dharmakīrti's insights. The first thing that might be meant by *anyāpoha* is the unique particular that is the object of perception. This is "excluded from the other" in the sense that its being unique is just its being distinct from everything else. A second possible use of the expression is to denote the universal or object-in-general that is the intentional object of conceptual cognition (including inference and all other forms of linguistically mediated cognition). This involves "exclusion of the other" in the sense that all thought of kind membership involves differentiation. Dharmottara adds (no doubt with Dharmakīrti's psychological model in mind) that the particular is the direct object of perception and the indirect object of conceptual cognition, while the object-in-general is the direct object of conceptual cognition and the indirect object of perception. This leads to the question how there can be coordination between these two forms of cognition. The answer is that the mental image copied from a perception is of the nature of the "exclusion of the other" in the sense that, due to the activation of impressions from past experience, it is such as to be incompatible with representations of things taken as dissimilar. And it is the excluding nature of this mental image that is the third sense of *anyāpoha*. This third sense is clearly related to the "Janus-faced" nature of Dharmakīrti's concepts to which Dunne refers.

In assessing the apoha theory as a theory of human cognition, Katsura points out that even after Dharmakīrti's emendations and Śākyabuddhi's clarifications, there remains the problem of explaining how things can appear to naturally fall into kinds (Dignāga's use of the taxonomic tree) or resemblance classes (Dharmakīrti's judgment of similarity). He suggests that in the end the apoha theorist must invoke human conventions. But in defense of this he cites the "Ugly Duckling" theorem of the theoretical physicist Satoshi Watanabe. This theorem concerns the intuitive idea that natural kinds are constructed out of resemblance classes: that, for example, two lotuses belong to a single kind because they share more predicates in common than do a lotus and a mugwort plant. But as Watanabe proved, this is false; there being infinitely many predicates that any two particulars (including a lotus and a mugwort plant) share, this way of understanding the sense that a lotus resembles another lotus more than it does any other entity cannot be made to work. So, the suggestion is, natural kinds must involve factors pertaining to the human subject; resemblance nominalism is a nonstarter.

Hugon showed that a deeper understanding of the apoha theory can be obtained by starting with its critics—in the case of her contribution, Kumārila and Uddyotakara. In his contribution to this volume, Masaaki Hattori investigates the deep and subtle critique of apoha developed by another Naiyāyika, Jayanta. One objection of particular interest is that the apoha theory is unable to account for the fact that in the expression "blue lotus," the denotations of the two words must share a common locus and be in the qualifier-qualified relation. This is said to be a problem for the apoha theorist because the Buddhist denies that absences are real entities, so that if what words denote are exclusions or absences, they can have no locus and enter into no relations. If on the other hand one holds that absences are reals, there will be the difficulty that two absences must have separate loci and likewise cannot be in the qualifier-qualified relation. The problem here is the one raised by question (4), only put in ontological form. Dignāga and Dharmakīrti both answer it by in effect making the particular the indirect object of the individual word. So the particular denoted by a use of "lotus" is differentiated from all those particulars that are nonlotuses, but among these are some that are nonblue; the use of "blue" in this use context serves to differentiate the denoted particular from those that are lotuses but nonblue.

Question (6) also receives much attention from Jayanta. For the realist about absences, an absence requires an absentee or object of negation, something positive that gives content to the negation. This leads to an objection first formulated by Kumārila, that the difference between the object of negation for "cow" (namely, the class of lions, tigers, elephants, squirrels, horses, zebras, etc.) and the object of negation for "horse" (namely, the class of lions, tigers, elephants, squirrels, cows, zebras, etc.) is vanishingly small. For each class differs from the other just in containing one kind not contained by the other, and there are potentially infinitely many kinds of animals. Here we have an interesting variant on Watanable's Ugly Duckling theorem, only used against the apoha theory.

Jayanta's discussion of the apoha theory includes what Hattori takes to be references to the views of Dharmottara, and Hattori outlines the contributions of this important thinker to the later tradition. Chief among these is the view that the denotation of a word can be neither existent nor nonexistent. This odd-sounding view is the consequence of what would become a kind of master argument for the apoha theory in the hands of Jñānaśrīmitra and Ratnakīrti: that since a word may be used in a particular context of use to denote either an existent or a nonexistent, the denotation of a word type can be neither and consequently must be a mere mental construction. Dharmottara's positive view is that word meaning is something mentally constructed and superimposed on the reals. Jayanta takes this to be significantly different from Dharmakīrti's view, but it is not entirely clear that the views cannot be reconciled.

Parimal Patil's first contribution to this volume is a discussion of what was probably the last Indian formulation of the apoha theory, that of Jñānaśrīmitra and Ratnakīrti. (His second contribution, available at the companion website for this volume, is a translation of Ratnakīrti's *Apohasiddhi.*) One of their aims, he claims, was to deploy the apoha theory to develop a general theory of mental content—of the intentional object of such mental states as perception, inference, and verbal cognition. He sees the theory as involving a kind of paradigmatism (like that of prototype theory), according to which one's command of the concept associated with the word "cow" involves the ability to recall a particular mental image on the basis of which one forms the appropriate exclusion class (*non-cow*) and its complement (the class of things like the paradigm, and hence cows). But, Patil claims, the success of the theory depends on both the failure of all the competing realist and nominalist theories and a number of questionable assumptions.

Among the latter is the assumption that there will be intersubjective agreement among the exclusion classes formed by the utterance of "cow" for English speakers, given that according to the theory these are formed through recall of a particular mental image. Since the image that occurs to each speaker will be distinct from that of any other speaker, Patil is here raising the "private language" horn of the dilemma in question (2). Now Dharmakīrti's response to this challenge (as Dunne and Hattori both discuss) is to invoke some common interest (such as the desire for milk) that all members of the exclusion class fail to satisfy. (Remember that failure can be construed as a lack or absence, something that Buddhists claim involves no ontological commitment.) Patil raises the question why this functionally determined exclusion class should agree across subjects, and he suggests that a modern and naturalistically inclined apoha theorist who appeals to natural selection to answer the question will quickly exhaust their explanatory resources and be reduced to saying that is just how it is. Presumably this is because a selectionist explanation will make reference to similarities across individual members of a species—for instance, the fact that all humans are mammals and so must share a taste for milk. And any such similarities that are appealed to in the explanation will themselves require further explanation. So the modern apoha theorist must either dogmatically assert that this is how things are or else embark on an infinite regress. Patil clearly finds the latter alternative uninviting, but one might wonder whether Jñānaśrīmitra necessarily agrees. As Patil discusses at the end of his paper, Jñānaśrīmitra does not think that any theory (including the apoha theory) can be ultimately true, but he nonetheless holds that a given theory might be more useful than its alternatives and so stand as a better formulation of the conventional truth than the other options. Moreover, Jñānaśrī also holds that there is a potentially infinite hierarchy of such theories, each standing as a better approximation to the ultimate truth than its predecessors.[10] So here too, as in the appeal to "beginningless ignorance," there may be more tolerance for infinite regresses than one might expect.

Patil also raises a number of questions concerning the account of mental content he finds implicit in Jñānaśrīmitra and Ratnakīrti. Among these is the question whether a contemporary apoha theorist would want to take on board Jñānaśrī's claim that perceptual and inferential/verbal cognitions have distinct phenomenal characters. He takes this to be the claim that while perception and inferential/verbal cognition of, for example, a fire have the same conceptual content, they differ by virtue of their qualitative character as experiences, their "what-it-is-like-ness." But it is not entirely clear that any apoha theorist says this. Dharmakīrti does say that a perception and the immediately following mental image that forms the basis for conceptual cognition differ in that the first is vivid while the second is indistinct, and this sounds like a difference in phenomenal character. But this is, according to all apoha theorists, a difference between a cognition without conceptual content and one with conceptual content—between a perception properly so called and what is sometimes called a perceptual judgment. Now there is an interesting problem in the neighborhood, one that stems from the fact that on Dharmakīrti's formulation there must also be a phenomenal difference between the perceptual judgment and a cognition that is entirely inferential in nature. When your sense of vision comes in contact with the fire, there is first the mental image, M_1, the vivid and unconceptualized presentation of color and shape, and then a moment later the perceptual judgment that employs an indistinct copy, M_2, of this presentation to form the concept *fire*. When you later tell me about your experience, your use of the word "fire" arouses in me a mental image, M_3, that I use in working out what the fact is that you are reporting. Now we are told that while M_1 and M_2 differ in terms of relative distinctness, the latter is routinely mistaken for the former, so they must resemble one another in phenomenal character. On the other hand, no one ever mistakes M_3 for either M_1 or M_2, yet it still must resemble them in phenomenal character. So there does seem to be a puzzle here. But current cognitive science tells us that subjects are able to make more fine-grained discriminations on the basis of a presented image than on the basis of a recalled image (Metzinger 2003, 43–62). Since M_2 and M_3 are recalled images, with the former being the product of immediate recall, it is possible that differences in phenomenal character among the images are the result of such functional differences. Patil is of course right to wonder whether this is of strictly semantic significance. But an adequate theory of concepts must account for more than just the semantic facts.

Patil also makes much of Jñānaśrī's claim that there can be no adequate account of mental content insofar as nothing that will serve as representational content (something that is necessarily conceptual in nature) can fully capture what is immediately given in raw experience. One might wonder, though, whether this is not just another way of making the point that Dunne has in mind when he refers to the "Janus-faced" nature of Dharmakīrti's concepts. One point it is not always easy to keep in mind is that the representationalist theory of perception was not the consensus position in classical Indian philosophy. Modern theories of concepts tend to presuppose such a view and with it the idea that mental content is necessarily "in the head." Externalist views of content and a direct realist view of perception are still minority views. In the Indian context, on the other hand, non-Buddhist philosophers consistently held the direct realist view, and even among Buddhist philosophers representationalism was controversial (Dhammajoti 2007, 136). In this context it seems plausible that the insistence one finds in Dharmakīrti and his successors that ordinary people routinely mistake the object of inferential cognition for the object of perception might reflect an older tradition of representationalist insistence that perceptual judgments cognize the external object only indirectly through a kind of automatic inference. It is that older tradition that seems to lie behind Dignāga's insistence that there are two distinct kinds of objects of cognition. Perhaps it is likewise at work in Jñānaśrī's claim that no single account can be given of cognitive content.

P. K. Sen's paper begins with a survey of some of the more important critiques of apoha theory by such thinkers as Uddyotakara, Kumārila, Jayanta, and Vācaspati. He then develops his own set of objections to the theory, along lines not unlike those of its Nyāya critics. Unlike Pind, Sen does not interpret Dignāga as having started with an argument against the existence of universals. This makes an important difference to how we assess his critics such as Uddyotakara and Kumārila, many of whose objections otherwise seem to miss the point. If all Dignāga were doing was developing an exclusion-based semantics that is formally equivalent to the realist semantics of Nyāya or Mīmāṃsā, and there were no independent reason to think that the meaning of a kind term (something that is distributed across a range of particulars) could not involve real universals, then we would be inclined to agree with Uddyotakara, Kumārila, and the like that the realist theory is preferable to the more cumbersome and counterintuitive apoha theory.

As Sen's discussion brings out, once the Buddhist argument against real universals was on the table, the debate over the apoha theory centered largely on issues of ontology. Uddyotakara's objections, for instance, all presuppose that whatever serves as the semantic value of an expression must be some real thing; he quite fails to grasp that apoha theorists are struggling to articulate an alternative vision according to which abstract objects are mental constructs. Of course Sen himself sees this, and so he neatly turns the tables on the apoha theorist, who sees the twin problems of causal efficacy and instantiation as the Achilles's heel of a realist semantics. To the apoha nominalist who asks how an eternal universal can cause anything and how a single universal can be present in many locations simultaneously, Sen asks how something imaginary can do either one. He concedes that Nyāya realism comes at an ontological price, but he insists that the Buddhist nominalist alternative has its costs as well. And he warns against dismissing the realist view based on a caricatured understanding of realism. Sen points out, for instance, that Nyāya does not posit a real universal for every general term; their semantics makes do quite well with a basic stock, through the combination of which other meanings are "cooked up." And the apoha theory is, he says, vulnerable to the problem of typicality effects common to prototype theory, proxytype theory, and other exemplar-based theories of concepts. If giving milk and serving as a draught animal are the functions that a Dharmakīrtian cites as the basis for the construction of the exclusion "not non-cow," then what are we to make of buffaloes and of aged and infirm cows? While Sen does not point this out, a sophisticated Nyāya theory of universals does contain resources for handling this problem: although the basic stock of real universals might be inflexible in their application, the problems associated with typicality effects might arise only with kind terms whose semantic values involve combinations of that basic stock, with flexibility built into the principles of combination. But then the apoha theorist might use a similar strategy to answer Sen's buffalo objection.

Through his examination of accounts of the apoha theory in critical Nyāya and Mīmāṃsā texts, Sen often manages to cast the theory in an interesting light. One instance of this is Śrīdhara's account of Dharmottara's arguments for apoha. One such argument is, according to Śrīdhara, that since it is agreed that goats, horses, camels, and squirrels have nothing in common, they can be considered as constituting a class only under the expression "non-cows." But according to the apoha theorist, since there are no real universals such as cowness, the things called "cows" likewise have nothing in common. So it stands to reason that these too should be taken as constituting a class only under the expression "not non-cow." This interesting and quite sensible-sounding argument is one that is more readily apparent in Śrīdhara's description than in any number of more "friendly" sources.

Georges Dreyfus's paper concerns Dharmakīrti's formulation of the apoha theory, only this time as seen through the eyes of some of his Tibetan commentators. Dreyfus holds that central to Dharmakīrti's overall theory is the role played by concept formation. But he begins by examining whether Dharmakīrti may be thought of as a resemblance theorist, someone who holds that we judge particulars to belong to a kind because they all resemble one another. Dharmakīrti does say that we naturally judge certain particulars to resemble others. Dreyfus points out, though, that he carefully refrains from asserting that such judgments are based on real resemblances among the particulars, an assertion that would lead to a failed nominalism. Instead, he seeks to explain such judgments in terms of the fact that the particulars involved perform a similar function (in the case of the stock example of the medicinal herbs, that of reducing fever). This makes it clear that Dharmakīrti wishes to bring out the interest-relative nature of perceived resemblance. It also leaves room for appeal to either karmic imprints or the forces of natural selection to account for a widely shared primitive similarity space, something that can then be employed to explain how we might construct more complex, language-mediated concepts. Crucial to this account, according to Dreyfus and some Tibetan commentators, is once again the "Janus-faced" nature of Dharmakīrtian concepts, which are at once mental particulars with unique and inexpressible phenomenal content and exclusions that are multiply instantiable. As such they are meant to perform the crucial function of mediating between perception and thought, thereby explaining how thinking can be constrained by reality despite its being incapable of capturing reality.

Dreyfus is nonetheless concerned about what he takes to be the seeming paradox at the heart of Dharmakīrti's system, that the (presumably correct) judgment "that is a cow" is as much in error as "that is Santa Claus" or "that is a garland of sky-flowers." And of course this worry is not misplaced; non-Buddhist philosophers regularly beat up on the apoha theorists on just these grounds. To dispel the paradox, Dreyfus recommends the distinction certain Tibetan commentators drew between precritical application and critical examination. In the former context the judgment "this is a cow" is deemed correct. In the latter it is erroneous, not only because there is no such thing as the property of being a cow but also because the subject of predication, the "this," is likewise a fictional construction. Critical examination nonetheless helps us see how such constructions as *cowness* and the spatiotemporally extended substance denoted by "this" are causally connected to real particulars, thereby explaining how a judgment that could not possibly be veridical can nonetheless contribute to successful practice. This, it might be added, could be usefully compared to our present understanding of color perception. While we perceive physical objects as colored, we know that mind-independent reality contains no such thing as color. It is nonetheless possible to explain how the mind-independent nature of physical objects causes us to perceive color and through this causal link also explain why judgments of color should have pragmatic value.

The gap between sensation and thought is also the subject of Jonardon Ganeri's essay. But Ganeri's contribution is among those that stand back from the historical details of the apoha theory and attempt to evaluate it on strictly philosophical grounds. In this case the assessment centers on the question whether the apoha theory has anything to offer to current efforts to close the gap between conceptual and nonconceptual representation. Taking Austen Clark's work on sentience as a model of such efforts, he spells out several places where the notion of exclusion or apoha might do important work. One such place arises because there is no straightforward correlation between stimuli and the qualities presented in sensation. In the case of color vision, for instance, any number of distinct combinations of wavelengths will all produce the experience of seeing red. What this is taken to show is that presented qualities are sorted into kinds on the basis of their place in a discrimination ordering. This, Ganeri points out, is tantamount to claiming that being red is just a matter of being excluded from what would be deemed nonred. To this it might be added that this is just the insight that seems to have been behind Dignāga's claim that determination under a determinable is fixed by contrastive relations.

A second place where Ganeri sees an opening for the apoha theory is in explaining the ability of sentient creatures to assign perceived features to places. This ability is crucial to the "bundling" of qualities that bundle theorists like modern empiricists and Buddhist reductionists employ to bridge the gap between sensation and thought. If space were given as just another quality alongside red and sweet, it could not be used to bundle red and sweet together in the construction of quality clusters (an ability that is crucial in turn to the construction of substances such as tomatoes). Ganeri suggests that the key difference lies in the fact that a quality like red has an incompatibility range—the instantiation of red at a particular place is incompatible with the instantiation of such other qualities as blue and yellow—while there is nothing comparable for places. So the ability to do something akin to referring to a particular place, an ability that a sentient creature of any degree of complexity must have, is likewise to be grounded in a kind of discriminatory exclusion. Ganeri worries, however, that Dharmakīrti's idealism might get in the way here. If, as most scholars now believe, Dharmakīrti's final position is that of Yogācāra, then he must deny the existence of space and consequently must explain the appearance of space as a matter of spatial qualities being given as part of the flux of sensory impressions. But this last bit need not follow. There is much evidence from cognitive science that the spatial organization of sensory data proceeds on the basis of action simulation on the part of the sentient creature: "here" is distinguished from "there," for example, on the basis of what can be done with and without locomotion. And it might be possible for the idealist to simply appropriate this way of individuating apparent places, given that actions come with their own presented qualities.

A third place where Ganeri thinks exclusion might help explain how sensory capacity could give rise to something like our conceptual scheme is in the ability to see the world as containing enduring, reidentifiable physical objects. This ability involves resources that are clearly unavailable at the level of pure sentience. How is the gap crossed? Ganeri points out that a creature with the ability to assign features to places will also be able to detect places that lack the absence of a feature. And, he points out, "no absence of pot here" is transformable, through the canceling of the two negations, into "this is a pot." The latter judgment clearly involves the concept of a reidentifiable particular. The suggestion is thus that our scheme of enduring substances can be constructed out of perceptual contents through the judicious application of exclusion. Having thus indicated places where the notion of exclusion might play a useful role in a modern gap-closing exercise, Ganeri goes on to explore ways in which specific details of Dharmakīrti's account can be tied to such an account. On this view the apoha theory is more than an interesting historical artifact.

Amita Chatterjee is similarly sanguine about the prospects for a cognitive science account of understanding that is informed by insights from the apoha theory. One current debate in cognitive science concerns whether cognition should be understood as a computational process (akin to the operations of a digital computer) or instead should be seen in terms of the dynamic interaction of the perception and action systems of an ecologically situated organism. One question for dynamicists or noncomputationalists is whether any theory of cognition can dispense entirely with mental representations (these being something that computational accounts seem well suited to account for). Chatterejee suggests that Dharmakīrti's formulation of the apoha theory might be used to sketch a noncomputational theory of mental representation, thereby enhancing the plausibility of the moderate form of noncomputationalism, which agrees with computationalism that mental representations play a role in higher-order cognition. This immediately suggests an answer to question (2), the question whether concepts are mental particulars or rules. A computational approach to representations sees them as rules or schemata (namely, the algorithms involved in computation), so if the apoha theory suggests a noncomputational approach to representations, then this option is ruled out. But the dynamicism of noncomputational approaches likewise rules out the possibility that a representation is a kind of mental particular. On a noncomputational approach, there is nothing that a concept actually is; our notion of a concept merely picks out one facet of a complex dynamic process.

This likewise suggests an answer to question (1), whether concepts are innate or learned. Chatterjee makes use of the notion of a Gibsonian affordance, which is a perceivable possibility for action. Affordances are relational properties that obtain between an organism and its environment, in this respect resembling the interests that Dharmakīrti claims shape the formation of an exclusion class (such as the interest in load-bearing animals that helps shape the concept *cow* as what is other than those things that fail to satisfy this interest). Now some affordances are sufficiently stable features of the organism-environment relation that natural selection can operate to foster the appearance of resemblances among the particulars involved. This process yields concepts that are innate for the species, the equivalent of Dharmakīrti's imprints due to "beginningless ignorance." But other affordances are more ephemeral. These will be the source of learned concepts, some of which are then shared through the medium of language. This account thus appears to resolve the old debate between rationalists and empiricists in favor of rationalism, since it holds that at least some concepts are innate. But as the notion of an affordance makes clear, the innateness at work here holds only at the level of the individual organism, not that of the species. From the perspective of the species, all concepts are acquired through interaction with the environment. Completely dissolved is rationalism's mystery of a preestablished harmony between mind and world.

The deepest source of affinity between the dynamicist approach and the apoha theory lies, for Chatterjee, in their shared conviction that "Objects in the world are created in stages by dynamic interaction between organisms and the world." What Dharmakīrti's formulation offers is a way of spelling out how these stages lead from the raw, unconceptualized perception of unique particulars, through the (protoconceptualized) perception of primitive resemblances such as of color, to full-fledged perceptual judgments about enduring substances such as pots and cows. Because this is a stage-wise process, representations are required for the account to work. Dharmakīrti offers a means of understanding in a dynamicist way how one set of interactions might come to be treated as perceptual input—serving as representations—at the next higher level. On this understanding of Dharmakīrti's bottom-up approach, apoha is not a logical operation we perform on sensory input, but a node in the interactive interplay of organism and environment.

A final point worth remarking on is Chatterjee's suggestion that the apoha theory might be made to work without the scheme-content dichotomy that was built into it from its inception in Dignāga's thought. This would require restricting the theory to what Buddhists call the conventional truth. Such an apoha theory would then be, like dynamicist theories of cognition, fully compatible with Putnam's internal realism. And it would thereby avoid all the difficulties that come with a strict separation between scheme and content, while at the same time avoiding the problem of relativism. To this it might be added that Madhyamaka appropriations of Dharmakīrti's thought aim to accomplish just this. Mādhyamikas routinely deny that there is such a thing as the ultimate truth; yet Dharmakīrti's formulation of the apoha theory makes full use of the distinction between conventional and ultimate truth. So when a Mādhyamika such as Śāntarakṣita claims that the apoha theory gives the best available account of human cognition, this is tantamount to saying that the ultimate-conventional distinction holds only conventionally.

The next two papers, by Bob Hale and Brendan Gillon, take up a characterization of the apoha theory (first developed in Siderits 1982) that includes the claim that the characteristic apoha expression "not non-cow" involves two distinct kinds of negation: verbally bound negation ("that *is-not* non-cow") and nominally bound negation ("that is not *non-cow*"). On this understanding of apoha, the use of two distinct negations is meant to answer the nongeneralization difficulty first raised by Kumārila (see Hugon, this volume): how, in a world of particulars, can reference to a real yield something general in nature? If the two negations involved in the apoha expression are of the same sort, then it seems they should cancel out, in which case we return to the particular with which we began. The suggestion is that the combination of two different types of negation does not obey the classical principle of double negation, so the generalization required for the apoha theory to work is logically feasible.

In his critical assessment of this strategy, Hale begins by raising what he calls the problem of compositionality, but this is a different problem than that of "blue lotus" and "pet fish" discussed under the rubric of question (4). He takes the central claim of the apoha theory to be that "*x* is *P*" is to be analyzed as "*x* is not non-*P*," which leads to the difficulty that since the latter involves a complex expression in which "*P*" is a constituent, the composition principle (the principle that the meaning of a complex expression is a function of the meanings of its constituents) must be violated. This is in fact the circularity objection of Kumārila and Uddyotakara discussed by Hugon. In exploring possible responses to this objection, Hale describes what he calls a nominalistic compositional semantics, which has as primitive expressions only logically proper names, the basic logical operators (including verbally bound negation), plus a distinctness predicate that functions like nominally bound negation for proper names. Hale grants that this strategy might be made to work if we equip the language user with something like Dharmakīrti's "indistinct" mental images. But he points out that then it is the mental images that do all the nominalistic heavy lifting (functioning as they do much like Locke's "abstract ideas"), in which case the resort to the two-negations strategy seems unnecessary.

Hale goes on to point out that the appeal to interests that is at the heart of Dharmakīrti's causal story appears to require that interests be repeatable kinds. In this case the nominalist is back to the usual dilemma of either building in real resemblances at ground level or else embarking on an infinite regress. In the end, Hale thinks the best the apoha nominalist might be able to do is simply claim that while "*x* is not non-*P*" is not the correct analysis of "*x* is *P*," it is ontologically less committed than "*x* is *P*" and so can be used to replace "*x* is *P*." But, it could be pointed out, this takes the apoha theory back to the most minimal characterization of Dignāga's formulation. What other resources the apoha theorist might summon to try to answer the many objections to which Dharmakīrti was responding, Hale does not say.

Brendan Gillon's contribution sets out to sketch a formal semantics for apoha nominalism interpreted in accordance with the two-negations idea. His result is negative: with the two negations interpreted as set-theoretic internal and external negation, no combination of the two yields the desired result that shows how cognition can proceed from the particular to the general. Other authors claim there is no evidence that any apoha theorist had the two-negations strategy in mind. Gillon is silent on this historical question. His conclusion is that if they had meant the apoha theory to work in this way, there is nothing in modern semantics that would vindicate their hypothesis.

Mark Siderits takes stock of his understanding of the apoha theory in light of the criticisms of others on both historical and theoretical grounds. To do so he spells out in considerable detail what he takes the bottom-up approach of Dharmakīrti and his followers to amount to when we stand back from the historical details and perform a rational reconstruction of the theory. This yields a response to question (2) that, like that of Chatterjee, rejects the dichotomy of concepts as either mental particulars or schemata. Thus, in response to the criticism that Dharmakīrti only gives us what Dummett calls a "code conception" of language (in which words serve as code for communicating ideas, which are private states of the speaker), the apoha theorist would say the concept of a concept is the reification of a dynamic process in which mental particulars play a role and which can be analyzed post hoc in terms of rule-governed behavior. This dynamic process is consistent over time for a cognizer, as well as across cognizers, because it leads to action that satisfies the interests of cognizers.

Siderits also suggests a response the apoha theorist might make to the common charge at the heart of question (5), that the theory inevitably smuggles in universals in the guise of real resemblances. This response involves what might be called a strategy of endless deferral, in effect conceding that the theory involves an infinite regress but denying that it is vicious. But it also tackles the question how there can be causation without universals, given that causal laws state relations between *kinds* of entities. The proposed response suggests that causation can be treated by the Buddhist nominalist as something to be reduced to more fundamental entities, in effect adding causation to the long list of things the Buddhist calls mere conceptual fictions. But this move will no doubt appear unsettling to those who see causation as crucial to the bottom-up account: how is the account to work if there really are no such things as causal laws?

Also addressed is question (6), the question whether the two negations involved in the apoha theory are any more than a gimmick. Siderits concedes that the two-negations strategy does not by itself explain how one can come to exhibit mastery of a concept, be it one that is "innate" (such as *yellow*) or "acquired through experience" (such as *pot*). Apoha theorists of a Dharmakīrtian stripe explain such mastery in terms of the interest-guided employment of mental images. He suggests, though, that the appeal to two distinct negations does respond to a very different question: not the epistemological question of how one masters a concept, but the metaphysical question of how there can be exploitable patterns in a world of unique particulars. The suggestion is that apoha theorists hit upon the idea of deploying a combination of verbally bound and nominally bound negations as a way of showing how Nyāya's taxonomic tree could be retained in a nominalistic universe. Of course Siderits must now concede that there is no historical evidence directly supporting his claim. But he does point to an episode in later Nyāya history that includes all the major features of his model of apoha, including a two-negation strategy linking particulars and universals and an objection based on the nongeneralization problem.

Finally, Parimal Patil's translation of Ratnakīrti's *Apohasiddhi*, or "Demonstration of Exclusion," is available at the companion website for this volume, www.cup.columbia.edu/apoha-translation. Patil has chosen to give his translation the title "Without Brackets." In most translations of Indian philosophical texts, one encounters a great deal of material enclosed in square brackets. This material is there for a good reason. Classical Indian philosophers placed such value on conciseness of expression that their writings are all but unintelligible to those not well versed in the tradition in which they worked. Consequently, a literal translation of the text is likely to be of little or no use unless supplemented by a considerable amount of background material, and many scholars have elected to use square brackets to indicate that supplementary material is being inserted. But this device interferes with comprehension of the argument, particularly for nonspecialists. Patil (like some other scholars) has chosen to avoid their use while still providing the material necessary to make the text intelligible. It will still prove challenging. But the hope is that those with little or no prior exposure to Indian philosophical texts and traditions will still be able to follow the argument after having read the essays that precede it in this volume.

Notes

1. The Sanskrit term *apoha* means "exclusion." The apoha theory is the theory that the meaning of a general term is "the exclusion of what is other." We will be using the word in unitalicized form, just as the word *karma*, which is Sanskrit in origin, is now used without italics.
2. "sarva eva hi bhāvā svarūpa-vyavasthitayaḥ. te nātmānaṃ pareṇa miśrayanti. . . . na hi sambandhināpi anyenānye samānā nāma tadvanto nāma syuḥ. bhūtavat kaṇṭhe guṇena" (PVSV ad PV I.40–41, ed. Gnoli, 24.25–25.6). Karṇakagomin explains that the reference here is to the popular religious practice of stringing together idols of the planets and other divinities.
3. At least not on the commonsense view. Buddhists maintain that strictly speaking what we call an apple is just a bundle of quality particulars such as the sweet taste, the red color, and so on. For them the commonsense view that there is no sweet taste without the apple that has it is only conventionally true.
4. Whether such "cross-cutting" disqualifies both the properties or only one of them, and whether the neat ontological hierarchy that is presupposed by this universal blocker is integral to a realist metaphysics, have been the subject of much contemporary debate (see Shastri 1964; Mukhopadhyaya 1984; and Ganeri 2001).
5. PV 1.152 (Gnoli): "na yāti na ca tatrāsīd asti paścān na cāṃśavat / jahāti pūrvaṃ nādhāram aho vyasanasaṃtatiḥ" (Gnoli 1960).
6. See *Nyāya-kumuda-candra*, vol. 2, 560–1 (Kumar 1939, 1941). Versions of this theory were adopted by followers of Rāmānuja (Qualified Monist Vedānta) as well as by Maddhva (Dualist Vedānta) logicians. Vyāsatīrtha of the latter school clarified how a single resemblance can reside, as it were, with one leg in the resembler and with another leg simultaneously in numerous other similar particulars. The category of resemblance admitted by these philosophers is very different from the resemblance admitted by Prābhākara Mīmāṃsakas, for the latter were realists about universals, while the Jainas and the Maddhvas rejected, as logically redundant, both universals and inherence. The only difference between Prābhākara and Vaiśeṣika, as regards universals, centers on their conceptions of inherence.

7. This, by the way, is almost suggested by the Buddhist author Arcaṭa in his HBṬ commenting on Dharmakīrti's HB: that a *vikalpa* is nothing but an unclear copy of a sensory image. At HBṬĀ 287, Durvekamiśra, in his gloss on Arcaṭa on Dharmakīrti, says explicitly that the false impression of being determinate is given by the faint and unclear concept thanks to the nonconceptual inner perception which takes place at the same time, implying thereby that the conceptual awareness borrows its seeming clarity from the nonconceptual awareness.
8. Compositionality, incidentally, was also a concern for Dignāga, whose discussion of the expression "blue lotus" is motivated by anxiety to distinguish coreferential terms from same-sense-bearing synonyms. Apoha theory could not strictly speaking be a theory theory of concepts, since it does not appeal to ontologically intrinsic essences at all; but the functional and dynamic aspects of a set of concomitance rules and a pragmatic-predictive capacity are features that apoha theorists also ascribe to concept employment.
9. For Frege as a major source of modern nominalist bias, see Bergmann 1958.
10. For more on Jñānaśrīmitra's contextualist semantics, see McCrea and Patil 2006.

1

How to Talk About Ineffable Things

DIGNĀGA AND DHARMAKĪRTI ON APOHA

Tom Tillemans

If one tries to tell something of a unifying narrative as to why the philosophy of language became what it did in the system known as "Buddhist logic," the central thread must be the Yogācāra separation between language and concepts, on the one hand, and ineffable, unconceptualized, real particulars on the other. To adopt a frequently used philosophical notion, Yogācāra, and logician apoha theorists, were unwavering subscribers to a "scheme-content" distinction. They held that we can and should make a clear and radical separation between what our linguistic-conceptual schemes create and impose upon an uninterpreted content and that content itself, which is real, accessible only to perception, outside the conceptual scheme, and hence free of the scheme's distortions and coloring.[1] The seemingly intractable problem in making a radical scheme-content separation, as Yogācāra wants to do, is that once you've insisted that things in the world exist in a way completely outside the natures imposed by the conceptual scheme, it becomes very difficult to backtrack and still somehow account for a linkup between language, concepts, and the world. Prima facie at least it becomes extremely difficult to say why one conceptual scheme or one description of the world would be better or more likely to be true to the world than another. After all, all languages and concepts would seem to end up on an equal footing, that is, having no connection with how things are.

APOHA THEORY AS A YOGĀCĀRA PHILOSOPHY

Now before we take up the ways Buddhist logicians, or Apohavādins, nonetheless manage to talk nonarbitrarily about the world, let us briefly look at the scheme-content distinction in Yogācāra and assure ourselves that the Apohavādins *are* in a significantly similar philosophical predicament. The Yogācāra scheme-content distinction figures in their account of things having three natures (*trisvabhāva*), a classical theme found in texts such as Vasubandhu's *Trisvabhāvanirdeśa* and *Triṃśikā* and in many others (including texts of Dharmapāla and at least implicitly Asaṅga's *Bodhisattvabhūmi Tattvārthapaṭala*). In brief, linguistic-conceptual natures are said to be "thoroughly imagined" (*parikalpita*) and hence are not real, but are projected or "superimposed" (*samāropita*) on "dependent nature" (*paratantasvabhāva*) of things, which is real but ineffable and is in itself outside and beyond the imagined.

Although the terminology used in Vasubandhu and other authors is that of "imagined nature" and "dependent nature" (*parikalpitasvabhāva, paratantrasvabhāva*), the substitution of "universals" (*sāmānyalakṣaṇa*) for the former and "particulars" (*svalakṣaṇa*) for the latter is easily and naturally made. Thus, a universal X taken as an apoha, that is, an exclusion of non-X, has all the scheme-like characteristics of the imagined natures. Being an absence, it is unreal and is a kind of conceptual fiction, thus supposedly enabling us to avoid acceptance of *real* universals; the system remains nominalist in the sense of recognizing only particulars (*svalakṣaṇa*) as real. Equally, particulars are the perceptual content upon which the scheme comes to bear, just like dependent natures in classical Yogācāra.

I mentioned that this rapprochement between three-natures theory and the apoha theory could be easily made, philosophically speaking. Indeed, it *was* explicitly made by more than one major thinker in the logicians' branch of the Yogācāra school. This is much in evidence in Dharmapāla, a sixth century granddisciple of Dignāga, who in his *Guang bai lun shi lun* on the *Catuḥśataka*, alternates between the "imagined-dependent" (*parikalpita-paratantra*) and "universal-particular" (*sāmānyalakṣaṇa-svalakṣaṇa*) dichotomies (see Tillemans 1990, 162–164). But it is especially clear in the Yogācāra chapter of Bhāviveka's *Madhyamakahṛdayakārikās*, where an anonymous Yogacārin defends the three-nature theory by relying on Dignāga's philosophy of language and uses the latter's apoha theory to explain how the conceptual scheme is constituted and imposed on its ineffable perceptual contents. If ever there was a reasonable doubt about the close connection between the Buddhist logician's philosophy of language and the Yogācāra worldview, Bhāviveka's *Madhyamakahṛdaya* lays that doubt to rest.[2]

THE PROBLEM OF REFERENCE

What do we mean when we speak of "talking about" things in the world that are somehow ineffable? To put it another way, "talking about" for Buddhists will eventually, but not immediately, lead to a problem of reference for Yogācāra-inspired philosophers and their Apohavādin allies. The problem doesn't arise immediately because Buddhist writers, indeed many Indian philosophers, regularly use terms like "what is expressed/talked about" (*vācya*), "what is signified" (*abhidheya*), and "the meaning of a word" or "things-meant-by-words" (*padārtha*) without having to make a distinction between the fact of words expressing properties as a sense or meaning and their referring to particular things in the world via those properties that are their senses.

As is well known, Indian philosophers of language, from early on, formulated their semantic theories in terms of a dichotomy: either words express general properties or they express particular things. There are then subsequent attempts to split that difference so that (senselike) properties and (referencelike) particulars are both expressed together. The Nyāya solution of having words express "particulars possessing the universal" (*jātivat, sāmānyavat*), for example, allows one to avoid the absurdities of words *just* expressing one or the other. (See the introduction, part 1, on classical Nyāya-Vaiśeṣika realism about universals.) Apohavāda, that is, the position that words *express* some type of double-negative property, is a similar attempt to split the difference, in that it supposedly allows one to express *both* a quasi universal and particulars.[3] The ambitious project of chapter 5 in *Pramāṇasamuccaya* (PS) is to show that apoha not only splits the difference but that it is the only philosophically satisfactory way to do so in that (contrary to Nyāya et al.) it avoids commitment to real universals.

Now once it has been decided rightly or wrongly that apoha is the best answer to that longtime semantic controversy, another and potentially different question arises for a Buddhist logician, implicitly in Dignāga, but much more clearly in Dharmakīrti. That question is the following: how is it, via the intermediary of these fictitious double-negative quasi universals or in some other fashion, that somehow the Buddhist logician can come to *pick out* real individual particulars that are in themselves supposedly ineffable? This is, in brief, the problem of bridging the scheme-content gap. It is a problem of reference and it obviously becomes especially intractable when reference is supposedly going to involve ineffable particulars that are somehow only perceptible and whose intrinsic natures are not actually conceivable or describable.

TWO APPROACHES IN APOHAVĀDA

I would maintain that there are, broadly speaking, two approaches in apoha theories to bridging the scheme-content gap: what we could call "top-down" approaches and "bottom-up" approaches. By "top-down" I mean a position that would somehow maintain that it is because of some specific—and perhaps even very ingenious—features of the logical operators of negation in the exclusion that the apoha does pertain to particular things, even though it does not have the ontological baggage of a real universal. In short, on a top-down approach the apoha would behave like a property, a sense, or a meaning, which belongs to the conceptual scheme but nonetheless qualifies and thus serves to pick out the real particulars in the world; because of some feature of double negation, we are spared commitment to real universals in addition to real particulars. On a bottom-up approach, causal chains and error are what serve to bridge the scheme-content gap, rather than the logico-metaphysical features of a special sort of double negation. The way words link to things is thus primarily explained through the existence of a causal chain from things to thoughts and then to the utterances of words.

The usual way in which Apohavāda has been presented in the West and in classical India (at least in the writings of major non-Buddhist critics) is via a top-down approach, that is to say, Buddhists think that somehow some nominalist mileage is to be gained by seeing double negation, a fiction, as serving to pick out real entities that are in their intrinsic nature ineffable and purely particular. Thus Potter (1963, 188) on Buddhist nominalism's elimination of commitment to universals (taking the standard Indian example of a universal, i.e., cowness): "Although it falsifies reality to describe it as having a certain positive character (e.g., cowness), it does not falsify [reality] to describe it as lacking a certain negative character (e.g., noncowness)." And Matilal (1971, 44): "Meanings, for Dignāga, are fictional constructions and they have a negative function . . . to exclude the object from the class of those objects to which [the name] cannot be applied."

Here is what I think would be a plausible summary of Dignāga's position. In the fifth chapter of his *Pramāṇasamuccaya* (PS), we find the recurrent idea that real things as they are in themselves, that is, as they appear to direct perception, are indivisible unities. They can nonetheless be regarded in terms of several conceptually separable "facets" (*aṃśa,* literally "portions") or properties, and no single word can express the totality of these properties. Thus, the word "cow" expresses one such property and "impermanent" expresses another, each word conveying some mind-created property in a purely negative fashion, differentiating the *X* (e.g., the cow or the impermanent thing, and so forth) from non-Xs. It thus looks like Dignāga more or less allowed the essentials of what I am calling a "top-down" approach: an apoha, a non-non-*X* expressed by a word, would enable us to pick out particulars in that the property would apply to things simply because of its logical features of double negation and not because it is a positive feature that would be present in the particulars themselves.[4]

Dignāga's adversaries in India, for example, Kumārila and Uddyotakara, saw him as adopting this top-down approach, replacing universals with fictional constructions in the form of double negations. And not surprisingly, one of their main objections was to question whether a Buddhist could ever hope to gain any advantage for his nominalist project by substituting double-negationese for positive universals. (For the details of the non-Buddhist challenge and the Buddhist response, see the papers by Hugon, Hattori, and Sen in the present volume.) Contemporary philosophers reconstructing the theory of apoha, such as Hans Herzberger and others, have generally sought to answer that type of question in the context of the top-down approach. In the case of Herzberger, he used some possibilities offered by Emil Post's theory of twofold propositions to come up with what he termed a "resourceful nominalism," one that would explain how predicates applied nonarbitrarily to individual things account for our naive semantic intuitions but nonetheless avoid ontological commitment to universals. For Herzberger (1975) every proposition would be analyzable as an ordered pair of content and commitment—"apohist double negation" would affirm content but deny ontological commitment. In what seems to be at least partially a top-down approach, Mark Siderits, in several articles, has taken the relevant double negation as involving two different types of negation, so that it is the combination of the two that picks out a class of individuals, all the while staying nominalistically unengaged to universals.

Let us leave the top-down approach to take up what we are calling a "bottom-up" approach, one which, I would maintain, is that of Dharmakīrti. The major change that happens between Dignāga and Dharmakīrti is the use of a causal approach to link language to the world. This causal chain from particulars to perception and finally thought and language is entirely absent in Dignāga and so constitutes a substantial evolution in the theory: indeed, it is arguably a new Apohavāda that Dharmakīrti has devised. We will take up a few representative passages.

In *Pramāṇavārttika* (PV) III, Dharmakīrti writes: "If it said that [universals, that is, apohas] will lose their status of being properties of [real] entities (*bhāvadharmatva*), this is not a fault, for the cognition of the [universal] was preceded by an apprehension of the entity."[5] To this Devendrabuddhi comments:

> When conceptual thought (*rnam par rtog pa = vikalpa*) arises in dependence upon tendencies (*bag chags = vāsanā*) that were instilled due to one's having seen [particular] forms and so forth, it determines (*zhen pa = adhyavasāya*) apprehended images (*rnam pa = ākāra*) of its own as being the images of form and so forth and thus practically applies [to forms, etc.] In this way, [thought of form, etc., i.e., thought of the universal] arises [indirectly] due to the influence of seeing [particular] forms and so forth, and determines [its own images] to be those [i.e., real features of form], and therefore [for the combination of these two reasons] one does call [the universal] a property of the [real] entity.[6]

Here is the idea. An apoha universal U can be said to be a property of particulars p_1, p_2, p_3, etc., because (1) the thought of U is causally conditioned by tendencies imprinted on the mind by direct perceptions of p_1, p_2, p_3, etc., these perceptions being in turn causally linked to p_1, p_2, p_3, etc.; and (2) the mind cannot distinguish between its own invented universal U imputed to entities and the entities themselves (which are particulars and actually lack U). Devendrabuddhi cites these two points as the combined reason for claiming that a universal can be said to be a property of entities; the same combination of reasons is also amply attested in Dharmakīrti.

There are thus two parts to the Dharmakīrtian theory of reference, that is, the explanation as to how words link up with, or refer to, several particulars taken to be of the same kind. The first part of the theory is a version of what is usually called nowadays a "causal theory of reference," that is, a type of theory that explains what a person Jones is referring to with such and such a word by detailing a long and complex causal chain from the object, via Jones's language learning and concept acquisition, to the representation of the object and then to the use of the word on a specific occasion. Dharmakīrtian Apohavāda is not much different on that score in relying on causality for reference, although it is certainly a complicated chain that is being postulated. Here are the details: Jones sees particular things and has perceptual images (*ākāra*) of them; these images regularly cause, due to "tendencies" imprinted on Jones's mind, the same type of judgment, "this is an instance of *U*," a judgment to which a generic image (i.e., a type of apoha/*vyāvṛtti*) appears; because the particular perceptual images all have the same effect in leading to the same judgment, they are all the same in their causal power and can be grouped together. The link from the apoha to the specific word is made by speaker's speech intention (*vivakṣā*): Jones wishes to use a specific word to express such and such an apoha, so that it is the intention which causally conditions the utterance of the word.

The second factor in this Dharmakīrtian theory of reference is error. This is in fact a supplementary feature of the causal account. The Apohavādin maintains that at a key stage in the causal chain of reference, namely, when the mind makes a judgment that such and such a thing is an instance of *U*, it must be seduced by its own inventions of generic images to think that they are real and pertain to the world. To put things in other terms, in order for us to apply language and concepts to things in the world, we need to somehow ignore the differences that there are between all particulars and think and talk in terms of common properties to whose reality we are unreflectively and naively committed, even if in our more sophisticated theoretical reflections we might be persuaded that they are merely our own inventions.[7]

That said, although error does have a role in apoha, it needs to be stressed that the rapprochement with a typical modern error theory is only partial. Indeed, error theories have, in one way or another, often been made to account for commitments to specific types of properties considered dubious by the propounders of eliminativist theories. Hartry Field has applied such a theory to mathematics; Paul and Patricia Churchland's eliminativist physicalism is arguably an error theory about "folk-psychological" language and concepts. J. L. Mackie, in his *Ethics: Inventing Right and Wrong*, argues that "good" and so forth actually refer to invented properties (goodness, etc.) but that nonetheless our actual ethical thinking and discourse does, and indeed will invariably, proceed as if ethical properties were genuinely present in, or intrinsic to, actions and states of affairs.

Returning to apoha, one obvious difference from such typical error theories is that of scope. The Apohavādin is not just theorizing about *some* specific kinds of properties: he in effect maintains that we act, think, and talk unreflectively as if all common properties were real, as if realism about universals were right, even though as Buddhist theoreticians we may be convinced nominalists and maintain in our more sober moments that only particulars are real. But it should also be noted that the apoha theory is not in fact explaining the linkup between scheme and content in a typical error-theoretic fashion, that is, as *merely* due to us *wrongly thinking* that there is such a linkup. And if that is so, then apoha is probably best *not* seen as a typical error theory. The point is that, contrary to error theorists' diagnoses of such and such ideas or practices being purely due to "metaphysical superstition,"[8] for the apoha theorist there actually *is* an important connection between thought, language, and particulars via a complex causality, even if in our subjective representations of that causal process we might invariably distort and misapprehend many of its key features. Significantly enough, as the passage from Devendrabuddhi makes clear, error by itself is by no means the whole story and presupposes the causal account. Indeed, with one notable exception, that is, certain indigenous Tibetan apoha theories, it generally is causality that does the main explanatory work in Apohavādins' accounts of reference.[9]

This dominant causal direction in Dharmakīrti's account also seems to result in a naturalistic explanation of reference. In what follows, I'll be speaking of a "naturalized theory of reference" in much the same way as, since W. V. Quine, we can speak of "naturalized epistemology," a theory that places the emphasis on what human beings do in knowing, referring, etc., rather than on philosophically *certifying* the rationality and justification of what they do.

Dharmakīrti largely focuses on describing the cognitive events that supposedly happen when people refer to things with words, and this he does by specifying a chain of events in which each event is causally linked with the next.[10] Notable is that at crucial stages in his Apohavāda he actually relinquishes the quest to ground or certify reference and seems to say, in effect, that we do such and such types of things and make such and such judgments, but at a certain point no more philosophically satisfying justification or certification can or need be given. For example, various particular X images are grouped together because they each in fact cause the same effect (*ekakārya*)—that is, a judgment (*pratyavamarśa*) like "this is an X" that is the same (*eka*) for all individual X images—and not because there is something "in them" in common.[11] The Apohavādin's critic then objects that to certify that the instances of the judgment "this is an X" are indeed the same, the Apohavādin would have to say that they all cause the same metajudgment, and in that case an obvious regress would loom. Dharmakīrti is aware of that regress and refuses to give a further justificatory account in terms of same metajudgments; instead, he just appeals to the fact that this is how the judgments appear to us, that is, as all seeming to have the same content.[12] Certification comes to a clear end at this stage, replaced by a mere pointing out of some complex facts.

EVALUATING THE NOMINALIST MERITS OF THE TWO APPROACHES

Apoha theory, as time goes on, has ever-expanding uses: for example, it provides a Buddhist account of concept formation, of the transition from perception to conceptualization, and gives an attempt at a solution to logical problems like substitutivity of identicals for identicals in opaque contexts. Some of this looks like psychology, even a kind of seventh-century cognitive science, and may well turn out to have considerable interest; other interesting aspects touch on familiar themes in the philosophy of logic (see Tillemans 1986). But irrespective of whatever else Buddhist apoha theory is used for and might help elucidate, does it have anything that we might find promising for an intelligent nominalism? The short and swift answer: it depends on which apoha theory you take.

First of all do top-down approaches, or modern reconstructions of them, actually respond to the (usual) Indian and Western anxieties about nothing of nominalist interest being gained by resorting to double-negationese? Herzberger himself acknowledged that his account would only work if a much cruder version of nominalism worked, namely, a so-called happy nominalism according to which universals are just *flatus vocis*, so that "Socrates is ill" is true just in case "is ill" is predicable of Socrates. If that is so, then the actual avoidance of commitment to universals would, in the end, not be due to apoha, but rather to more usual crude/happy nominalist strategems. What is needed, then, if this top-down Apohavāda is to be more than a (slightly) nonobvious *flatus vocis* nominalism is to find some ingenious features of double negation that themselves genuinely do the work a universal is supposed to do in grouping entities, but without the ontological baggage that nominalists abhor.

It seems that Dignāga, in the fifth chapter of his *Pramāṇasamuccaya*, did indeed think that apoha provided that type of stand-in for a universal.[13] But this is a hard route to go and it is far from clear that it will, in the end, deliver what is sought after, especially given Kumārila-style arguments. True, we might well have more or less *simultaneous* understandings of both X and non-non-X, but difficulties become acute when the apoha nominalist wishes to accord some type of privileged place to understanding the double-negative stand-in, taking it as what words primarily express.[14]

One might well therefore say, "so much the worse for trying to make top-down apoha nominalism work," and explore the merits of a Dharmakīrtian theory, or elements of such a theory, as the best prospect for nominalism. I think that, broadly speaking, this is indeed the most philosophically promising way to go. However, there are major consequences in transforming Apohavāda into a bottom-up theory in this Dharmakīrtian way, for it will no longer be a theory that ensures nominalism via double-negative stand-ins. The point is this: much or all of the nominalist mileage that Dignāga supposedly gained by resorting to double negation is now going to be gained in a quite different manner, that is, via a naturalistic explanation invoking causal chains. After all, if Dharmakīrti succeeds in linking up scheme and world via his naturalistic account and avoids commitment to real universals in such an account, why should he bother with the Dignāgan approach concerning double negation? He would have his nominalism not because of ingenious features of a double-negative stand-in, but because he would supposedly have given an adequate naturalistic account of the linkup between scheme and world, an account in which real universals played no role. There is, thus, whether acknowledged or not, a significant rupture with pre-Dharmakīrtian positions.

Now, of course, Dharmakīrti and later writers, like Śākyabuddhi et al., would not acknowledge that rupture and indeed made recurrent attempts to find a place for *both* double negation and causal chains in their nominalistic account of how scheme and world link. Thus, the logico-metaphysical aspects of double negation come in when Dharmakīrti's commentators wish to explain how a number of particular mental images are all images of the same color, blue, or how we can say that the resultant judgments, "this is blue," are all the same judgments of something being blue, for the problem is that if there really is something like sameness we would again seem to arrive ineluctably at real universals. Sameness of apoha/*vyāvṛtti* (exclusion) will thus be invoked to render talk of sameness innocuous.

There seems to be an air of déjà vu about this. Why would double negation fare better here to guarantee that one and the same *X*-ness could apply to many images? If it worked here, then why not in a top-down theory without any causal chain at all? Fortunately, apoha qua ingenious double negation is only at most a limited part of Dharmakīrti's account of how scheme and world link and is not, I would maintain, the main theme at all. Indeed, from Dharmakīrti and his commentators on, apoha theory expands its concerns, all the while taking on considerable hybridness due to holdovers from previous authors. This is, alas, what makes later apoha theories often impossible to summarize in an easily digestible fashion.

In any case the main theme in Dharmakīrti's establishment of nominalism, I would maintain, is a causal account culminating in a strategic refusal to justify metaphysically samenesses that we do in fact recognize. And arguably this *is* a promising tack. After all, there are attractive nominalisms that, in effect, take samenesses or resemblances as *primitive*, needing no further explanatory postulates, be they real universals or their ingenious replacements. As we saw previously, taking sameness as primitive seems to be what Dharmakīrti actually did in his bottom-up theory, although only when we were near the end of the causal chain and dealing with sameness of judgments. The basic standpoint, then, seems to be that at the end of the causal story, the apparent similarity between certain judgments is not an *analysandum* needing a further *analysans*; it is primitive and all the other samenesses (e.g., between perceptual images or between individual things) are explicable in terms of that sameness of judgment.

Of course, an easy and uncharitable reading of Dharmakīrti would be that he just simply failed to provide an account for the linkup between scheme and world because (after numerous frustrating arguments in dense Sanskrit) he had no adequate logico-metaphysical analysis that would certify or justify why things should be grouped together as we think they should. At a crucial point in the theory, so the argument might go, Dharmakīrti just closed his eyes on the problem and took recourse in a type of "ostrich nominalism" (the phrase is that of Armstrong 1980) that refused to face up to explanatory duties. I think, however, that this would be wrong: I am not convinced that *this* refusal to enter the metaphysical fray was a bad thing.[15] I think we can get an inkling of why it was even a good thing by making a distinction—as David Lewis has done in answer to Armstrong—between giving an *account* of sameness and giving an *analysis* of it. In the latter case, we would demand that sameness be grounded in something else, an *X* that is more fundamental. A satisfactory account of sameness, however, need not be that type of analysis. As Lewis (1983) points out, sameness can well be accounted for by giving it a determinate place in a well-developed theory, but as itself a primitive notion that is not analyzable into further facts. Even realist theories have to at some point appeal to primitive notions (like instantiation, participation, etc.). So why shouldn't nominalism?

In short, a fact like sameness of judgment might well figure as primitive in a Dharmakīrti-style causal theory of how scheme and world are linked without the whole theory being charged with disingenuous shirking. If that is right, then arguably the Dharmakīrtian approach may well go in the direction of what Lewis maintains is an "adequate nominalism," providing an alternative to the usual game whereby a realist uses "one over many" conundrums to force supposedly responsible thinkers to either accept real universals or come up with some kind of stand-in that would satisfy metaphysical scruples. The interesting feature of this version of bottom-up Apohavāda, if the theory is carried out consistently, would be Dharmakīrti's enlightened refusal to play a metaphysician's game that was best put aside.

Notes

1. Such scheme-content separations, with a few variations here and there, are actually extraordinarily widespread in numerous domains of intellectual reflection, East and West. The now classic critique of various attempts to make scheme-content separations in science, anthropology, linguistics, and philosophy of language is Donald Davidson's essay "On the Very Idea of a Conceptual Scheme," that is, Davidson 1984.
2. The point is developed in Tillemans forthcoming.
3. The history of how Apohavāda evolved on this matter is complicated. The broad outline is as follows. While Dignāga accepts that Apohavāda accounts for both the universal and particular being expressed, he argues at length in PS 5 against the Nyāya "thesis of words expressing the particulars possessing the universal" (*tadvatpakṣa*). Kumārila, in *Ślokavārttika* (Apohavāda 120, ed. Shastri), then objects that Dignāga's view that words express a (universal qua) exclusion and particulars will itself be a form of *tadvatpakṣa* and that the problems Dignāga invokes will also apply to Apohavāda. Dharmakīrti in PV I.64 and PVSV rehabilitates *tadvatpakṣa* and makes it compatible with Apohavāda by stressing that the apoha universal is fictional and that there is no genuine separation to be made between it and the particulars. For the historical and textual details, see Tillemans forthcoming.
4. Cf. the succinct account of M. Hattori's ed. of chapter 5 of PS, that is, Hattori 1982, 103. See also PS V.12: "bahudhāpy abhidheyasya na śabdāt sarvathā gatiḥ / svasaṃbandhānurūpyāt tu vyavacchedārthakāry asau" (Though that which is expressed has many facets, one does not understand them all from a word. The [word], however, isolates objects according to its connection). Cf. PSV V cited in PVSV 62–63 (ed. Gnoli): "śabdo 'rthāntaranivṛttiviśiṣṭān eva bhāvān āha" (A word talks about entities only as they are qualified by the negation of other things). On this passage and similar passages from Dignāga's *Sāmānyaparīkṣā*, see Pind 1999.
5. PV III, k. 53 (ed. Tosaki): "bhāvadharmatvahāniś ced bhāvagrahaṇapūrvakam / tajjñānam ity adoṣo 'yam."
6. PVP, P. 167b8–168a1: "gzugs la sogs pa mthong bas bsgos pa'i bag chags la brten nas rnam par rtog pa skye ba na / rang nyid kyi gzung ba'i rnam pa la gzugs la sogs pa'i rnam pa nyid du zhen pas 'jug pa de ltar na gzugs la sogs pa mthong ba'i stobs kyis skye ba'i phyir dang / der zhen pa'i phyir dngos po'i chos yin no zhes tha snyad du byas pa yin pa yin no."

7. Elsewhere I have dubbed this the "theory of unconscious error." For discussion and textual sources, see Tillemans 1999, 209–211. Tibetan Buddhist thinkers, such as Sakya Pandita (sa skya paṇḍita), elaborate upon the difference between the unreflective, practical perspective (*'jug pa'i tshe*) and the theoretical perspective (*'chad pa'i tshe*), maintaining that the latter cannot be confused with, nor take the place of, the former. See Tillemans 1999, 235–236n19. See also the article by Dreyfus (in this volume).
8. "Metaphysical superstition" is a phrase coined by John Haldane and Crispin Wright in discussing error theories. See Haldane and Wright 1993, 9. On error theories, see Hale 1999, 288–291.
9. The post-Dharmakīrti school that placed the *least* emphasis on causality in its version of apoha is the Geluk (*dge lugs*). See Tillemans 1999, 211–212. My thanks to Mark Siderits for helping me to better perceive the dissimilarities between Dharmakīrtian *Apohavāda* and typical error theories.
10. There is often an attempt to invoke "beginningless tendencies to do something," but this is again another natural fact, albeit a strange one, a little like Plato's anamnesis in the *Menon*.
11. Cf. Dharmakīrti's PVSV (ed. Gnoli) to PV I.109: "naiṣa doṣaḥ / yasmāt / ekapratyavamarśasya hetutvād dhīr abhedinī / ekadhīhetubhāvena vyaktīnām apy abhinnatā" (This fault [i.e., that perceptual cognitions of individuals would also all be radically distinct] does not occur. This is because the [perceptual] cognitions are the cause of the same judgment [about them] and thus the cognitions are not different; due to them being causes of the same [perceptual] cognition, the individual things too are not different). The translation of *ekapratyavamarśa* poses some serious problems. Most authors in the present volume (with the exception of Hattori and Tillemans, as well as Dunne 2004, 121, 344) have rendered it along the lines of "judgment [of individuals] as being the same," "judgment of sameness," etc., rather than simply "same judgment." Taking *ekapratyavamarśa* as "judgment [of individuals] as being the same" is admittedly in keeping with the Tibetan rendering *gcig tu rtogs pa*. It also seems to reflect the understanding of Karṇakagomin, PVSVṬ p. 227.13: *ekapratyavamarśasyeti svaviṣayasyaikākārapratyayasya*. If one takes the compound in this way, however, k. 109's iterated reasoning that causes are the same because effects are the same becomes less clear. And while it is no doubt true to Dharmakīrti's philosophy that judgments are of sameness/identity amongst individual things, the interpretation of the *eka* compound as "same judgment" remains much simpler philologically.
12. For the details in Dharmakīrti, Śāntarakṣita, et al., see Dunne 2004, 121–126.
13. See PSV to PS V.36d, where Dignāga says that the exclusion of other is what has the features usually attributed to real universals, namely, unity, permanence, and application to each individual (*ekatvanityatvapratyekaparisamāpti*).
14. On difficulties in assigning such a priority to double negation while preserving a principle of compositionality of meanings, see the paper by Bob Hale in the present volume. Revealing too are the later Apohavādins' attempts to reply to the charges of "interdependence" (*anyonyasaṃśrayatva*) levelled by Kumārila and others, for these attempts work only if the privileged place of double negatives is sacrificed. In brief, Kumārila in *Ślokavārttika* V.83ff. had invoked a version

of compositionality of meanings to argue that understanding non-non-X depended upon understanding non-X and that this latter understanding in turn would be dependent upon understanding X; if this X is to be understood as a non-non-X, we would arrive at a vicious circularity where nothing could be understood. Dharmakīrti and later Apohavādins, like Śāntarakṣita, tried to break out of this interdependence by saying that what X-ness is can be independently understood, just as non-X-ness can be independently understood, because we understand which things produce the judgments "this is an X" and which things do not. See, for example, *Tattvasaṃgraha* 1063: "gāvo 'gāvaś ca saṃsiddhā bhinnapratyavamarśataḥ" (cow and non-cow are well established because there are different judgments). In short, the causal account of how individual animals can be thought to have the same cowness is again invoked, this time to find a way out of the otherwise vicious circularity. However, if that and the tu quoque strategy used in PV and PVSV (see Pascale Hugon's paper in this volume on Dharmakīrti's way out of circularity) were to work as an answer to Kumārila et al., it would seem that there could be nothing fundamental about the apoha, non-non-cow, so that it would be what we must understand above all or first of all in using the word "cow" or in thinking about cows.

15. My reading of Dharmakīrti here is different from the Buddhist nominalism of Mark Siderits, who holds that Apohavādins do take very seriously the usual "one over many" arguments for universals and do enter the fray by proposing a nominalistically acceptable stand-in. See Siderits 2006. Whereas top-down Apohavāda does indeed take the "one over many" argument seriously and proposes a double-negative stand-in, it may well be the merit of Dharmakīrti that his approach is not primarily top-down but causal. At the end of the causal chain, it seems to unabashedly sidestep the metaphysical issues. Cf. Dunne (2004, 126): "On the other hand, one might suppose that we are engaged in a frustrating and fruitless enterprise when we yearn to specify in precise terms the metaphysical warrant for our use of the term 'red.' In that case, Dharmakīrti's answer is quite satisfactory, or perhaps even liberating."

2

Dignāga's Apoha Theory

ITS PRESUPPOSITIONS AND MAIN THEORETICAL IMPLICATIONS

• *Ole Pind* •

There is hardly a single aspect of Dignāga's philosophy that has generated as lively a controversy on the Indian philosophical scene as his apoha theory. Although it is possible to form an idea of the nature of this debate through the writings of Uddyotakara, Kumārila, and Mallavādin, who each wrote detailed refutations of Dignāga's views, most of the arguments against the apoha theory remain fairly obscure as long as they are not studied with reference to their proper philosophical context: Dignāga's own writings. Unfortunately, most of his works on epistemology and logic are no longer extant in their original Sanskrit versions, and the few that have survived can only be studied in their Chinese or Tibetan translations and a handful of Sanskrit fragments found scattered in the relevant literature. The loss of the greater part of Dignāga's works may seem paradoxical in view of his seminal influence upon the development of Buddhist epistemology (*pramāṇavāda*). As it is, the only extant Dignāgan work that makes it possible to form a comprehensive view of his epistemology and logic is the *Pramāṇasamuccaya* (PS), of which the fifth chapter is specifically devoted to an exposition of the apoha theory.[1]

As the title indicates, Dignāga composed PS as a compendium of his works on epistemology and logic. The main idea was evidently to provide scholars and students with a summary of his theory of the means of knowledge (*pramāṇavāda*), assuming that if needed they would refer to the more detailed expositions of his other works. The PS is thus marked by extreme economy of exposition and tantalizing ellipsis. Although it has been assumed that it records the final stage of development of Dignāga's thought, we cannot a priori exclude the possibility that he composed other works after PS. Whatever may have been the case, the fact remains that the major part of his works is irretrievably lost and that the "Examination of Apoha" (*apohaparīkṣā*) chapter of PS remains the only extant exposition of his philosophy of language.[2] Before attempting a discussion of the basic theoretical presuppositions of this chapter, it would be interesting to address the question of its sources.

Judging from the numerous parallels between PS and the *Nyāyamukha* (NM), Dignāga appears to have written PS in the form of a patchwork of more or less edited text from works he had written earlier. In the introduction to PS, he mentions NM as one of the sources he exploited, and in the concluding chapter he refers readers who want a more detailed criticism of the doctrines advocated by the philosophical systems of Nyāya, Vaiśeṣika, and Sāṃkhyā to three "Examinations" (*parīkṣā*): (1) *Nyāyaparīkṣā*; (2) *Vaiśeṣikaparīkṣā*; and (3) *Sāṃkhyaparīkṣā* (the last of which is also mentioned in NM). It is evident, however, that Dignāga did not exploit only these works. The "Examination of Apoha" chapter is probably largely based upon the *Sāmānyaparīkṣā* (SP), fragments of which are quoted by Siṃhasūri in his commentary, *Nayacakravṛtti* (NCV), on Mallavādin's *Nayacakra* (NC) (v. NCV p. 628, 7–8), thus indicating that Mallavādin, at least to some extent, based his criticism of Dignāga's philosophy of language upon the *Sāmānyaparīkṣā* (SP), as did probably Uddyotakara. It thus seems reasonable to assume that the scope of the problems dealt with in the SP is to a large extent identical with that treated in the "Examination of Apoha." It is possible to form an idea of some of the questions that Dignāga addressed in SP from a reference in Dharmakīrti's *Pramāṇavārttikasvavṛtti* where he deals with the question of whether a term denoting a particular feature (*viśeṣaśabda*) such as "cow" (*go*) at the same time applies to general features like existence (*sattā*), substanceness (*dravyatva*), etc., that are concomitant features of the entity cow. Dharmakīrti answers that this assumption has already been rejected by the Ācārya (*nirloṭhitaṃ caitad ācāryeṇa*; PVSV, ed. Gnoli, 89, 6). According to Karṇakagomin, he is referring to *Sāmānyaparīkṣā* and other Dignāgan works.[3] As one would expect, Dignāga deals with the same problem in the "Examination of Apoha." The idea is that a term denoting a particular feature only applies to this particular feature. However, since a word like "cow" denotes an object that is defined by a hierarchy of concomitant general features, the latter are by implication indicated by the former. This analysis is based upon the principle that it is possible to deduce a set of general features from a term denoting a particular feature if they have a well-defined place, according to their extension, in a conceptual hierarchy of terms. Questions relating to taxonomy were of great interest to Dignāga; in fact, a substantial part of PS V is devoted to the analysis of the possible relations between terms denoting the general features of a thing and their subextensions. Another work dealing with the apoha theory is *Dvādaśaśatikā*, which appears to be quoted by Dharmakīrti in the *Pramāṇavārttikasvavṛtti*.[4] It is clear from the few quotations ascribed to Dignāga's *Hetumukha* that he also addressed the question of *anyāpoha* in this work. Other Dignāgan works dealing with *apohavāda* are not known.

According to Dignāga's general formulation of the apoha theory at the very beginning of the "Examination of Apoha," any given word (*śabda*) expresses its meaning (*svārtha*) through exclusion of the meanings of other words (*anyāpoha*). In this regard the sign function of the word is said to be analogous to the function of the inferential indicator. Elsewhere Dignāga claims that a word expresses its meaning through exclusion of other words (*śabdāntarāpoha*), thus emphasizing the functional symmetry between the word (*śabda*) itself—formally a configuration of phonemes—and its corresponding reference (*artha*).[5] Dignāga's apoha theory thus stands out as a unified semantic theory dealing with the word not only in terms of its content (*śabdārtha*), but also in terms of its being an expression (*śabda*) invested with meaning (*vācaka*). Before addressing the problem of the theoretical implications of the apoha theory, it would seem necessary to briefly review a number of features that are fundamental to Dignāga's epistemology.

Dignāga's thinking evolves, as is well known, from a fundamental dichotomy: the opposition between the realm of particulars (*svalakṣaṇa*) and the realm of universals (*sāmānyalakṣaṇa*). This dichotomy constitutes the basic theoretical presupposition of his epistemology, thus defining its nature and scope. The realm of particulars consists of any given object as it is reflected in sensation (*pratyakṣa*). Sensation, as such, is restricted to individual occurrences of any given entity (*svalakṣaṇa*), which, by definition, is beyond linguistic representation and thus inexpressible.[6] In contrast to the realm of particulars, the realm of universals is exclusively defined in terms of abstract types. It consists of those generalized objects (*sāmānya*) that are indispensable for making correct inferences (*anumāna*) or—structurally amounting to the same thing—for obtaining knowledge through verbal communication (*śābda*). The sign, whether it is the inferential indicator (*liṅga, hetu*) or the word (*śabda*), does not primarily concern that particular indicator and indicated or that particular word and signified object, but the invariable relationship (*avinābhāva, sahabhāva, sambandha*) that holds between any occurrence of, for example, smoke and fire, or of substance (*dravya*) and existence (*sattā*), or between any occurrence of, for example, the word "cow" (*gośabda*) and the signified object cow (*go*). Thus, the indicator or the word is the type and not the token or occurrence. Things are only definable in relation to their type. The bare individuals, that is, particulars (*svalakṣaṇa*), remain outside the reach of signs.[7] This means that the word or the inferential indicator cannot convey a concept of the individuals in a form that accounts for their individuality, but it can do so in a general form, that is, through the types that are instantiated through individual occurrences, for example, of smoke and fire or of the word "cow" and the denotation cow. Although types are recognized through their realizations in concrete instances, they are not definable in terms of their realizations: they can only be defined in terms of what they are not, that is, through negation (*nivṛtti, pratiṣedha*) or exclusion (*apoha, vyavaccheda, vyāvṛtti*) of their complements (*anya*). Thus, the sign function of the word or the logical indicator is constituted by a relation between two generalized types—in the case of the word between the signified object's generalized type (*arthasāmānya*) and the word's generalized type (*śabdasāmānya*)—the natures of which are established through exclusion of the other. Dignāga appears to regard *arthasāmānya* as a cognitive image (*ākāra*) having the characteristic of an abstract type, that is, a universal (*sāmānyalakṣaṇa*), being located in the mind. It contrasts with the word's individual reference (*arthaviśeṣa*), which belongs to the domain of individuals and therefore is, by definition, inexpressible.[8]

Although Dignāga never touches on the question of the properties of the word type (*śabdasāmānya*), we must assume that it is characterized by the same abstract cognitive features as the object type (*arthasāmānya*). The word thus unites two abstract images, that is, an acoustic image and a representational image, which together constitute the sign function, whereas the corresponding relation between individual signified objects (*arthaviśeṣa*) and individual words (*śabdaviśeṣa*) belongs to the domain of individuals and thus cannot constitute a sign function. This is the idea underlying the following verse, which Indian writers on Dignāga's apoha doctrine often quote when discussing his view on words (*śabda*). It stems from one of his lost works, presumably the *Sāmānyaparīkṣā* (SP):

> It is not claimed that there is a signifier-signified relationship between an individual signified object and an individual word (*arthaśabdaviśeṣa*) because the [individual signified object and individual word] have not previously been observed [together]; their common feature [i.e., the *arthasāmānya* and the *śabdasāmānya*], however, can be taught.[9]

Although it is not absolutely clear to what extent Dignāga's formulation of the apoha theory is indebted to contemporary schools of philosophy, this much is certain: that it represents Dignāga's solution to the epistemological problems raised by his rejection of the idea that universals (*jāti* or *sāmānya*) are real entities.[10] They were conceived by the Nyāya-Vaiśeṣika school as ubiquitous entities inherent in substances (*dravya*), thereby qualifying them (*viśiṣta*) as belonging to a certain class of things having certain distinctive features. Indeed, the scope of the apoha theory only becomes fully understandable when we realize that Dignāga used exclusions of others (*anyāpoha*) as a substitute for universals, in contexts where the Nyāya-Vaiśeṣika school of philosophy would formulate its own theories with reference to real universals. This hypothesis is confirmed not only by the "Examination of Apoha" chapter itself but also by the writings of Dignāga's main critics. In fact, most of the fifth chapter of PS is concerned with analyzing the theoretical problems that follow from the assumption that the ground of application (*pravṛttinimitta*) of any given word is a universal. Although Dignāga and subsequent Buddhist philosophers could easily show that the assumption that universals are real entities has absurd consequences, it is nonetheless clear that the rejection of universals must have caused a serious epistemological problem, which they were forced to address. However, the moot point is what motivated Dignāga to substitute exclusions of others (*anyāpoha*) for universals.

Commenting upon the introductory verse of the "Examination of Apoha," Dignāga writes,

> The word (*śabda*)—which is connected, by virtue of being invariably concomitant (*avinābhāvitva-sambandha*), with some attribute (*aṅga*) of the object (*viṣaya*) to which it is applied (*prayujyate*)—indicates (*dyotayati*) this

> [attribute] by excluding other signified objects (*arthāntaravyavaccheda*) just as [the inferential indicator] "the quality of being produced" [indicates its proper signified object through exclusion of other signified objects], etc.

This paragraph introduces a number of theoretically important concepts, of which the concept of the word's connection with its signified object in terms of their being invariably concomitant is crucial because a correct understanding of its implications throws light on one of the most important aspects of the apoha theory: the question of how to justify the existence of a universally valid connection between any given word and its reference or any given inferential indicator and the indicated property. As appears from the introductory statement and numerous parallel instances in PS, Dignāga claims that the function of the word is identical with that of the inferential indicator, in the sense that knowledge deriving from verbal communication (*śābda*) is inferential like knowledge stemming from an inference (*anumāna*). The condition of its being correct knowledge, however, is that there be an invariable connection between the sign—the word or the inferential indicator—and the signified. It is among other things this question to which the apoha theory, according to Dignāga, is a solution. If we understand his solution to this problem, it becomes easier to understand other features of the theory.

How is such an invariable connection established? There is good reason to believe that the tradition that Dignāga opposed referred to universals as a means of establishing such connections. It appears indirectly from a revealing passage in the *vṛtti* ad PS II 16, in which Dignāga shows the absurd consequences of the assumption that universals are real entities, that certain philosophers attempted to solve the problem of how to justify the existence of universally valid connections between properties (e.g., between smokeness and fireness) by claiming that knowing a universal to be resident in a single substratum is equivalent to knowing it to be resident in all.[11] This claim is only understandable in a philosophical context in which it was assumed that universals always instantiate in the same way. Hence, they could serve as a means of establishing universally valid connections of the kind that were required by the logical theory of the period. However, if one rejects the idea of the universal as untenable, one is left with the problem of accounting for the possibility of universally valid connections. Dignāga evidently solved this fundamental epistemological problem by relying upon exclusion of others (*anyāpoha*). In the context of Dignāga's reductio ad absurdum of the Nyāya-Vaiśeṣika and Sāṃkhyā view of what constitutes a universal, it is interesting to notice that he rejects, in the immediately preceding paragraph, the possibility of affirmation (*vidhi*)—as opposed to negation (*pratiṣedha*) or exclusion (*apoha*)—which would seem to indicate that the question of affirmation traditionally was linked up with the assumption of real universals.[12] The reason why affirmation is impossible is, as Dignāga explains in the commentary ad PS II.15,[13] that individual occurrences are always context bound and therefore cannot assume the role of being a type. If one were to establish an invariable connection between, for example, smoke and fire, on the basis of their individual occurrences, it would never be universally valid, since its form would be restricted to the perception of the properties of that particular smoke and that particular fire, in the same way that sensation (*pratyakṣa*) by necessity is restricted to individual objects. Therefore, he concludes that such relations cannot be formulated in an affirmative form that is universally valid (*vidhi*)—which by implication involves the joint presence (*anvaya*), that is, concordance, of indicator and indicated and therefore is restricted to individual occurrences of things—but they can be formulated in terms of exclusion of other, which basically generalizes joint absence, or difference (*vyatireka*). Thus, concordance and difference do not have the same force, as noted by a number of Dignāga's critics: difference is primary (*pradhāna*).

In PS V.34 (q.v.)[14] he addresses in greater detail the epistemological question of what constitutes exclusion of other:

> Since it is not observed [to apply] to the signified objects of other words, and since, moreover (*api*), it is observed [to apply] to a member (*aṃśa*) of [the class of] its own signified objects, the word's connection [with its signified object] is easy to make (*sambandhasaukarya*) and there is no uncertainty (*vyabhicāritā*). (34)

For (*hi*) *anvyaya* ("concordance," "joint presence") and *vyatireka* ("difference," "joint absence") are what enables a word to denote its signified objects. And these two are: its occurrence (*vṛtti*) in the homologous cases (*tulya*) and its nonoccurrence (*avṛtti*) in the heterologous cases (*atulya*). Now, its occurrence in the homologous cases certainly cannot (*nāvaśyam*) be stated (*ākhyeyā*) for all (*sarvatra*) [the homologous cases], but [it can be stated] for some (*kvacid*), because, as the [number of] signi-

> fied objects (*artha*) is infinite (*ānantya*), such a statement is impossible (*ākhyānāsambhava*). It is possible, however, to indicate its nonoccurrence in the heterologous, even though they are infinite, merely through its not being observed [(*adarśanamātra*) in the heterologous]. And precisely therefore (*ata eva*) it is explained that since it is not observed [to occur] in any other cases (*anyatra*) but its proper relata (*svasambandhin*), the fact that [a word] denotes its own signified object is an inference based on its exclusion of those [other cases] (*tadvyavacchedānumāna*). Indeed, if the inference were by means of concordance (*anvayadvāreṇa*), the word "tree" (*vṛkṣaśabda*) would not lead to any doubts about whether the same (*ekasmin*) entity (*vastuni*) is a "Śiṃśapā" [tree] or the like. Just like that doubt (*saṃśayavat*), there would also be doubts about whether it has *earthenness* (*pārthivatva*) and *substanceness* (*dravyatva*), etc. However, since the word "tree" is not observed [to apply] to things that are nonearthen (*apārthiva*), etc., the inference is only through difference (*vyatireka*).

This passage contains Dignāga's answer to the problem of how to justify a valid connection between types. The idea is that to establish a connection between the word and its denotation, one has to proceed by way of induction, which in the Indian philosophical context means through the observation of concordance (*anvaya*) and difference (*vyatireka*) of the two objects—the word and its signified object or the inferential indicator and the indicated—through which the types are realized. This procedure entails a division of things into two sets: a set of similar things (*tulya*) and a set of dissimilar things (*atulya*). Thus, for instance, the word "tree" is only observed to apply to any member of the set of trees, whereas it never applies to things that are members of the dissimilar set, that is, nontrees. Complete induction through *anvaya*, however, is ruled out a priori because it is not possible to observe the connection, in time and space, between all individual occurrences of, for example, the word "tree" and individual trees because they are infinite. Dignāga therefore suggests that one can establish a connection with reference to the mere fact that the word "tree" is not observed to apply to what is not a tree. The mere fact that the word is not observed (*adarśanamātra*) to apply to things that are dissimilar to the things through which its meaning is realized can be generalized so as to hold for everything dissimilar to the object to which the word "tree" is applied. The meaning of a term thus becomes equivalent to an inference based upon the exclusion from its scope of what it does not denote (*vyavacchedānumāna*). Dignāga illustrates his point by recalling the fact that if the meaning of the word "tree" were established through concordance (*anvaya*) there would be no doubt about the mental image (*ākāra*) it would evoke in a given case.

There is doubt, however, because the usage of the word "tree" does not evoke an image of a particular kind of tree, it only conveys a general notion of *treeness* that applies to all kinds of trees. However, since the word "tree" applies to an entity that is defined by concomitant features like *existence* (*sattā*), *substanceness* (*dravyatva*), *earthenness* (*pārthivatva*), etc.—its so-called relata (*sambandhin*) or adjuncts (*anubandhin*)—it is clear that they are indicated as well, provided that they have a well-defined place in the hierarchy of terms defining the entity in question. Thus, for instance, *earthenness* (*pārthivatva*), which is a subextension of *substanceness* (*dravyatva*), indicates the latter, which in turn indicates *existence* (*sattā*) because whatever is earthen (*pārthiva*) is also a substance (*dravya*), and whatever is a substance is also existent (*sat*). The underlying idea is that if the terms in a systematic hierarchy giving the essential attributes of a certain entity are all coreferential, they are logically related according to their extension, so that it is possible to infer other attributes from any given term denoting one of their subextensions. It is this fact to which Dignāga refers in the introductory paragraph of the "Examination of Apoha" where he introduces the term *aṅga*, the Dignāgan term for any given attribute. It is obvious that Dharmakīrti's concept of reasons that are essential properties (*svabhāvahetu*) is indebted to Dignāga on this point. In fact, the whole question of *svabhāvahetu* centers on the coreferentiality (*sāmānādhikaraṇya*) of a systematic hierarchy of terms defining a particular entity, in other words, their nondifference (*abheda*), that is, their syntactical agreement.[15]

Dignāga's view that mere nonobservation (*adarśanamātra*) is constitutive of the invariable connection between the indicator and the indicated or of the connection between the word and its reference raises a number of complicated epistemological issues. These can be followed from Pāṇini's definition of elision (*adarśanaḥ lopaḥ*; Astdh I.1.60) through Dignāga's alleged pupil Īśvarasena's theory of mere absence of perception (*upalambhābhāvamātra*)[16] to Dharmakīrti's theory of nonperception (*anupalabdhi*). Dharmakīrti, however, breaks with the Dignāgan tradition, because in contrast to Dignāga, who for theoretical reasons takes difference (*vyatireka*, i.e., "joint absence") to be the principal factor in establishing the universal concomitance, Dharmakīrti clearly regards concordance (*anvaya*, i.e., "joint presence") and difference (*vyatireka*) as having the same force.[17]

It is obvious that Dignāga's treatment of the problem of the feasibility of establishing an invariable connection between the word and its signified object is indebted to his logical theory. Indeed, the way in which he deals with this vital problem is analogous to his treatment of the second and third members of the logical canon: the so-called triply characterized logical reason (*trilakṣaṇahetu*). In the context of Dignāga's logical theory, the question of the invariable connection naturally belongs in the context of the exemplification (*dṛṣṭānta*), the so-called elucidation (*pradarśana*), which was used for stating the invariable connection. It is therefore not at all surprising to find that Dignāga claims in the fourth chapter of PS, which is devoted to a discussion of the nature of the exemplification, that pervasion (*vyāpti*) can only be stated in terms of exclusion of other (Kitagawa 1973, 518). In fact, the Dignāgan use of the delimitative particle *eva* as a means of clarifying the character of the formulation of the triply characterized reason is clearly a corollary of exclusion. This is confirmed by the concise but highly interesting *vṛtti* on 38c–d of the "Examination of Apoha," where Dignāga discusses the relationship between concordance and difference:

> Again, if the word were to denote its signified object without relying upon (*anapekṣya*) a negation of other signified objects (*arthāntaranivṛtti*), in that case
>
> > it would be established exclusively (*eva*) through *anvaya* ("concordance," "joint presence") [with its signified object]. (38c)
>
> Rather when a word denotes its signified object, it would not be through *anvaya* as well as *vyatireka* ("difference," "joint absence"). Now, this is claimed to be the case. However, since the denotation (*abhidhāna*) works by restricting either [a word] or both [words in a proposition] (*anyatarobhayāvadhāraṇa*), denotation of a signified object (*arthābhidhāna*) is also by means of *vyatireka*, as, for instance, [in Pāṇini's Astdh I.4.49]: "That which the agent (*kartṛ*) wants to obtain most of all (*īpsitatama*) is termed *karman* (i.e., 'direct object')."
>
> Suppose, however, that the word's signified object is merely exclusion of other [signified objects] (*anyāpohamātra*), it would then denote its signified object (*arthābhidhāna*) exclusively (*eva*) through *vyatireka*.
>
> This would be the case if we did not accept *anvaya*. However, (*tu*),
>
> > I do not claim that [the word's] pervasion [(*vyāpti*) of its signified objects] is with the principal (*mukhyena*) (38d),
>
> entity (*bhāvena*). For (*hi*) it is impossible, as I have already explained [in PS II.16], that there be universals in things, whether they are distinct (*bhinna*)

or not (*abhinna*) from [their substrata]. But let us grant that the signified object is qualified by the exclusion of other objects (*arthāntarapohaviśiṣṭe 'rthe*) without there being any [real] general property, in accordance with [PS V.34a] "since it is not observed amongst the signified objects of other words (*adṛṣter anyaśabdārthe*)." Then the *anvaya* ("joint presence") of the word [and its object] and the *vyatireka* ("joint absence") do not pertain to different objects.

There is thus no doubt that the restrictive/delimitative value ascribed by Dignāga to the particle *eva* is equivalent to the value of apoha. This is also made abundantly clear from Dignāga's treatment of the distribution of the restrictive (*avadhāraṇa*) *eva* elsewhere in PS. Apart from the fact that *eva* and apoha belong in the same context, there is a direct line of development from Dignāga's statement in 38c to Dharmakīrti's expression "a statement results in an exclusion" (*vyavacchedaphalaḥ vākyaḥ*) in *Pramāṇavārttika* IV.192 (= *Pramāṇaviniścaya* II.11) and his description of the use of *eva*. (Indeed, one only has to refer to the examples "Caitro dhanurdharaḥ" and "Pārtho dhanurdharaḥ, nīlaḥ sarojaḥ"[18] used by Dharmakīrti [PV IV.192] to illustrate the use of *eva* in order to understand the scope of Dignāga's remarks and thus to interpret the quotation from Pāṇini, distributing the restriction [*avadhāraṇa*] accordingly.) The succeeding paragraph (38d) would seem to corroborate the hypothesis that originally the problem of pervasion (*vyāpti*) was addressed in the context of the Nyāya-Vaiśeṣika theory of universals. Dignāga, however, denies the possibility of substantial pervasion, that is, pervasion that implies the assumption of the pervaded being real universals qualifying their proper substrata, for in Dignāga's view it is not inherent universals that qualify things, but rather the fact that words define things as excluded from what they are not.

Dignāga's non-Buddhist critics directed a substantial part of their criticism of the apoha theory against the assumption that an absence as such could have a qualifying force, and in this connection they also addressed the question of the value of the excluded. Since this side of the apoha doctrine is among its more controversial aspects, it would seem natural to address these criticisms in this connection. Dignāga does not discuss the role of apoha as a qualifier in PS, but merely restricts himself to ascribing to the apoha, in a well-known passage in PS, the value of the

Nyāya-Vaiśeṣika universal.[19] There is, however, an interesting Sanskrit fragment ascribed to Dignāga in which he deals with this question in some detail, although one cannot say that it throws light on all aspects of this idea. It was known to all post-Dignāga scholars discussing the apoha theory, and there is reason to believe that it stems from the *Sāmānyaparīkṣā* (SD). The largest portion of this text is found in *Tattvārthabhāṣyavyākhyā* V.24. It reads:

> For the word is said to designate [its signified object] while effecting, for the sake of its own signified object (*svārtha*), its exclusion from other signified objects (*arthāntarāpoha*). (The word "for" is used in the sense of "because.") For instance, the word "tree," while effecting, for the sake of its own signified object, the negation of the word "nontree" (*avṛkṣaśabdanivṛtti*), indicates that its own signified object is characterized as tree (*vṛkṣalakṣana*). And thus the word's signified object is a thing qualified by negation (*nivṛttiviśiṣta*), but it is not mere negation (*nivṛttimātra*), for mere negation would be completely indefinable (*alakṣanīya*) because it is a nonentity (*avastutva*), like, for instance, descriptions of such things as horns of an ass, and blunt sharpness.[20]

Textual evidence thus seems to indicate that Dignāga conceived of the opposition between the excluded nontree and the thing being qualified by the negation of nontree, that is, tree, in terms of a privative opposition between tree and nontree. Thus, the word "tree" expresses the presence of a particular distinctive feature, whereas nontree expresses its absence. Paraphrasing Dignāga's statement, we may say that the word functions as a limitation operator in that it delimits its own signified object from other signified objects by establishing a boundary between its own referent, tree, and its nonreferent, nontree. This boundary is the result of a conjunction of the presence and absence of a particular distinctive feature. However, according to Dignāga the negation of nontree is the qualifying property of tree, and this entails a peculiar logico-semiotic aporia which all Dignāga's critics, and first of all Kumārila, did not hesitate to point out: if any given word and its signified object are defined in terms of a privative opposition in which the presence of a term of the type *A* necessarily implies the absence of a term of the type non-*A* and vice versa, the implication becomes tautological. Hence, we may conclude that within the structure of the privative opposition the distinctive feature coincides with the opposition itself: the term *A* at the same time identifies its signified object as *A* and differentiates or excludes it from non-*A*. There is reason to believe that this is what Dignāga had in mind when talking about negation of other as the qualifier of any given signified object, although the formulation as such is analogous to the idea that Dignāga rejects, namely, that a thing is qualified by the universal inherent in it.

It is obvious that Dignāga did not consider a term of the type non-*A* to be without content: it denotes in a general form the absence of the particular distinctive feature that determines the signified object of the positive term *A*. He addresses the question of the type non-*A* in PS 43b, introducing the crucial notion of the "single property" (*ekadharman*), which Kumārila made subject to a detailed discussion in *Ślokavārttika*. (Apohavāda 61ff.). Dignāga writes:

> Nor is the objection that no cognition can occur justified,
>
> > because [the word] excludes by means of [the single] general feature (*sāmānya*).(43b)
>
> For (*hi*) it does not exclude a different universal (*jāti*) for each individual substance (*pratidravyam*), but rather (*kiṃ tarhi*), [it excludes] with the single property of their general feature (*sāmānya*) due to the intention of expressing the [objects] to be excluded (*vyavacchedyavivakṣayā*). And on this point (*atra*) we have explained that [the signified object] is inferred merely through [its] not being observed in the heterologous [instances] (*vijātīye adarśanamātreṇānumānam*). Yet, this problem (*doṣa*) [that no cognition can occur] concerns only (*eva*) you; for if [the word] were to apply (*varteta*) by universally pervading (*vyāptyā*) its proper homologous [objects] (*svasajātīya*), the pervaded (*vyāpya*) would be infinite (*ānantya*). Therefore, as in the statement "it is a nonhorse because it is horned" (*viṣāṇitvād anaśva iti*), the inference is an exclusion of this [namely, horse] (*tadvyavacchedānumānam*)[21] because of not observing hornedness in a horse (*aśve viṣāṇitvādarśanena*), but [hornedness] does not exclude the white mares, etc. (*karkādīn*), each separately (*pratyekam*), nor does it apply to every single cow individually, etc. (*ekaikeṣu gavādiṣu*). Also you maintain a theory of cognition based upon concordance and difference (*vyāvṛttyan uvṛttibuddhimatam*). And the principle (*nyāya*) is the same in this context.

The problem that Dignāga addresses in this text is the objection that each exclusion would seem to imply the exclusion of innumerable entities. Consequently, definite knowledge would seem to be impossible. However, as Dignāga explains, entities are not excluded each individually, but rather they are excluded collectively, according to the general theory of exclusion, as instantiating an absence of the single distinctive feature defining the excluding term. The fact that the excluded term is defined by the absence of a single distinctive feature—the Dignāgan "single property" (*ekadharman*)—does not mean that it is without reference and thus not interpretable: it is merely used to define collectively all those entities in which a particular feature is absent. Thus, for instance, the term "nonhorse" (*anaśva*) of the inference "it is a nonhorse because it is horned" only conveys the idea of a horned animal that is not a horse, without reference to the specific nature of the animal that is denoted by the term "nonhorse."[22] Dignāga's final reference to the fact that the opponent also agrees that cognition proceeds by concordance and difference (*vyāvṛttyanuvṛttibuddhi*) is interesting because it gives us a hint of the ideas that he attempted to amalgamate, which thus become important for the assessment of the historical background against which Dignāga worked out his own apoha doctrine. I shall return to this point.

One of the most remarkable features of the apoha theory is the fact that Dignāga, according to a prima facie reading of his description of the inferential character of apoha, would seem to consider verbal knowledge equivalent to an inference from difference (*vyatireka*). His critics were not slow in pointing out this apparent violation of the canonical rule of the triply characterized reason (*trilakṣanahetu*), which does not admit of this type of inference. Kumārila, for instance, closes his criticism of the apoha theory by criticizing inferences based upon difference. The target of his criticism is probably Uddyotakara and his school, whom Kumārila apparently accuses of not having the right to reject Dignāga's view because they accept inferences through difference (cf. also NC[V] p. 666,12ff.). This controversy thus shows that Dignāga's contemporary critics took his remarks about inference based upon exclusion (*vyavacchedānumāna*) to be equivalent to an inference from difference. It is difficult not to agree with Dignāga's critics, and this apparent theoretical inconsistency perhaps explains why Dharmakīrti seems to consciously avoid the issue: he prefers to reinterpret the Dignāgan doctrine about the inferential nature of verbal knowledge (*śābda*) by taking it to mean that *śābda* indicates the presence of intention (*vivakṣā*) in the speaker, rather than interpreting it along the lines of Dignāga's own theory. In this regard he seems to fall back on Bhartṛhari's view that spoken words manifest the intentions (*vivakṣā*) of the speaker.

The invariable concomitance between word and object naturally presupposes learning the scope of the word in question, in other words, the word-meaning connection, that is, the *vyutpatti*. Dignāga deals with this question in an interesting paragraph ad k. 50b toward the end of the "Examination of Apoha." He writes:

> But how can the knowledge of the signified object (*arthapratipatti*) that someone who has not yet been shown [the word's] connection [with its signified object (*akṛtasambandha*)] gets from a word be an inference (*anumāna*), as, for instance, the one that is expressed [in the proposition]: "this is a breadfruit tree" (*ayaṃ panasaḥ*')?
>
> In this case there is no knowledge of the signified object through the word *panasa*. Why?
>
>> Because the signified object is shown by someone to whom [the connection] is known. (50b)
>
> Since a [word's] signified object is established by an authority (*vṛddha*) to whom the connection [of the word with its signified object] is well known (*prasiddhasambandha*) via the demonstrative pronoun "this" (*ayaṃśabda*) and ostension (*hastasaṃjñā*), there is no knowledge of the signified object through the word *panasa*, but it is rather (*kiṃ tarhi*) the scope of the name (*saṃjñāvyutpatti*) ["panasa" that is taught]. On the other hand, coreference (*sāmānādhikaraṇyam*) between this [that is, the word "panasa"]—whose purpose is [to teach] a name—and the demonstrative pronoun "this" serves to show the connection (*sambandhapradarśanārtham*), the assumption being (*iti kṛtvā*) that the [connection] is what is expressed by (*abhidheya*) both [terms]. And since the word *panasa* does not [yet] have this [namely, the breadfruit tree] as its signified object, its purpose is that of [teaching] a name.

Dignāga thus assumes that ostensive definition is at the basis of the learning of the connection between the word and its signified object. This view, of course, should be interpreted from the perspective of his statement about concordance (*anvaya*) and difference (*vyatireka*), which shows that in Dignāga there is no inductive assumption in the proper sense of that word: the connection between word and its signified object (*artha*) or between the logical indicator and the indicated is taken to be invariable as long as it is possible to claim that the word or the indicator excludes everything in which the particular feature expressed by the excluding term is absent. Thus, the excluded represents nothing but a generalized, hypothetically posited absence of the feature that defines the scope of the excluding term.

In the immediately following paragraph (ad k. 50c), Dignāga discusses the nature of the connection, which he claims is only due to representations, that is, imaginary, depending on the connection made by the mind between the word and its signified object. The connection is not an object of knowledge conveyed by the word in question.

> [Objection:] Then it is precisely the connection that will be the word's object of knowledge (*prameya*).
>
> > The connection is not [the word's object of knowledge] because it is imagined (*vikalpitāt*). (50c)
>
> For (*hi*) the connection is imagined when the signified object and the word *Panasa*, after having been apprehended (*upalabhya*) through separate instruments of knowledge (*pramāṇa*), are connected (*sambaddha*) by the mind (*manasā*) that thinks: "this [word denotes] this [signified object]" (*ayam asyeti*), in the same way as the inference-inferendum connection (*anumānānumeyasambandhavat*). Therefore, verbal knowledge (*śābda*) is not a separate instrument of knowledge.

Dignāga's epistemology, logic, and philosophy of language are no doubt indebted to his contemporaries, although it is far from clear to what extent they influenced him. One sometimes gets the impression that he tried to amalgamate ideas that would seem at least prima facie to be incompatible. The influence of Bhartṛhari on Dignāga is one instance. Thus, for example, it is not entirely clear, in spite of Dignāga's unusually explicit exposition in the "Examination of Apoha," how he would defend adopting Bhartṛhari's view of the sentence as the primary source of verbal knowledge and still remain consistent with the apoha doctrine, which basically is an extension of his logic.

There is one aspect of Dignāga's apoha doctrine that seems to point toward another possible inspiration and probably the most important one. The fact is that Dignāga worked out the apoha theory on the basis of a conceptual tree that ultimately stems from Vaiśeṣika taxonomy. This explains his claim that exclusion is not a universally pervasive feature, but only operates under certain conditions. If we take our point of departure in the conceptual tree that he received from the Vaiśeṣika tradition, the point in question becomes clear. The Vaiśeṣika tree is basically constituted by the supreme universal (*sāmānya*) *existence* (*sattā*) and its subextensions, the particular universals (*sāmānayaviśeṣa*) *substance* (*dravya*), *quality* (*guṇa*), and *action* (*karman*), which each ramifies into innumerable subextensions on various levels. According to Dignāga the particular universals exclude each other, whereas the supreme universal *sattā* only excludes *asattā* (cf. the interesting quotation from *Hetumukha* concerning the use of the term *asat*), but not the particular universals with which there is concordance. The same principle is extended to, and remains in force for, all the different subextensions: they exclude each other provided that they belong to the same level in the hierarchy, but they do not exclude their possible subextensions, just as they are not excluded by the relevant term in the hierarchy whose subextensions they are. To generalize: there is concordance in the tree *vertically*, but exclusion *horizontally*. In short, the principle is a Dignāgan version of the type of tree that is delineated briefly in *Praśaṣṭapādabhāṣya* 7, which describes the relationship between the terms constituting the tree in terms of compliance (*anuvṛtti*) vertically and distinction (*vyāvṛtti*) horizontally. It seems obvious that Dignāga has adopted the same principle of analysis (cf. his reference supra to *anuvṛtti* and *vyāvṛtti*). This general principle, of course, becomes ontologically untenable under certain circumstances where mutually exclusive terms go together in defining a single entity, which thus would seem to be in internal contradiction with itself. Dignāga addresses this problem in a fairly complicated way that can only be described as a politics of terms, individual terms allying themselves with other terms in much the same way as kings ally themselves with other kings according to the rule of the *cakras* laid down in a political treatise like Kauṭilya's *Arthaśāstra*. Ultimately, this part of the apoha theory would only make sense if we assume that it represents a Dignāgan version of problems entailed by Vaiśeṣika taxonomy. If this assumption is true, it also becomes understandable why Dignāga would name the fifth chapter of PS "Examination of Apoha." In the perspective of his other "Examinations" (*parīkṣā*), this is only understandable as an indication that he is subjecting current views on exclusion to a critical analysis while arguing at the same time for the necessity of his own view: the Dignāgan theory of exclusion. The chapter as a whole would seem to corroborate this assumption, and it would thus seem necessary to reconsider the historical background of Dignāga's theory of the means of knowledge.

Notes

1. The bulk of this paper was written in 1991 at a time when the Sanskrit version of Jinendrabuddhi's PSṬ had not yet become accessible. For the Sanskrit text of PSṬ, chapter 1, as well as a description and history of the manuscripts, see Steinkellner et al. 2005. Readers are kindly referred to my forthcoming Sanskrit restoration and annotated English translation of PSV V, including an edition and English translation of substantial parts of PSṬ V.
2. Dignāga himself probably did not call the chapters of his PS "Examinations" (*parīkṣā*), but just numbered them as first, second, third, etc. The addition of "Examination" may have been made by a scribe. For one thing, the term usually suggests a critique or refutation, and clearly Dignāga is not doing that in the apoha chapter of PS. For more details, see the introduction to my forthcoming translation and study of PS V.
3. Cf. PVSVṬ 337, 13–14: "nirloṭhitaṃ caitad ācāryeṇa Diṅnāgena sāmānyaparīkṣādau yathā viśeṣaśabdānāṃ sāmānye vṛttir iti."
4. Cf. PVSV I 62, 26: "arthāntaravyāvṛttyā tasya vastunaḥ kaścid bhāgo gamyate"; cf. Siddhasenagaṇin's *Tattvārthabhāṣyavyākhyā* V 24 (quoted in NCV 548, 24–25): "yathā Dvādaśaśatikāyām—yady apy uktam aprasaktasya kim arthaṃ pratiṣedhaḥ? iti nāivāitat pratiṣedhamātram ucyate, kin tu tasya vastunaḥ kaścid bhāgo 'rthāntaravyāvṛttyā loke gamyate yathā viṣāṇitvād anaśvaḥ iti."
5. For a study of this feature of Dignāga's apoha theory, see Pind 1991.
6. The idea that individuals are inexpressible is also presupposed by the objection quoted in Bhartṛhari's alleged commentary on VP I.69: "pratiniyatasvarūpabhedā vyaktayaḥ. na hy asaṃvedyam avyapadeśyam avidyamānaṃ vā vyaktīnaṃ rūpam." It probably represents a view, that is, the so-called *tadvatpakṣa*, that was current among contemporary grammarians and Nyāya-Vaiśeṣika philosophers. They held that the signified object of a word is the particular thing as endowed with a universal (*jātivān arthaḥ*).
7. As is well known, the realm of particulars is exclusively accessible to sensation (*pratyakṣa*), which by definition is devoid of representation (*kalpanā*). The scope of universals, however, is defined by the sign function, whether it be the linguistic (*śābda*) or the inferential (*liṅga*) sign, and is thus characterized by representation.
8. Cf., for example, PS II.3; cf. Kitagawa 1973, 450; for an English translation, see Hayes 1980, 248–249.
9. "nārthaśabdaviśeṣasya vācyavācakateṣyate / tasya pūrvam adṛṣṭatvāt; sāmānyaṃ tūpadekṣyate." The verse is inter alia quoted in NCV 615, 12–13; cf. TSP ad TS 961

(= ŚV Apohavāda 102). For an analysis of the implications of this verse, see Pind 1991.

10. Dignāga's concise refutation of the assumption that universals are real categories is found in PS II 16; cf. Kitagawa 1973, 464–465; for an English translation, see Hayes 1980, 257–258.
11. Cf. Kitagawa 1973, 464: "gal te rten gcig bzung bas kyang thams cad gzung ba yin na ni, de yang rten bzhin du du mar 'gyur ro" (in Vasudhararakṣita's translation); "ci ste spyi gcig la brten par gzung na yang thams cad gzung ba yin no zhe na ? de la brten bzhin du du mar 'gyur ro" (in Kanakavarmin's translation); for an English translation of this passage, see Hayes 1980, 258.
12. This assumption seems to be corroborated by Śāntarakṣita, who quotes (in TS 1096) the following short phrase from Dignāga's *Hetumukha* (identified by Kamalaśīla ad loc.): "affirmation is impossible"(asambhavo vidheḥ); Śāntarakṣita then explains that affirmation (*vidhi*) is impossible "because universals, etc., are impossible" (sāmānyāder asambhavāt).
13. For an English translation, see Hayes 1980, 257. Note that Hayes interprets Tib. *sgrub/bsgrub* (cf. Kitagawa 1973, 463, 468ff.) as if it were equivalent to Sanskrit *sādhana*. In this context, however, *sgrub/bsgrub* = *vidhi*, which Kitagawa (1973, 114 [line 23]) accordingly translates "kentei teki na shikata."
14. For an English translation, see Hayes 1988 ad loc.
15. Cf. the use of *abheda* in the Sanskrit fragment from *Nyāyamukha* (NM) concerning the definition of *pratyakṣa*: "yaj jñānārtharūpādau viśeṣaṇābhidhāyakābhedopac āreṇāvikalpakaṃ tad akṣam akṣaṃ prati vartate iti pratyakṣam" (quoted in TSP ad TS 1236).
16. See Steinkellner 1966; cf. HB II 154ff.
17. Cf., for example, Dharmakīrti's implicit criticism of Dignāga's reference to *adarśanamātra* as constitutive of apoha in PV III *Pratyakṣapariccheda* 172a–c: "anyatrādṛṣṭyapekṣatvāt kvacit taddṛṣṭyapekṣaṇāt / śrutau sambadhyate 'poho." The criticism is implicit in the clause *anyatrādṛṣṭyapekṣatvāt,* for which Dignāga would have *sarvatra* (i.e., in the *atulya* in toto); see PVBh p. 264, 30ff. ad loc. cit.; note especially the following reference to Dignāga's view on 265, 23: "anye tu punaḥ sarvato vijātīyād vyāvṛttir kvacid vidheye vṛttim apekṣata iti vyatireke tātparyam anvaye tu neti, vyatireka eva prādhāyena pratyāyate"; see also Kumārila's criticism of Dignāga's view in ŚV, *Anumānapariccheda* 131cd–132: "aśeṣāpekṣitvāc ca saukaryāc cāpy adarśanāt / sādhane yady apīṣṭo vyatireko 'numāṃ prati / tāvatā na hy anaṅgatvaṃ yukti śābde vakṣyate." Kumārila's reference to *śābda* (i.e., to the chapter on verbal knowledge) is to Apohavāda 75 (q.v.). It is perhaps not a random mistake that Jñanaśrīmitra quotes in his *Apohaprakaraṇa* a slightly edited version of Dharmakīrti's verse, substituting *sarvatra* for *anyatra* (see op. cit. 207, 10–11).
18. This concerns the different scope of the restrictions that are supposedly present in all statements, whether *eva* is explicitly stated or not. To take the first two: "Caitra is an archer," that is, Caitra is only an archer = Caitra is not a nonarcher and there can be other archers too; "it is Pārtha [alone] who is the archer," that is, no one other than Pārtha is the archer = Pārtha (i.e., Arjuna) is the only excellent archer among the Pāṇḍava brothers. See Kajiyama 1973; Gillon and Hayes 1982.

19. Cf. the following Sanskrit fragment ad loc. cit.: "Now the qualities of a universal are characterized as oneness, permanence, [and] extension to each [particular]; they are present in the [apoha] alone" (PSV ad 36; "jātidharmāś cāikatvanityatv apratyekaparisamāptilakṣaṇā atraiva tiṣṭhanti"). See Kamalaśīla's version of this passage from the *vṛtti* in TSP 389, 9–11.
20. Cf. the Sanskrit fragment quoted in Nayacakra (NC), ed. Muni Jambuvijaya, vol. II 548, 13–16: "tathā cāha dattakabhikṣur eva: arthāntarāpohaṃ hi svārthe kurvatī śrutiḥ "abhidhatte" ity ucyate. hiśabdo yasmādarthe. yathā vṛkṣaśabdo 'vṛkṣaśabdanivṛttiṃ svārthe kurvan svārthaṃ vṛkṣalakṣaṇaṃ pratyāyayatīty ucyate, evaṃ ca nivṛttiviśiṣṭaṃ vastu śabdārthaḥ, na nivṛttimātram, alakṣaṇīyam eva ca syān nivṛttimātram, avastutvāt, kharaviṣāṇakuṇṭhatīkṣṇatādivarṇavat."
21. Cf. Jinendrabuddhi's exegesis of the term *tadvyavacchedanumana* at PSV V.34.
22. It is presumably the same problem Dignāga addresses in the only surviving Sanskrit fragment from the *Dvādaśaśatikā*, see note 4 above.

3

Key Features of Dharmakīrti's Apoha Theory

John D. Dunne

The apoha theory contains a number of occasionally technical and even counterintuitive elements, and the main purpose of this chapter is to present its most fundamental features in a straightforward fashion. At the outset it is critical to note that, while certainly unified in its overall scope, the apoha theory underwent historical development that led to divergent interpretations among its formulators, and any single, unified account of the theory would be problematic. Hence, this chapter will focus on a pivotal historical moment in the theory's development, namely, its articulation by the Buddhist philosopher Dharmakīrti (fl. 625), especially as interpreted by his immediate commentators, Devendrabuddhi (fl. 675) and Śākyabuddhi (fl. 700). To contextualize this particular layer of interpretation, I will begin with a brief historical overview and then present some contextual material under two headings: Dharmakīrti's causal model of cognition along with the minimalism about concepts that such a model encourages and the basics of his ontology. With these matters in place, I will then examine the fundamental points of Dharmakīrti's apoha theory.

HISTORICAL OVERVIEW

The apoha theory finds its first explicit articulation in the work of Dignāga (fl. 425), the first Buddhist philosopher to employ rigorously the style of discourse that we may call *pramāṇavāda*, or "*pramāṇa* theory." This style of discourse, which appears to arise primarily from the early efforts of the

Nyāya school, focuses on what constitutes a *pramāṇa*, that is, the reliable means or (literally) the "instrument" for arriving at a trustworthy or reliable cognition (*pramiti*). Verbal testimony and various forms of inference are considered important forms of *pramāṇa*, and these are understood to operate through the use of conceptuality (*vikalpa*). Thus, a major topic of discussion within *pramāṇa* theory is the issue of conceptual cognition (*savikalpakajñāna*), and since conceptual cognition is thought to necessarily take a universal (variously called *sāmānya*, *jāti*, *ākrti*, etc.) as its object, a discussion of conceptuality requires a theory of universals. Dignāga's apoha theory is an attempt to formulate a theory of universals—and, hence, a theory of conceptual cognition—that takes a nominalistic approach which rejects the realism about universals found in other, non-Buddhist philosophical traditions of classical South Asia.

Dignāga's formulation of the apoha theory was explicitly criticized by the Nyāya philosopher Uddyotakara (fl. 525) and by the Buddhist thinker Bhāvaviveka (fl. 530),[1] who developed a similar theory of his own. Dignāga's thought—including the apoha theory—receives a significant reworking at the hands of Dharmakīrti, and it is his reformulation that forms the basis for all subsequent treatments, whether Buddhist or non-Buddhist. Among Buddhist thinkers, the earliest commentarial layer consists of works by Devendrabuddhi and Śākyabuddhi, and while they propose some innovations, their interpretations do not range far from Dharmakīrti's works. Thinkers such as Śāntarakṣita (d. 787) and Kamalaśīla (fl. 765) incorporate Dharmakīrti's philosophy into their Mādhyamika perspective, but the details of his *pramāṇa* theories are not significantly revised. However, by the time of later commentators such as Jñānaśrīmitra (fl. 1000), Ratnakīrti (fl. 1025), Karṇakagomin (fl. 975), and Mokṣākaragupta (fl. 1100), a general trend toward ever-greater realism about universals becomes evident. In Tibet, realist interpretations gain momentum, and in some cases receive criticism, at the hands of numerous prominent thinkers, some of whom are considered in this volume. Since the presentation given in this chapter focuses on the earliest layer of interpretation, it may appear to conflict with the more realist approaches of some later Buddhist authors, but the general contours and mechanics of the theory will nevertheless remain the same.

COGNITIVE MODEL AND MINIMALISM ABOUT CONCEPTS

Dharmakīrti articulates the apoha theory within his commitment to a causal and descriptive model of embodied cognition and the minimalist approach that this commitment brings to concepts. As a way to frame this aspect of Dharmakīrti's philosophy, we might compare it to the contemporary notions of "naturalized" epistemology. Broadly, the term "naturalized" refers to the project of integrating phenomenological or epistemological theories into an explanatory framework that is in some fashion tightly bound with the natural sciences. Clearly, Dharmakīrti's approach cannot share a concern with the natural sciences of today, but it does share the impetus toward "an empirical psychological study of our cognitive processes," which is part of what it means for epistemology to be "naturalized" in our contemporary context.[2] More specifically, Dharmakīrti's work can be treated as an extension of the *Abhidharma* tradition upon which it is based, and it thus assumes a detailed account of matters including the various components of attention and metacognition, the workings of memory, the types and characteristics of emotional states, the physiological composition of the sense faculties, and so on. While the *Abhidharma* may diverge from our contemporary scientific understanding of such issues,[3] it similarly presents itself as engaged in a careful and allegedly empirical description of human psychology, including cognitive and affective processes. It is crucial to recognize that Dharmakīrti participates in this larger Buddhist project, and that his work is an extension of it.[4]

One expression of this feature of Dharmakīrti's work is that his epistemology, which encompasses his theory of concepts, is "event based" in that it aims to describe the way that particular mental events (*jñāna*) can be reliable in regard to successful human action. These events, moreover, arise in accordance with a causal model of cognition that includes physiological and psychological elements. The model presumes that any causally efficacious thing endures for only an instant; thus, the model involves the causal interaction of momentary entities. To the extent that any causally efficient entities appear to endure over time, they are actually a series of momentary entities that are causally related to each other in such a way that one moment in the sequence acts as the primary cause for the next moment in the sequence. Thus, if one is observing a patch of blue, the matter that constitutes that patch actually endures for only an instant; nevertheless, the patch appears to endure longer because the matter constituting the patch occurs in a sequence of moments of that matter, each instance of which arises from the previous moment of matter and perishes as it produces the next moment. Such is also the case with the matter that constitutes the body, including the sense faculties, and with consciousness itself, with the proviso that in the case of consciousness, the moments that constitute the flow of consciousness are mental—not material—in nature.

On this model, a cognitive event is a moment of consciousness under particular causal conditions, and for the purposes of understanding the theory of concepts involving apoha, perhaps the best example of such an event is the act of recognition (*pratyabhijñāna*) that occurs when a perceived object—for example, a patch of blue color—is conceptually labeled or recognized—that is, as "blue." Such an event presumes three causal streams: (1) the causal stream of the matter constituting the perceived object; (2) the causal stream of matter constituting the sense faculty; and (3) the immaterial causal stream constituting the mind. Each of these streams is reducible to discrete, causally efficient moments that endure only an instant. When the object comes into relation with the sense faculty, if other conditions are in place, it causes the mind to arise with a phenomenal form of the object in a subsequent moment. This phenomenal form is not a mere mirror of the object because its phenomenal appearance is conditioned by factors other than the object, including the state of the sense faculty and the various cognitive and affective features of the previous moment of mind. However, in epistemically reliable contexts, the phenomenal form does bear a "resemblance" (*sārūpya*) to the moment of the object that created it inasmuch as the relevant causal characteristics of the phenomenal form are restricted by the causal characteristics of the object. The phenomenal form that first arises through sensory contact is "nonconceptual" (*nirvikalpaka*) in that it has not undergone the apoha process, but under the right conditions, the apoha process occurs in yet another subsequent moment, and this moment of mind now has a phenomenal form that is "conceptual" (*savikalpaka*). In this third moment, the conceptualized phenomenal form loses the vividness or clarity that is characteristic of a perceptual form, and it is thus unclear (*aspaṣṭa*). That is, while the phenomenal form arising as a perceptual cognition through sensory contact is vivid or clear, the conceptual cognition of recognition following upon that nonconceptual cognition loses some degree of phenomenal clarity.[5]

This raises the question of what Dharmakīrti means by a concept, and without going into great detail at this point, we can note some features that relate to the event-based, causal model he employs. In one of his last works, Dharmakīrti writes, "a concept is a cognition with a phenomenal appearance that is capable of being conjoined with linguistic expression."[6] The qualification that the appearance—that is, the phenomenal form—is "capable" of being construed with a linguistic term is important because it is meant to point out that infants can have conceptual cognitions even though they do not yet have the ability to use language. As the commentator Dharmottara notes, "as long as the child who is seeing the nipple does not determine it to be what has been previously seen by thinking 'this is that,' he will not stop crying and direct his mouth to the nipple."[7] Thus, even infants have conceptual cognitions because, as will also be discussed later, a central feature of such a cognition is that the current contents of experience are construed as identical (*ekīkaraṇa*) with previous experience. Thus, in the case of recognition, the interpretation of the perceptual phenomenal form as "blue" involves construing that phenomenal form as the same as a previously experienced phenomenal form. And this is precisely what it means for a mental event to be conceptual, namely, that it involves identifying two things as the "same." This very minimal criterion for conceptuality means, among other things, that even animals can use concepts. Consider, for example, the scientific research conducted in the past decades on pigeons. It is now established that they have the capacity to recognize repeatedly pictures of fish, landscapes, and even the cartoon character Charlie Brown, despite their obvious lack of any evolutionary need to do so. This capacity to repeatedly identify Charlie Brown in various settings, sometimes with considerable distortion to his usual form, would indicate to Dharmakīrti that pigeons must be using concepts, even though they lack the capacity to mentally form an expression such as "that is Charlie Brown."[8]

Dharmakīrti's minimalist approach to conceptual cognition is also reflected in the examples that he uses to articulate his theory of concept formation through apoha. In general, the examples involve only single terms, often articulated in simple expressions such as "this is a cow" or "this is a jug." It seems clear that Dharmakīrti thus intends to examine not complex sentences, but simple predicative constructs that in at least some cases must be prelinguistic. These examples, along with the minimalist criteria that mark a cognition as conceptual, suggest that Dharmakīrti is not attempting to account for the formation of sentences or linguistically structured acts of predication, but rather that he seeks to articulate a more basic theory of what must be in place for more complex, linguistically structured cognition to be possible. If this view is correct, then we might best see his theory as explicating the very minimal form of conceptuality that is akin to "feature placement" (i.e., minimal cognitive events such as "blue here now") as discussed by Jonardon Ganeri elsewhere in this volume.

DHARMAKĪRTI'S ONTOLOGY

A second topic that must be addressed before examining the central features of the apoha theory is Dharmakīrti's ontology. In brief, Dharmakīrti follows the basic Buddhist rubric of the "two realities" (*satyadvaya*), the ultimate (*paramārthasat*) and the conventional (*saṃvṛtisat*). For Dharmakīrti, only causally efficient things are ultimately real. This claim rests largely on the notion that, to be known as real, a thing (or its effects) must impinge on the senses, for it is on the basis of sensory experience that we can assert the ultimate reality of a thing.[9] Thus, in a paradigmatic sense, the causal efficacy of an ultimately real thing consists in its ability to causally interact with the senses in such a way that a phenomenal form of the thing is created in the next moment of consciousness in accord with the model described earlier. Indeed, it is this production of a phenomenal form through contact with a sensory object that comprises what Dharmakīrti calls "perception" (*pratyakṣa*). Hence, this also means that any object of perception must be ultimately real because only a causally efficient thing can participate in the causal process that leads to the creation of a perceptual phenomenal form.[10]

By limiting ultimate reality to things that have the capacity to participate in a causal process, Dharmakīrti can deny ultimacy to universals—a denial that is directed at a large range of non-Buddhist thinkers who take real, extramental universals to be the objects of (or at least required for) conceptual thought and language.[11] Dharmakīrti's denial of universals appeals to various arguments, and one line of reasoning points to the incoherence of maintaining that a real universal is either ontologically identical to or different from the particulars in which it is allegedly instantiated. Another approach appeals especially to the notion that, to be ultimately real, an entity must be causally efficacious. In the context of causal efficacy, many of Dharmakīrti's arguments rest on two claims: causal efficacy requires change and change is incompatible with universals. If, for example, the universal "cowness" (*gotva*) were to change, then it would be something other than cowness, since to change is to become other. Thus, to change, "cowness" must become "noncowness," and this would mean that all the objects qualified by cowness would suddenly become noncows. If, however, cowness does not change, then it cannot be causally efficacious because it could not move from a state of not producing a specific effect (for example, a phenomenal form in a perceptual event) to a state of producing that effect.[12] Finally, alongside these technical arguments from causal efficacy comes a more commonsense approach: an assortment of particulars that we label "fire" actually produce heat, but that label or concept "fire" cannot boil our tea.[13] Thus, whether due to the incompatibility between causal efficacy and a universal's lack of change or the simple intuition that one cannot eat the idea of an "apple," universals cannot be causally efficacious. Hence, in Dharmakīrti's ontology, they can only be considered real in a conventional sense.

On this basis, Dharmakīrti denies the ultimate reality of universals, and that denial is consistent with his nominalist project. We should here note that Dharmakīrti's notion of a universal differs significantly from the realist theories of philosophers such as the Naiyāyika Uddyotakara. Specifically, for realists such as Uddyotakara a universal exists in distinction from the conceptual cognition that apprehends it, but for Dharmakīrti a universal does not exist independent of a conceptual cognition.[14] In rejecting the ultimate reality of universals, Dharmakīrti is denying that an expression or concept refers "in an affirmative manner" (*vidhirūpa*) to its referent by virtue of that referent's instantiation of a real universal to which that expression or concept is related. By denying the reality of universals, Dharmakīrti makes this realist account impossible.

While denying the ultimate reality of universals, Dharmakīrti must still account for how conceptual cognitions can guide action in the world. This obligation stems in part from Dharmakīrti's understanding of why we use concepts. As he puts it, we use concepts not simply out of some pernicious habit, but rather with a specific purpose or goal in mind.[15] We might, for example, seek to heat ourselves in front of a fire, and we might then use the conceptual recognition of a fire—one following upon the nonconceptual perception of fire—to walk over and reach a real, particular fire that has the capacity to fulfill our telos (*artha*), the goal that we seek. On this understanding of why we use conceptual cognitions, Dharmakīrti is obliged to show how our words and concepts yield useful information that enables us to act effectively in the world, even without the presence of any ultimately real universal.

APOHA: A SUMMARY

Dharmakīrti's problem, as we have sketched it, is to explain how concepts can provide useful information without any ontological commitment to the existence of universals, and his response to that problem is the apoha theory. This section presents a summary of the theory, and the following two sections examine some of its details: the notion of particulars having the same effect and the role of "imprints" (*vāsanā*). This chapter will conclude with a review of some key features of the apoha theory.[16]

Overall, Dharmakīrti maintains that three different types of universals can be constructed through apoha: those based upon the real (i.e., particulars), those based upon the unreal, and those based upon both. To simplify our task, let us consider only the type of universals that are based upon real things, most especially those relevant to the act of recognition sketched earlier. Dharmakīrti discusses this form of recognition in a key passage:

> Having seen that things, although different, accomplish this or that telic function (*arthakriyā*) such as the [production of a] cognition, one conjoins those things with expressions that take as their object the difference from things that are other than those [that accomplish the aforementioned telos]. Having done so, then when one sees another thing [with that telic capacity], one has a recognition of it [as being the same as the aforementioned things].[17]
>
> (PV I.98–99aB)

And in his commentary, Dharmakīrti explains:

> It has already been said [at PV I.75] that even though [some] things, such as the eye and so on, are distinct, they accomplish the same telic function. [A person] sees that among [things], some accomplish that same telic function, such as the [production of] a cognition; as such those things are [conceptually] distinguished from the others [that do not perform that function]. Those things thus produce, by their very nature as real things (*vastudharmatayā*), a false awareness in [that person]; that awareness is associated with expressions that have as their object the exclusion [of those things] from [the others] that do not perform that [aforementioned function]. This false awareness is [the recognition] "this is that." It arises because the imprint [placed in the mind by that person's previous experience] has been activated [by what s/he is presently seeing]. [In this cognitive act of recognition] the difference [among those unique things] is glossed over (*saṃsṛṣṭabheda*).[18]

In terms of the basic contours of the apoha theory, this passage is useful for understanding how, in the absence of real universals, a concept such as *fire* can be applied nonrandomly to only some objects. For Dharmakīrti, the explanation is that one constructs a sameness for a class of objects on the basis of their difference from other objects. The warrant for that construction is that every object is in fact completely unique in its causal capacities or "telic function" (*arthakriyā*). In the construction of a sameness that applies to certain objects, however, one focuses on a subset of causal capacities that are relevant to one's telos or goal (*artha*), and one thus ignores other capacities that distinguish even the objects we call "fire" from each other. The sameness that applies to all fires is thus, strictly speaking, a negation: it is the exclusion (*vyāvṛtti*) of all other things that do not accomplish the desired telic function. Since each individual fire is actually unique, the conceptual awareness formed through exclusion is "false" (*mithyā*) or "erroneous" (*bhrānta*) in that it presents those objects as the same. Nevertheless, since it is rooted in their causal characteristics, that "erroneous" awareness can successfully guide one to objects that will accomplish one's goals.

In presenting the apoha theory, the passage also draws on the causal model of cognition discussed earlier. As noted previously, the act of perception consists of a phenomenal form being generated in consciousness by the interaction of the senses with an object. When an act of recognition is to occur, a perceptual phenomenal form activates an "imprint" (*vāsanā*) such that the phenomenal form in a subsequent moment of consciousness is now construed in terms of an exclusion that forms a class of entities. The phenomenal form is thus conceptualized in an act of recognition whose minimal structure would be "this is that."

The successful act of recognition thus involves an appeal both to a regular causal process and to the conditioning preserved as "imprints" in the perceiver's mind. In terms of the appeal to causal characteristics, the causal model of perception requires that the phenomenal form that arises through contact with the object is an effect of the perceptual process, and since it arises through a causal process, that phenomenal form is a particular. As a particular, each phenomenal form is utterly unique, and it cannot be distributed over other particulars. Hence, phenomenal forms themselves *cannot account* for the universal—the "sameness" (*sāmānya*)—that enables us to see one object and then another and recognize that the two are the same, for example, that they are both "fire."[19]

Nevertheless, each phenomenal form, precisely because it is a unique particular, can be the basis for the construction of the appropriate universal. As noted earlier, a particular is necessarily causally efficacious, and this means that it arises from causes and produces effects. Moreover, the range of effects that it can produce is restricted by the causes from which it has arisen. A particular's uniqueness thus amounts to the fact that it has arisen from specific causes and that it therefore is capable of producing a restricted range of effects.[20] If we consider a phenomenal form that arises from what we would call "fire," that phenomenal form (a mental particular) is unique or "excluded" (*vyāvṛtta*) from all other particulars in that no other particulars arise from exactly the same causes or produce exactly the same effects. The phenomenal form, being the unique effect of the unique particulars that produced it, thus serves as the basis for excluding the phenomenal forms produced by other particulars.[21]

The view, however, that each phenomenal form excludes all others by virtue of its uniqueness is not in itself adequate to account for our use of concepts: we require a notion of sameness and not just difference. We must have some notion of sameness because we need to account for *anvaya*, the "repeatability," "distribution," or "continuity" applicable to any cognition that involves construing two or more things as the "same" (*eka*). When we reflect on the conceptual cognition of "fire," for example, it appears to assume a "fireness" that is present in multiple instances, and in this sense the concept of *fire* has *anvaya*. Here we encounter the relevance of factors occurring in the mind in which the concept will arise. One such factor is the imprint of previous experience, which we will discuss in greater detail later. Another factor is the set of expectations that arise from having a particular goal, one that Dharmakīrti always frames as obtaining the desirable or avoiding the undesirable. These essentially behavioral goals create a "desire to know" (*jijñāsa*), that is, a need for information about what will or will not accomplish that goal. This desire to know, in turn, places "limits" (*avadhis*) on the causes and effects upon which we focus. In other words, we have expectations about what we wish to obtain or avoid, and our concepts are constructed in relation to those expectations.[22]

In the case of the concept *fire*, some set of interests—such as the desire for warmth—or other such dispositions prompt us to construe the phenomenal form in question as distinct from entities that do not have the causal characteristics expected of what we call "fire." At the same time, we ignore other criteria, such as having the causal characteristics expected of that which is "smoky" or "fragrant," because these are not part of what we desire to know, so as to accomplish our goals. When we look at an object that we will call "fire," it produces a phenomenal form that, given the context of our expectations, activates the imprint of a previous experience. Both the current phenomenal form and the form that arose in the previous experience exclude all forms that we would *not* call "fire"; but suppose that the current fire is smoky, while the previously experienced fire was not. Indeed, from Dharmakīrti's ontological perspective the two fires really are not the same at all, but our desire to achieve a goal—such as warming our hands—that is accomplished by fire creates a context that compels us to *ignore these differences*. And since we have ignored the differences between those two phenomenal forms—the current one and the one that caused the imprint—we can construe both of them as mutually qualified by a negation, namely, their difference from phenomenal forms that do not activate the imprints for the concept *fire*. That mutual difference, which Dharmakīrti calls an "exclusion" (*vyāvṛtti*), thus becomes their nondifference. In short, that exclusion or nondifference pertains to all things that are different from those that do not have the expected causal characteristics—in this case the causal characteristics expected of that which we call "fire."[23] In this way, exclusions, being formed on the basis of the phenomenal forms in conceptual cognitions, are construed as negations that qualify those forms. Thus, while the phenomenal forms themselves are completely unique—they do not have *anvaya* and thus are not distributed over other instances—they can be construed as qualified by a negation that *does* have *anvaya*, inasmuch as that negation applies to all the instances in question because they exclude what is not a "fire."

Dharmakīrti thus arrives at a theory of universals (*sāmānyalakṣaṇa*) that requires both the phenomenal form and the exclusion. That is, strictly speaking, a universal is a combination of that which is not distributed (i.e., lacks *anvaya*) and that which is distributed. The phenomenal form, as a mental particular, is not distributed, but the exclusion (*vyāvṛtti*), as a negation applicable to all the phenomenal forms in question, is distributed. Lacking distribution, the phenomenal form alone cannot be the universal. But in Dharmakīrti's theory of qualities, a negation cannot exist in distinction from that which it qualifies; therefore, the negation alone also cannot be the universal. The universal must therefore be a phenomenal form that we construe in terms of a particular type of negation, namely, the exclusion of that which does not have the expected effects. However, to apply this negation to all the phenomenal forms in question, one must construe all the phenomenal forms in question as having the same effect. Let us examine this important issue in greater detail.

CONCERNING SAMENESS OF EFFECT

Dharmakīrti claims that a universal is constructed on the basis of the exclusion of all the entities in question from those that do not have the expected causal characteristics. Dharmakīrti recognizes, however, that if certain things—such as those called "jugs"—are excluded from others because those others do not have the expected causal characteristics, one is also asserting that all the things we call "jugs" have the *same* causal characteristics, namely, those expected of a "jug." For Dharmakīrti, this amounts to the claim that, in the case of all jugs, we may identify at least some of their causes as the "same" (*eka*), and most importantly, we may likewise identify at least some of their effects as the "same."[24]

Dharmakīrti's focus upon sameness of effect becomes particularly salient when he presents his apoha theory in terms of the act of recognition discussed earlier. As noted previously, when an object is perceived, it produces a sensory cognition containing a phenomenal form that, being a (mental) particular, is no less unique than the object that produced it. If Dharmakīrti were to claim that objects are the same because those effects—the phenomenal forms they produce—are the "same" (*eka*), then it seems that he must contradict his ontology of particulars: if he says that two phenomenal forms, which are mental particulars, are the same, then how can he say that all particulars are unique? Speaking in the voice of an objector, and using "cognition" to refer to the phenomenal form, Dharmakīrti puts the problem this way:

> *But each cognition is an effect of those individuals, and cognition is different* [PV I.108cd] for each real thing. That is, as with the individual that caused the awareness, the cognition in which it appears is distinct; therefore, how can all those specific individuals have the same effect? For the cognition is their effect, and it is different in each case. In other words, the single effect of jugs and so on, such as bearing water, is different for each substance because the substances are different. Hence, those individuals, being different, do not have the same effect.[25]

To avoid this problem, Dharmakīrti maintains that the cognitions—that is, the cognitions with phenomenal forms related through causality to their objects—are not what account for the sameness of those objects' effects. Instead, those cognitions themselves act as causes for another cognition, a "judgment" (*pratyavamarśajñāna*) in which the thing in question is construed as the "same" (*eka*) as other things. He explains:

> This is not a problem, *because the cognition produced by each individual in question is nondifferent since each cognition is the cause of a judgment [of the individual as] the same [as the other individuals in question]. And since they are the causes of the same cognitions, the individuals are also nondifferent.*[26]
>
> (PV I.109)

Dharmakīrti admits that the cognition—or more precisely, the phenomenal form—produced by each object is indeed unique. Hence, one cannot *directly* use those phenomenal forms as the warrant for the claim that the objects are the same because they have the same effect. If those phenomenal forms are the basis for the construal of the objects that produced the phenomenal forms as the same, then Dharmakīrti must first show how those phenomenal forms—the effects of the objects—are themselves the same. To do so, he once again turns to the principle that entities are the same if they produce the same effect. That is, he maintains that those phenomenal forms are all the same because they all produce the same effect, namely, a judgment (*pratyavamarśajñāna*) that presents the aforementioned phenomenal form in such a way that it appears to be the same as the others. Thus, all the phenomenal forms can be the same in that each leads to a judgment, such as "this is fire," that presents its content—the phenomenal form—as the same as the content of the other judgments, inasmuch as in each case the content is presented as "fire." With this point in place, Dharmakīrti can then maintain that, if those phenomenal forms are the same because each leads to a judgment in which the phenomenal form is presented to be the same as the others, then one can also say that the objects that produced those phenomenal forms in the first place are all the same because they too produce that effect. Thus, the warrant for the sameness of the objects is that they produce the same effect: the phenomenal forms. And the warrant for the sameness of the phenomenal forms is again that *they* produce the same effect: a certain type of judgment in which each phenomenal form is presented to be the same as the others.

As we have described it so far, this theory leaves itself open to an obvious rebuttal: what warrants the sameness of the judgments? That is, Dharmakīrti's initial problem is that objects are unique, so the sameness required by language and concepts must be accounted for by sameness of effect. But if he turns to the phenomenal forms produced by those objects, he has the same problem because those cognitions, like the objects themselves, are unique. If he now turns to the claim that those phenomenal forms are the same because they produce the same judgment, then he appears to have fallen into an infinite regress. In other words, it would appear that we need, once again, to warrant the sameness of those judgments by appealing to the sameness of *their* effects; and of course, the sameness of the judgments' effects will once again require the same warrant, and so on.

Dharmakīrti's response to this problem is expressed, if somewhat elliptically, in his commentary on the verse cited earlier. Note that here he uses the metaphor of an "overlap" or "mixing" (*saṃsarga*) of objects whereby the nature of one is somehow partially present in the nature of the other. For Dharmakīrti, such an overlap is impermissible in the case of causally efficient things, since causally efficient entities are particulars, and they must be unique. At the same time, what it means for two objects to be *conceptually* construed as the "same" in the relevant way is precisely that the conceptual cognition presents them as overlapping in some fashion—for example, overlapping in that they are both "fire." With this and other such issues in mind, he comments on the aforementioned verse:

> It has already been explained[27] that the natures of things (*bhāva*) do not overlap, and that a cognition of a thing in which the phenomenal form presents a thing as if its nature overlapped with other things is an error. However, those distinct things indirectly (*krameṇa*) become the causes for concepts; as such, they produce a conceptual cognition in which they seem to overlap, and they do so "by their nature." Moreover, this is called their "nondifferent difference"—namely, their exclusion (*viveka*) from other things that by nature do not cause that effect; they are understood to be excluded in this fashion because they cause some same effect, such as a cognition.
>
> In terms of the cognition that each individual produces, even though it is different for every substance, the cognition appears nondifferent from the others in that by its nature the cognition causes a judgment [of its content] as the same [as the others]; that is, the judgment overlays the phenomenal form in the awareness with a nondifference. The individuals in question cause a thing (*artha*) such as an awareness [in which the phenomenal form] appears nondifferent [from the others] and which in turn causes that kind of judgment. Therefore, those individuals through their nature produce the same cognition with a phenomenal form that presents them as overlapping whose ultimate [object] is their difference in nature (*svabhāvabheda*) from all other things, as has been repeatedly stated. Therefore, the nondifference of things consists of the fact that they have the same effect.[28]

Dharmakīrti's solution to the problem of infinite regress is that sameness of effect does *not* act as the warrant for the sameness of the judgments in question. Instead, he cleverly shifts what he means by being "the same" (*eka*). The judgments are the same not because they have the same effect, but because they phenomenally present their content as the "same": by overlaying the phenomenal forms in the cognitions that produce them with a "nondifference" (*abheda*), each judgment presents its phenomenal content to be the same as the previously experienced phenomenal content. This amounts to an appeal to some unspecified combination of experience and mental dispositions: when we look at certain things, we just interpret them all as "fire," in the context formed by previous experience, certain dispositions, and the way that we use the term "fire." This appeal to experience and dispositions highlights the importance of mind dependency or "subjective factors" in the process of constructing exclusions. That is, Dharmakīrti maintains that when we construct exclusions, we do not do so haphazardly or out of some pernicious habit; rather, we have some purpose in mind, and that purpose provides expectations and interests that form the context of our concept formation. An apple and a strawberry, for example, will be different if we are concerned with their distinctive effects, but if we are only concerned with their coloration, we ignore that difference in light of the sameness constructed in terms of color. And of course, our use of "apple" and "strawberry" is dependent on our habituation to certain linguistic practices. And, as we will see later, questions of habituation and dispositions relate closely to Dharmakīrti's notion of "imprints" (*vāsanā*).

While Dharmakīrti's appeal to experience and dispositions reflects the mind-dependent aspects of the exclusion process, it is coupled with something more: an appeal to the nature (*prakṛti* or *svabhāva*) of things themselves. That is, when several objects produce cognitions that in turn produce the same judgment, "this is fire," it is not *just* my own expectations, conditioning, and other relevant dispositions that go into the construction of that exclusion. Rather, beyond my own subjectivity, the entities in question *by their nature* (*svabhāvataḥ*, *prakṛtyā*, etc.) produce cognitions whose content is capable of being construed as "fire." In combination with mind-dependent factors, this assertion of the nature of things puts an end to any infinite regress. We can pose the question "but *why* do those objects all produce cognitions that can lead to the same judgment?" And Dharmakīrti can answer, "because it is their nature to do so."[29]

Some interpreters may feel rather dissatisfied with Dharmakīrti's appeal to nature. In effect, he is saying that when we can call all fires "fire," for example, it is not that they all instantiate the universal "fireness"; nor that they all possess some real, specifiable similarity; nor even that they all have the "same" effect in a way that we can ultimately specify in objective terms. Rather, all those things are just different from nonfire things, and the reason for their difference is simply that *by their nature* they appear that way to us when we attend to what we mean by "fire." Even the seeming objectivity of this appeal to nature may disappoint some, for a thing's "nature" (*svabhāva*) is also conceptually constructed through the apoha theory.[30] On this interpretation of what Dharmakīrti means by nature, Dharmakīrti's talk about the nature of things that we call "fire" is best understood as a way of saying that, in ultimate terms, there is no metaphysically defensible reason for the fact that we call them "fire." Thus, if one is hoping for an ultimately defensible metaphysical reason, then Dharmakīrti's answer to the problem of sameness is dissatisfying. On the other hand, one might suppose that we are engaged in a frustrating and fruitless enterprise when we yearn to specify in precise terms the metaphysical warrant for our use of the term "fire." In that case, Dharmakīrti's answer is quite satisfactory, or perhaps even liberating.

IMPRINTS

We have seen that, in a context such as recognition, a universal is constructed on the basis of a thing's causal characteristics, and the universal is not arbitrary because it is constrained both in terms of the object and the subject. Objectively, it is the appeal to a thing's nature—that is, its causal characteristics—that places constraints on the formation of the concept. Since there is an infinite number of things from which an object may be differentiated, an unlimited number of universals may be constructed for that thing. Nevertheless, to be formed "on the basis of a real thing" (*bhāvāśraya*), the universal must be constrained by the causal characteristics of the thing in question; hence, even though an infinite number of universals can be properly constructed for any thing, an infinite number of universals also cannot be properly constructed because they would not conform to the thing's causal characteristics.

Elsewhere, I have argued that this appeal to an object's nature includes a covert reliance on regularities in the features of minds because even though seemingly objective, an object's nature is in part mind dependent for Dharmakīrti.[31] An even more obvious appeal to some kind of regularity in minds, however, is his explicit reliance on imprints (*vāsanā*) as playing a key role in the construction of concepts through apoha. Imprints clearly lie on the subject side, and they stand alongside other subject- or mind-located constraints, such as expectations, context, conventions, and so on. All of these may be thought of as involving conditioning (*saṃskāraṇa*), but imprints play a special role in this regard.[32]

In general, "imprints" (*vāsanā*) are a mechanism for the expression of karma, especially on the Yogācāra model, and Yogācāra idealism is Dharmakīrti's final view.[33] In the Yogācāra system, these imprints are stored in a type of subliminal or implicit consciousness known as the "storehouse" or "receptacle" consciousness (*ālayavijñāna*), the theory of which Dharmakīrti explicitly refers to and adopts, at least in part. According to the Yogācāra system before Dharmakīrti, imprints continue to add seeds to the storehouse, and these imprints in turn can be later activated, thus causing effects on experience. One way of understanding the theory of the storehouse is that it thus solves a central problem of the Buddhist notion of karma, namely, how it is that past intentions and actions can yield significant effects in the (continually reborn) mindstream, even over vast periods of time. To some extent, the theory of imprints and the storehouse probably serve similar purposes for Dharmakīrti, though his exact relationship to the Yogācāra tradition that precedes him remains unclear. What is clear is that, in employing the technical vocabulary of imprints, Dharmakīrti also bends it to his specific purpose, namely, accounting for the way that conceptuality operates.[34]

A complete account of Dharmakīrti's notion of imprints not only is beyond the scope of this chapter, but also may not be possible. Although he refers repeatedly to imprints, the precise mechanism of their operation receives no attention. Nevertheless, as is clear from the he passage on recognition cited earlier, he clearly distinguishes two basic forms of imprints: those that are "placed" (*āhita*) in the storehouse by experiences and those that are innate or "beginningless" (*anādi*). Both types of imprint play crucial roles in the apoha theory. As noted previously, in the act of recognition, a phenomenal form arises when, with other cognitive conditions in place, an object comes into relation with a sense faculty; when the act of recognition ensues, that phenomenal form activates an imprint from a previous experience, and the object that caused the current phenomenal form is construed to be the same as the object that caused a phenomenal form at that time. It is the placement of this imprint that crucially allows for the fundamental "unification" or "construal as the same" (*ekīkaraṇa*) which is the principal marker of conceptual cognition in Dharmakīrti's system. Without an imprint of previous experience that could be activated by the present experience, there would be no possibility of recognizing the contents of present experience as identical with what has been perceived before.

The imprint placed by previous experience plays a crucial role, but it cannot account in itself for all that is necessary for a concept to be formed through apoha. Consider a phenomenal form of, for example, an object that can be correctly called a "fire." That object should be capable of creating a phenomenal form that leads to concepts other than "fire"; depending on the specific characteristics of that object, the mind might form any number of other concepts, such as "light" or "visible object." There is something about the phenomenal form that performs the "exclusion" (*vyāvṛtti*) of these other possibilities—that is, there is something about the phenomenal form that inhibits the activation of imprints that would lead to the construal of the object not as a "fire" but rather as light and so on. Obviously, the imprint of previous experience itself cannot be what inhibits the activation of other imprints, because if the imprint of the previous experience of "fire" excludes the imprint for "light," then the object could only be construed as "fire" and never as "light." For Dharmakīrti it is clear that one of the major factors in inhibiting these other imprints is precisely the goal-oriented expectations that are present in the moments of mind prior to the arising of the phenomenal form. That form thus arises as conditioned by such expectations, which themselves are generally articulated in terms of obtaining the desirable or avoiding the undesirable. Conditioned by expectations focused on such a goal, the phenomenal form that arises is thus one that is primed for the imprint of "fire" to be activated, or perhaps more accurately it is biased against the activation of irrelevant imprints.

The imprint of a previous experience must thus be supplemented by some other cognitive factors such as expectations, but even this is not enough to explain how the concept arises. Recall that for Dharmakīrti the previously experienced object and the presently experienced object are actually *not identical in any way at all.* Ontologically, they are entirely distinct, and the raw, uninterpreted phenomenal forms initially arising in perception are also not actually identical even in their phenomenal appearance. So why would they ever be construed as the same? How would one first learn, even before the acquisition of language, the capacity to identify past experience with present experience if, in fact, the two experiences are not actually the same? According to Dharmakīrti, the answer is that one does not need to learn to do so because one's mind already has a powerful imprint—which might be better called a disposition—that causes one to identify objects as the same in an automatic fashion. This disposition is not learned; indeed, on Dharmakīrti's view it would be *impossible* to acquire it through experience because this would require an experience of two objects that are in fact the same, but for Dharmakīrti all perceptible objects are necessarily different in all ways.

This imprint is termed "beginningless" by Dharmakīrti in that it is a fundamental disposition in the mind of any sentient organism that has not acquired the capacity to eliminate it. In this sense the imprint might best be called "innate" in that it is not acquired, but is rather part of a sentient being's cognitive architecture just by virtue of being a sentient being. More specifically, it is the innate capacity to form concepts, defined in the minimal sense which even pigeons are capable of. If one adopts the perspective of evolutionary psychology, this fundamental disposition would provide a key evolutionary advantage in avoiding previously experienced dangers or approaching previously experienced opportunities—or, in Dharmakīrtian terms, it is necessary for the acts of recognition involved in obtaining the desirable and avoiding the undesirable. At the same time, however, this fundamental disposition radically distorts our experience of the world, such that we treat things that are actually different—the person I met a year ago and the person I am seeing now—as if they were the same. This distortion is so pervasive and it leads to such a dysfunctional engagement with the world that Dharmakīrti calls it "ignorance" (*avidyā*).[35]

Even with this fundamental capacity for (erroneously) identifying an object with a previously experienced object, one other aspect of the apoha process needs to be accounted for by an appeal to yet another kind of imprint. In brief, the difficulty here is a discontinuity between phenomenal content and reference. What is actually presented in the conceptual cognition of *fire* is a phenomenal form, but that phenomenal form lacks the capacity to produce heat and so on; only an actual fire has such capacities. Hence, for the concept to guide effective action, its phenomenality must be ignored and it must be treated as if it were the object, such as an actual fire, that it represents. Another innate imprint—and again, the best translation here would be "disposition"—accounts for this feature of conceptual cognition. As with the innate disposition that provides the automatic ability to see different things as the same, this imprint is obviously useful. If one needed to reflect upon the phenomenal representations in conceptual cognitions so as to understand that they are actually pointing to something in the world, conceptual thought would become hugely inefficient. One would lose much of the advantage in being able to recognize that what one is hearing now is the same lion's roar that one heard previously. Instead, to sustain the practical efficiency of conceptuality, the mind must automatically mistake the concept for the object that it represents. As with the other innate disposition, however, this erroneous feature of conceptuality comes at a price because it also involves a distortion. Perhaps the problem here is best illustrated by moving beyond Dharmakīrti to contemporary psychology where, according to one prominent variation on cognitive behavior therapy, the tendency to mistake thoughts for reality can become so dominant and rigid that it leads to the psychopathology of "cognitive fusion" and the behavioral dysfunction (such as pathological avoidance) that ensues.[36]

CONCLUSION: THE POWER OF CONCEPTS

Dharmakīrti's formulation of the apoha theory stems from his nominalist rejection of real universals, and one might suppose that he would therefore view concepts as somehow weakened or unimportant. After all, concepts are not ultimately real. Yet the need to provide a thorough account of concept formation points to the opposite conclusion, namely, that Dharmakīrti believes conceptual cognition to be crucial to his project. The centrality of conceptual cognitions becomes especially apparent when Dharmakīrti explains a key feature of Buddhist contemplative practice, namely, that the repeated contemplation of key concepts can itself transform an individual. To conclude this chapter, let us consider briefly the way that Dharmakīrti explains this transformative role of concepts.

We have seen that the phenomenal form plays an especially important role in the apoha theory. In the case of recognition, for example, it provides the causal link between the concept and the particular. Because it is an appearance in the mind, the phenomenal form also explains how conceptual cognition can have a positive content, even though what accounts for the "sameness" in the cognition is actually the negation that Dharmakīrti calls an "exclusion" (*vyāvṛtti*). Likewise, since the phenomenal form is construed in terms of a beginningless imprint that makes one mistake it for the actual object to which it refers, a conceptual cognition can provoke one to act on an object in the world, even though the phenomenal form that is actually appearing in the cognition is not actually that object.[37]

Phenomenal forms have to do with the content of experience, and in the context of contemplative practice, Dharmakīrti is especially concerned with the notion that one can have a direct, nonconceptual experience that is transformative. This direct experience, however, must somehow be of specific concepts, such as selflessness and impermanence, that are the objects of the contemplative's practice. The question then is, on Dharmakīrti's theory of concepts, how is it possible to move from a vague, conceptual cognition to a vivid, nonconceptual one? Dharmakīrti alludes to the theoretical basis for such events:

> A cognition that apprehends a linguistic object (*artha*) is a conceptual cognition of that [object] which it is cognizing. The actual nature [of any cognition qua mental event] is not a linguistic object; therefore, any [awareness of awareness itself] is direct [and hence nonconceptual].[38]
>
> (PV III.287)

Previously we noted that in the formation of a concept through the apoha or exclusion process, a phenomenal form is presented in such a way that it becomes vague—not vivid like a perceptual phenomenal form. In short, the image is vague in that it is not a phenomenally clear depiction of the object that it represents. Nevertheless, even though the judgment's image is vague as a representation, it is nevertheless an image. In other words, the judgment does contain some type of phenomenal content. And as a mental event, that phenomenal content is a real mental particular that can be known in its nature as a mental event through reflexive awareness (*svasaṃvitti*). In relation to that reflexive awareness, however, the content no longer appears to stand for something else; that is, it is no longer conceptual. In other words, as that which is known through reflexive awareness, every cognition—even every conceptual cognition—is a mental particular.

In this way, Dharmakīrti proposes what might be called a "Janus-faced" theory of concepts, which he explains most succinctly in another passage. There, an objector says, "if [a universal] is also a real object (*artha*) in terms of having the nature of awareness, then you would have to conclude [that it is a particular]" (PV III.9cd).[39] In other words, if the phenomenal content of a conceptual cognition can be known reflexively as a mental event, then it would seem that universals, the objects of conceptual cognitions, must be ultimately real, since they would be known through perception, albeit the unusual form of perception that is reflexive awareness. Dharmakīrti responds, "since we do indeed assert [that a universal is a particular], your statement poses no problem for us. But it is a universal because it [is imagined to have] the same form for all [the objects that it seems to qualify. It has that same form] because it is based upon their exclusion [from other objects that do not have the expected causal characteristics]" (PV III.10).[40]

Thus, when construed as a sameness distributed over a class of particulars, a concept is a universal; but considered as a mental event, the concept is a particular. In this way, inasmuch as it is distributed over a class of things, a universal is actually a negation, since on Dharmakīrti's view, only a negation formed through exclusion can be distributed in this way. But inasmuch as the negation is not ontologically distinct from the mental image that occurs in the conceptual cognition, that cognition is a particular qua mental event. The one proviso that must be added is that, when considered as a mental event, the conceptual cognition loses its distribution, and when it loses its distribution, it is no longer a concept.

This Janus-faced aspect of concepts gives us a means to explain how a contemplative can focus repeatedly on a concept and, through that contemplation, eventually arrive at a cognition that is now clear and vivid—

a cognition that is now a "yogic perception" (*yogipratyakṣa*). Dharmakīrti compares this process to the hallucinations of a lovesick man: When the man focuses intently and repeatedly on the memory (a concept) of his beloved, he is in part focusing on a mental event, which is a particular. With sufficient and intense repetition, he will have a clear experience—a perception—of that event itself. The contemplative follows the same type of process, and her efforts also result in a nonconceptual knowledge of a concept qua phenomenal form.[41] Hence, Dharmakīrti says, "therefore, that to which one meditatively conditions oneself, whether it be real or unreal, will result in a clear, nonconceptual cognition when the meditation is perfected" (PV III.285).[42]

In this way, for Dharmakīrti, concepts can become a powerful means of inducing experiences that transform—or distort—the mind. It would seem that although concepts are ultimately unreal, they are still potent and, in some cases, dangerous.

Notes

1. For Bhāvaviveka's dates, see Eckel 2008, 25. For Bhāvaviveka's critique and response to the apoha theory, see the translation of V.60–68 in Eckel 2008, 265–273.
2. In his influential article on the topic, Kim 1988 uses this phrase ("an empirical psychological study of our cognitive processes") to gloss a central element in Quine's groundbreaking work on what it means for epistemology to be naturalized. Kim also sees Quine's project as requiring the abandonment of normativity; it is "to go out of the business of justification" (380), and in this respect the "naturalization" may not suit Dharmakīrti's project. Nevertheless, Kim also notes that, on his reading of Quine, naturalized epistemology should be a "law-based predictive-explanatory theory, like any other theory within empirical science; its principal job is to see how human cognizers develop theories (their 'picture of the world') from observation ('the stimulus of their sensory receptors')" (389). It does not seem implausible to treat Dharmakīrti's approach in a similar manner—that is, as a project whose goal is a careful and largely empirical description of embodied human cognition.
3. It is worth noting, however, that there is also considerable convergence. For an accessible work on the emotions in this regard, see the Dalai Lama and Paul Ekman 2008.
4. A thorough account of the *Abhidharma*, especially in terms of its analysis of cognitive and affective states, is not yet available in a Western language. A helpful work in this direction is the second chapter in Waldron 2003. See also Potter et al., 1998. *Abhidharma* metaphysics has received more attention, and a good starting point is the chapter on *Abhidharma* in Siderits 2007.
5. The notion that the phenomenal forms in conceptual cognitions lack phenomenal clarity arises in various contexts for Dharmakīrti, but it is especially

evident in his notion of yogic perception, as discussed in the conclusion of this chapter.

6. NB I.5 (ed. Shastri): "abhilāpasaṃsargayogyapratibhāsā pratītiḥ kalpanā."
7. NBT 25 (ed. Shastri): "bālo 'pi hi yāvad dṛśyamānaṃ stanaṃ sa evāyam iti pūrvadṛṣṭatvena na pratayavamṛśati tāvan noparatarudito murkham arpayati stane."
8. The earliest publications in this area include Hernstein and Loveland 1964; Siegel and Honig 1970; Hernstein, Loveland and Cable 1976; and several others.
9. The crux of the matter is stated in perhaps its most succinct form by Dharmakīrti when he says, "to be is to be perceived" (PVSV ad PV I.3, ed. Gnoli, 4.2: "sattvam upalabdhir eva"). The corollary is that, minimally, a real thing must have the capacity for the "projection" (*arpana*) of its own phenomenal form into awareness. See PVSV ad PV I.282 (ed. Gnoli, 149.21ff.). Note that all the passages from PV and PVSV cited in this chapter are translated in Dunne 2004.
10. For a more detailed account, see Dunne 2004, 84–89.
11. For the general contours of this debate and references to recent work on the topic, see Dunne 2004.
12. For an extensive treatment of momentariness and the varieties of arguments in its favor, see Oetke 1993. A list of relevant passages in PV and PVSV is given in Dunne 2004, 97n68. For the question of permanence as applied to universals, a representative passage is found in PVSV ad PV I.144a (ed. Gnoli, 69.21ff.).
13. Dharmakīrti makes this point at a few junctures in his works, but perhaps the most amusing is his remark that, if one is interested in accomplishing a goal, one is not concerned with the reality or unreality of the universals appearing phenomenally in awareness, just as a lustful woman would not bother to inquire whether a eunuch is handsome or not! For a translation of the relevant passages, see Dunne 2004, 310–312.
14. The most concise statement of the causal inefficiency—and hence irreality—of universals is found in the Svavṛtti (PVSV ad PV I.166, ed. Gnoli, 84.10): "tasmāt sarvaṃ sāmānyam anarthakriyāyogyatvād avastu." See also PV III.1–3 and the amusing metaphor of the eunuch (PVSV ad PV I.210–211, ed. Gnoli, 106.27–107.9).
15. See PVSV ad PV I.93 (ed. Gnoli, 45.32–46.9).
16. PVSV ad PV I.68–75, translated in the appendix to Dunne 2004, contains a number of the arguments discussed in the remainder of this chapter. My discussion builds on the work of many others (see Dunne 2004 for references), as well as more recent contributions by Patil (2003) and Arnold (2006).
17. PVSV (ed. Gnoli, 49.16ff.): "jñānādyarthakriyāṃ tāṃ tāṃ dṛṣṭvā bhede 'pi kurvataḥ / arthāṃs tadanyaviślesaviṣayair dhvanibhiḥ saha / saṃyojya pratyabhijñānaṃ kuryād apy anyadarśane."
18. PVSV (ed. Gnoli, 49.19ff.): "uktam etat bhede 'pi bhāvās tulyārthakriyākāriṇaś cakṣurādivad iti / tām ekām jñānādikām arthakriyāṃ teṣu paśyato vastudharmatayaivānyebhyo bhidyamānā bhāvās tadvyāvṛttiviṣayadhvanisaṃsṛṣṭaṃ tad evedam iti svānubhavavāsanāprabodhena saṃsṛṣṭabhedam mithyāpratyayaṃ janayanti."
19. On the three types of universals, see PVSV ad PV I.191 (Gnoli, 95.19ff.) and PV III.51cd (ed. Sāṃkṛtyāyana): "sāmānyaṃ trividhaṃ tac ca bhāvābhāvobhayāśrayāt."

20. See, for example, PVSV ad PV I.166 (ed. Gnoli, 84.22–85.2) and chapter 3 in Dunne 2004.
21. The notion that uniqueness of particulars is ultimately the basis for constructing universals through exclusion is expressed at numerous points, including: PVSV ad PV I.70 (ed. Gnoli, 38.17ff.); PV I.72cd and PVSV ad cit. (ed. Gnoli, 49.16ff.); PVSV ad PV I.64 (ed. Gnoli, 35.2–3); and PV III.169 (ed. Sāṃkṛtyāyana).
22. The basic role of expectations in the construction of universals is indicated by the repeated use of the term *abhimata* ("expected") with relation to the causes and effects on the basis of which a universal is constructed. See, for example, PVSV ad PV I.93. See also the reference to *abhiprāya* in PV I.68–70 and PVSV ad cit. (ed. Gnoli, 39.8). The notion of the negative "limit" (*avadhi*) in opposition to which an exclusion is constructed appears to occur only once in the Svavṛtti (i.e., PVSV ad PV I.185).
23. The notion that certain entities may be considered nondifferent because they are all different from all other entities is emphasized at several places, including PVSV ad PV I.75d (ed. Gnoli, 42.6ff.), PVSV ad PV I.95cd (ed. Gnoli, 48.4) and especially in PVSV ad PV I.137–142 (see the appendix to Dunne 2004).
24. Although Dharmakīrti specifically discusses the construction of universals in terms of entities having the same types of causes, he tends to focus upon sameness of effect. Passages that mention both ways of constructing sameness include PVSV ad PV I.137–142 (ed. Gnoli, 68.24–69.2). Other examples include PVSV ad PV I.40–42 (ed. Gnoli, 25.19–23), PVSV ad PV I.64 (ed. Gnoli, 35.2–4), and PVSV ad PV I.75d (ed. Gnoli, 42.5–8).
25. PVSV ad PV I.108cd (ed. Gnoli, 56.10–14): "*nanu dhīḥ kāryaṃ tāsāṃ sā ca vibhidyate* [PV I.108cd] pratibhāvaṃ / tadvat tatpratibhāsino vijñānasyāpi bhedāt / katham ekakāryāḥ / tad hi tāsāṃ kāryaṃ tac ca bhidyate / yad apy udakāharaṇādikam ekaṃ ghaṭādikāryaṃ tad api pratidravyaṃ bhedād bhidyata eveti naikaṃ bhedānāṃ kāryam asti."
26. PVSV ad PV I.109 (ed. Gnoli, 56.15–57.7): "naiṣa doṣaḥ / yasmāt / *ekapratyavamarśasya hetutvād dhīr abhedinī / ekadhīhetubhāvena vyaktīnām apy abhinnatā.*"
27. Although Dharmakīrti could be referring to a number of passages, a likely candidate is PV I.68–75 and PVSV ad cit.
28. PVSV ad PV I.109 (ed. Gnoli, 56.15–57.7): "niveditam etad yathā na bhāvānāṃ svabhāvasaṃsargo 'stīti / tatra saṃsṛṣṭākārā buddhir bhrāntir eva / tāṃ tu bhedinaḥ padārthāḥ krameṇa vikalpahetavo bhavanto janayanti svabhāvata iti ca / sa tv eṣām abhinno bheda ity ucyate jñānādeḥ kasyacid ekasya karaṇāt atatkārisvabhāvavivekaḥ / tad api pratidravyaṃ bhidyamānam api prakṛt yaikapratyavamarśasyābhedāvaskandino hetur bhavad abhinnaṃ khyāti / tathābhūtapratyavamarśahetor abhedāvabhāsino jñānāder arthasya hetutvād vyaktayo 'pi saṃsṛṣṭākāraṃ svabhāvabhedaparamārthaṃ svabhāvata ekaṃ pratyayaṃ janayantīty asakṛd uktam etat / tasmād ekakāryataiva bhāvānām abhedaḥ."
29. This appeal to the nature of things becomes apparent when, in a related context, Dharmakīrti remarks: "Indeed, it is not correct (*na . . . ahati*) to question (*paryanuyoga*) the natures of things, as in "why does fire burn? why is it hot, and

water is not?" One should just ask this much, "from what cause does a thing with this nature come?" [PVSV ad PV I.167ab (ed. Gnoli, 84.19–21): "na hi svabhāvā bhāvānāṃ paryanuyogam arhanti kim agnir dahaty uṣṇo vā nodakam iti / etāvat tu syāt kuto 'yaṃ svabhāva iti."

30. See Dunne 2004, 158–173.
31. Ibid.
32. The term "imprint" (*vāsanā*), including the notion of "beginningless" (*anādi-*) imprints, occurs numerous times in Dharmakīrti's works, often in connection with the apoha theory. In the PVSV alone, it occurs repeatedly in these discussions related to apoha: PVSV ad PV I.58, PV I.64, PV I.68–70, PV I.72, PV I.75, PV I.98–99, PV I.106–107, PV I.151–152, PV I.161, PV I.205, PV I.238, and PV I.286.
33. Dharmakīrti assumes a Yogācāra stance at various points in his works, but perhaps the most sustained treatment is found beginning at PV III.333.
34. For more on the notion of the imprints and the storehouse consciousness, see Waldron 2003 and Schmithausen 1987.
35. See PVSV ad PV I.98–99 (ed. Gnoli, 50.20): "vikalpa eva hy avidyā."
36. See Hayes et al 1999. Cognitive fusion is first clearly articulated in chapter 3.
37. For more on these features of the phenomenal form, see Shōryū Katsura's article on the three forms of apoha in this volume.
38. PV III.287 (ed. Sāṃkṛtyāyana): "śabdārthagrāhi yad yatra taj jñānaṃ tatra kalpanā / svarūpaṃ ca na śabdārthas tatrādhyakṣam ato 'khilam." See also the remarks by Woo 2003, 441–444.
39. PV III.9cd (ed. Sāṃkṛtyāyana): "jñānarūpatayā arthatve sāmānye cet prasajyate."
40. PV III.10 (ed. Sāṃkṛtyāyana): "tathaiṣṭatvād adoṣo, 'rtharūpatvena samānatā / sarvatra samarūpatvāt tadvyāvṛttisamāśrayāt."
41. For more on yogic perception and for other references to recent work on Dharmakīrti's theory, see Dunne 2006.
42. PV III.285 (ed. Sāṃkṛtyāyana): "tasmād bhūtam abhūtaṃ vā yad yad evābhibhāvyate / bhāvanāpariniṣpattau tat sphuṭākalpadhīphalam."

4

Dharmakīrti's Discussion of Circularity

Pascale Hugon

In the mutually critical dialogue about the import of words between Indian realists and the proponents of the theory of apoha, a number of shared ideas related to facts of language figure as touchstones when putting the respective opponent's theory to the test. Many of the disputations thus concern grammatical notions such as the use of subject and predicate terms, gender and number, qualification, and coreference. On a more general level, both parties are equally challenged to account for the practical success of language in transactional usage within a social community, that is, for the fact that we do set verbal conventions and successfully apply them. Along these lines, detractors of the apoha theory contest the possibility of establishing a verbal convention if the meaning of a word were to be the "exclusion of what is other" (*anyāpoha*). One famous argument to this effect points out that the apoha theory involves a problematic circularity, insofar as the understanding of what is excluded from the other would have to rely on what is "other than that," which in turn is contingent upon knowing what is "that." This objection is probably also the first that comes to mind when considering a classical example like the word "cow": if the meaning of "cow" is the "exclusion of noncow," how can one know what is not a cow if one does not know what a cow is in the first place?

In the following, I will first consider the two sources that are most often cited in relation to this objection, namely, Uddyotakara's *Nyāyavārttika* (NV) and Kumārila's *Ślokavārttika* (ŚV),[1] and then turn to Dharmakīrti's discussion of this argument in the *Pramāṇavārttika* (PV).[2] Dharmakīrti's discussion involves an interesting feature, namely, he does not offer a straightforward answer in defense of the apoha theory, but proceeds through a counterargument alleging that the realist himself has to face a problem of circularity. I will present the various stages of this discussion and attempt to evaluate to what extent Dharmakīrti's response resolves this difficulty in the apoha theory.

UDDYOTAKARA'S ARGUMENT

Although Kumārila's wording stands out as the "standard form" of the circularity argument, Uddyotakara's critique of apoha in the NV already includes a similar objection that is interpreted as a circularity argument by the NV's commentator Vācaspatimiśra.[3] Uddyotakara's line of argument relies on two premises that spell out his own scenario: (1) one cannot negate what has not been previously understood; (2) an initial understanding occurs when there is a positive denotation.[4] Thus, according to Uddyotakara, a word having a positive object is a condition for an initial, affirmative understanding, which in turn is a condition for a potential subsequent negation. Conversely, without a positive object of the word, there can be no initial affirmative understanding, and hence no subsequent negation. For this reason, proponents of the apoha theory who do not admit, for example, that the word "cow" denotes a positive cow are denied the possibility of an affirmative understanding (of what is a cow) and furthermore of a negative understanding (of what is not a cow).[5]

According to Vācaspatimiśra, Uddyotakara's argument brings to the fore an "unanswerable interdependence" (*duruttaram itaretarāśrayatvam*) between the understanding of *x* and the understanding of non-*x*: not only is the latter, as a negation, dependent on the former, but the understanding of *x* is also, insofar as it amounts to the understanding of what is excluded from non-*x*, contingent upon the latter.[6] This interpretation reflects the outline of Kumārila's criticism of apoha that I will consider later. However, one should note that Uddyotakara does not introduce the notion of "interdependence" in his formulation of the argument, whose focal point is, I think, the necessity of a positive denotation as a condition for any understanding. In other words, for Uddyotakara, the apohavādin's failure to understand *x* is not so much a consequence of the dependence of the understanding of *x* on that of non-*x* as an upshot of his refusal to admit something positive as being a word's denotation.

KUMĀRILA'S ARGUMENT

Kumārila devotes a whole section of the ŚV to criticizing the apoha theory.[7] Notably, the specific formulation "exclusion of what is other" generates an inquiry into what this "other" is. For instance, what is the "noncow" (Skt. *a-go*) that is to be excluded to understand the meaning of the word "cow"? Two options are considered: (1) noncow is that which is not expressed by the word "cow" (*gośabdānabhidheyaḥ*); (2) noncow is that which is other than (a) cow (*gor anyaḥ*).[8] Each option is shown to give rise to a major problem for the apohavādin: in the first case, one that I will term "nongeneralization," in the second, that of circularity.

Nongeneralization

Let us briefly consider the first issue, which will be met again in Dharmakīrti's discussion of circularity. Asserting that noncow is that which is not expressed by the word "cow" raises the question how one is able to know what is not expressed by that word. Were the Buddhist to contend that it can be identified as that to which the word "cow" is not applied at the time of setting the convention, he would incur the consequence that the word "cow"—which is meant to be a class word—does not express something general. Indeed, since the Buddhist does not admit real universals, the only thing present at the time of setting the convention is a particular animal. Therefore, what the word "cow" is not applied to is absolutely everything else. And what is excluded from everything else is, in turn, only that particular animal. Hence, for the Buddhist, the word "cow" would end up denoting just this single instance.[9]

This argument points out that the experience of a single entity does not enable an understanding of an "other," the exclusion of which, in turn, would generate something like a "class," because what is "other" would unavoidably be everything else. The nongeneralization objection can also be seen to some degree as representative of a type of circularity problem insofar as the "exclusion of other," if it is to be construed with a particular as its starting point, can only bring us back to where we started, namely, to a particular. The "exclusion of what is other" is therefore unfit to fulfill the function of a universal.

Circularity

In the second option considered by Kumārila, "noncow" is analyzed as the negation of "cow" conveyed by the particle "non-" (in Sanskrit by the initial prefix "*a-*" in "*a-go*"). Following the precept (already met in Uddyotakara's argument) that "one cannot negate what is not known,"[10] the understanding of the negation "noncow" requires that of the negandum "cow." At this point the apohavādin is offered two options: (1) What is (a) cow is established positively, which allows for the understanding of the negation "noncow" but makes the whole theory of apoha useless; if cow is established positively, what is the need of understanding it through the exclusion of noncow? (2) Cow is to be established as the exclusion of noncow. Although "exclusion" is not rephrased as "negation," the object to be excluded, just like the negandum before, needs to be ascertained. Hence, the establishment of what is a cow requires the understanding of "noncow." This second option thus makes the understanding of "cow" dependant on the understanding of "noncow," and since the latter is already considered to be dependent on the former, any attempt to understand either one or the other generates a problem of circularity. In Kumārila's words this problem is termed "interdependence" (*anyonyasaṃśrayaḥ*):

> This noncow that is to be eliminated [should be] established, and it has the nature of the negation of cow. There, one should say [what is] precisely this cow that is negated by the negative prefix "non-." If this [cow] has the nature of the exclusion of noncow, there is interdependence. And if [one concedes that] cow is established in order that what is to be excluded [is established], the idea of apoha would be vain.
>
> When cow is not established, there is no [established] noncow. And in the absence of that [established noncow], how [could one understand] cow [if it is to be the exclusion of the latter]?[11]

The circularity involved here is not one that brings us back to our starting point; on the contrary, it is a consequence of the absence of a starting point, either in the form of the understanding of what is x or of what is non-x. Like nongeneralization, circularity threatens the possibility of establishing a fruitful convention (*saṅketa*).[12] In the first case, a convention could only generate an overrestricted application of the word insofar as it fixes the word for an individual instance rather than for a class. In the second case, the setting of a convention is made impossible by the fact that the cognizer knows neither what constitutes the domain to which the word should apply nor what constitutes the domain to which it should not apply.[13] This argument is to be understood as a reductio ad absurdum, because the possibility of establishing conventions that fix the use of words in this manner is taken for granted. The apohavādin must therefore admit that the object of a word is not an "exclusion of what is other," but something positive.[14]

DHARMAKĪRTI'S DISCUSSION OF CIRCULARITY

When dealing with the charge of circularity, Dharmakīrti, as mentioned earlier, does not provide a straightforward answer, but proceeds via a tu quoque argument.[15] In outline the discussion is as follows:

1. Statement of the objection (PV I.113c–114)
2. Dharmakīrti's counterattack and the realist's answer
   - 2a. Dharmakīrti imputes the same fault to the realist: the elimination of non-x when setting a convention generates a problematic interdependence (PV I.115–116)
   - 2b. The realist refuses the grounds for the imputation of the fault: there is no elimination of non-*x*; rather, there is mere positive understanding of *x* (PV I.117–118b_1)
   - 2c. Dharmakīrti refutes the realist's rejoinder: a restriction qua elimination of non-*x* is necessary to avoid indeterminacy (PV I.118b_2–d)
   - 2d. The realist gives his solution to the problem of circularity: that which is non-*x* is easily understood contrastively with regard to an experience of *x* (PVSV 60, 5–10)
   - 2e. The realist claims that his answer does not fit the Buddhist's theory: for the Buddhist, everything is different by contrast from the experience of a particular instance (PVSV 60, 10–13)
3. Dharmakīrti's position
   - 3a. Adaptation of the realist's solution: the judgment of sameness (*ekapratyavamarśa*) allows one to distinguish a positive and a negative domain of instances (PV I.119)
   - 3b. Establishment of the background conditions for the Buddhist's answer: although particulars are all distinct, they do cause judgments of sameness (PV I.120–121)

1. Dharmakīrti's restatement of the charge of circularity sums up the two elements that are said to generate a problem for the apohavādin: (1) the understanding of non-*x* and of *x* qua "exclusion of non-*x*" are interdependent; (2) neither that which is *x* nor that which is non-*x* is known initially. The consequence is the impossibility of knowing either of the two (let alone both) and hence of establishing a convention.[16]

2. The tu quoque argument deserves some attention, for one can indeed wonder how the realist can be charged with circularity, since it does not appear that he resorts to the exclusion of what is other, and he actually initially claims that he does not (2b). To this effect, Dharmakīrti invokes the necessity of a restriction in the setting of a convention to guarantee that the application of the convention will not be deviant; it is not enough to know what a word applies to, one must also make sure that it does not apply to anything else. The given rule is, nonelimination at the time of setting a convention prevents elimination at the time of application, as is illustrated by the fact that jackfruit trees and sandalwood trees not being eliminated at the time of setting the convention "tree" prevents their being set aside when applying the convention "tree."[17] Hence, the request "bring a tree" can result in getting a jackfruit tree, a sandalwood tree, or any other instance of tree. But another consequence of this rule is that the nonelimination, for instance, of cows or stones at the time of setting the convention "tree" could result in getting a cow or a stone—in other words, a nontree—with the request "bring a tree." In brief, without the said restriction, all actions based on linguistic conventions would have an uncertain outcome.

This restriction amounts to an elimination (*vyavaccheda* or *nirākaraṇa*) of what is other. Now, Dharmakīrti offers two versions of how the restriction can take place when setting a convention: (1) In the "strong version" (PV I.115), the elimination of what is other occurs before and as a condition for the understanding of what is *x*. (2) In the "weak version" (PV I.118), the restriction is inherent in the affirmative establishment (such as "this is a tree") in the form of an implicit negation. That is to say, "this is a tree" is to be understood as "*only this* is a tree."[18] The word "only" (in Skt., the particle *eva*) has the function of eliminating the possibility that something other than "this" could qualify as a tree.[19]

The realist's first reaction is to deny that any negation or restriction occurs when setting a convention (2b). But Dharmakīrti's argument for the necessity of a restriction to avoid uncertainty (2c) is compelling enough to make the realist settle for the weak version of the restriction thesis. Making the elimination of what is other a constitutive element of the setting of a convention unavoidably leads to the question of the interdependence between the understanding of what is *x* and what is other than *x*, because one cannot eliminate what is not known, and knowing what is non-*x* presupposes knowing what is *x*. But why would the realist have to concede that neither one is known, for he can indeed appeal to the understanding of *x* through the perception of the universal? The reason supplied here by Dharmakīrti is that before and even during the setting of a convention, the person for the benefit of whom the convention is made knows neither what is *x* nor what is not *x*; indeed, it is to establish this that the convention is set.[20] Thus whether in the strong or in the weak version of the restriction thesis, there is both interdependence between the two understandings and the impossibility of ascertaining either one of the two independently even for the realist. Hence, the latter has to confront the problem of circularity.

2d. The realist maintains that no such difficulty is involved in his system: the positive understanding of what is (an) *x* results from the perception of the universal, and the understanding of what is not (an) *x* is acquired by contrast. This contrast is described (1) from a subjective point of view—one understands something as "other than *x*" by experiencing a cognition different from the one that occurred when perceiving *x*—and (2) from an objective point of view—one understands that when the universal perceived at the time of setting the convention is not present, what is being seen is "other than *x*."[21]

The understanding of what is "other" is specified here as being something that the cognizer can acquire himself (*svayam*); the ability to differentiate situations on the basis of what is seen and experienced is not something he needs to be taught. Not only will any subsequent experience of something "other" be recognized as such, the cognizer does not have to wait until he actually experiences something different to understand that any subsequent different experience will be an experience of something other. Symmetrically, the present experience itself is ascertained as different from those presenting another aspect. The understanding of what is not (an) *x* thus prevents any overapplication of the convention as required. It remains dependent on the initial positive understanding, but as the latter can be ascertained independently, although the two are interdependent, no problematic circularity follows.

2e. The realist claims that such a solution is not available for the apohavādin.[22] Indeed, the realist's account relies on the existence of a real universal perceived at the time of setting the convention. But according to the Buddhist, only a particular is perceived at the time of the convention being set. Thus, we again meet the problem of nongeneralization already mentioned earlier: contrasting the perception of a particular with other perceptions only results in differentiating that particular from everything else. The Buddhist is thus left without a suitable positive starting point.

3. In response to both this last argument and the initial objection of circularity, Dharmakīrti proceeds to adapt the realist's explanation in compliance with the Buddhist's ontological commitments.[23] The apohavādin does not have at his disposal as an objective pole of contrast a real universal

whose presence and absence can be contrasted, and as we have seen, basing the contrast on a particular would bring up the problem of nongeneralization. Nevertheless, the subjective pole of contrast remains available in the form of a cognition termed the "judgment of sameness" (*ekapratyavamarśa*) that is seen to occur after experiencing certain particulars but which does not occur after experiencing others. Whereas for the realist the cognition of a certain aspect was caused by the universal, the judgment of sameness is not caused by any real existing sameness. It is caused by distinct particulars and takes the form of an aspect of conceptual thought superimposed upon them in the form of a unique nature. Just like the "aspect of cognition" invoked by the realist, the judgment of sameness can be the basis of contrast with other judgments and situations in which such a judgment is absent. Unlike the individual cognitions (conceptual and nonconceptual) qua mental events, the judgment of sameness is distributive; it does not occur only when experiencing the particular present at the time of the convention being set. The "judgment of sameness" can thus play the same role as the realist's universal: it allows for the determination of the positive domain and, through contrast, of the negative domain. As stated by Dharmakīrti, "the cognizer who relies upon the one thing that is a knowledge called the "judgment of sameness" himself distinguishes the objects that are the cause of that [judgment] and not that [judgment]."[24]

Dharmakīrti's commentators and followers are very explicit about the fact that the cognizer differentiates between particulars that have a certain effect and those that do not *before* the actual setting of the convention.[25] This claim contrasts with an argument Dharmakīrti sets forth twice in his discussion with the realist, namely, that the cognizer knows neither what is *x* nor what is non-*x* before (or even at the time of setting) the convention, and that the convention is meant to enable him to make that distinction adequately. We have seen that the realist succeeds in meeting this requirement: the understanding of *x* occurs at the time of setting the convention and is followed by the understanding of what is other. Moreover, the realist specifies that the understanding of non-*x* is something that the cognizer can acquire himself (*svayam*). He does not have to be taught either how to perceive a universal or how to contrast this experience with other experiences. According to the apohavādin's version, however, the "judgment of sameness" is unlikely to occur when experiencing the particular present at the time of setting the convention unless this particular is apprehended as being part of a similar domain. What Dharmakīrti's commentators thus make clear is that, indeed, similar and dissimilar domains have already been established and the convention merely fixes a word to refer to this specific dichotomy. In Dharmakīrti's account, this differentiation is also something that the cognizer is said to do "himself" (*svayam*). This spells out the fact that the capacity to distinguish particulars on the basis of their causal capacities is not something that is taught in convention setting, nor does the setting of a convention actually establish this distinction. On the contrary, it is the differentiation that prompts the establishment of conventions, and it is a condition for learning the use of words. Thus, notably, Śāntarakṣita sums up his answer to the circularity problem in the following verse: "Cows and noncows are well established due to different judgments. It is only the word ["cow"] that is not established, and it is applied according to the wish of the speaker."[26]

The difficulty that remains, however, for the Buddhist's account to be satisfactory lies in the possibility of such a judgment of sameness occurring in the absence of any similarity between the unique particulars. The question how distinct particulars can have the same effect was dealt with earlier by Dharmakīrti (see the example of the antipyretic plants in PV I.73–74), and the nature of the "judgment of sameness" has been explained in PV I.109. Thus, at that point in the discussion on circularity, the apohavādin takes the possibility of a judgment of sameness for granted. Still, in the second part of his answer (3b), Dharmakīrti repeats some of the elements of the previous discussions. Whereas the first half of PV I.119 offers a solution in the form of a "positive starting point" qua the judgment of sameness, the emphasis, in the part of Dharmakīrti's answer found in PV I.119d–121 and his prose autocommentary, resides in the notion of "difference"; in this account, the prominence of a positive experience gives way to the mutual differentiation of particulars in view of their respective causal capacities. Accordingly, the word fixed at the time of setting the convention is said to apply to the distinction.[27]

ANALYZING THE CHARGE OF CIRCULARITY AND DHARMAKĪRTI'S ANSWER

The Charge of Circularity

To ensure a fruitful transactional usage based on a common language, verbal conventions must be established in such a way that the scope of application of a word is defined both positively (the language user must know what the word applies to) and negatively (the user must know what it does not apply to). The two respective domains—whether they are considered as "what the word does (or, respectively, does not) apply to" or "what is (or, respectively, is not) *x*"—are disjoint and exhaustive and are interdependent insofar as each is the complement of the other. It is nonproblematic to *describe* one domain as being constituted by what is not in the complementary domain; it is not even completely tautological insofar as it indicates that the domain under consideration is not unlimited. It is possible as well to *understand* the scope of one domain as a function of the complementary one, provided the latter is known. This interdependence causes difficulties when the scope of each domain is not yet known and each domain is only understandable as the complementary domain of the other. In the texts under consideration, what we call the "circularity problem" is spelled out by the two conjoint conditions of (a) interdependent understanding (*anyonyāśraya, anyonyasaṃśraya*) and (b) neither of the two domains being understood independently (*ekagrahābhāva*).[28] In such a case, each domain is circumscribed only in terms of its complementary domain, so that the location of the boundary between the two domains cannot be fixed.

Kumārila presumes, when raising the circularity objection, that the apohavādin replaces the positive knowledge pertaining to the positive domain—in agreement with our intuition that the instances to which a word applies are apprehended as having something in common—by one that relies on the prominent understanding of the negative domain. Moreover, the understanding of the reference of a word on the basis of the latter is pictured as a mental operation similar or equivalent to an operation of negation. Kumārila's assumption—following the intuition that the instances of the negative domain do not have anything in common—is that a negative domain can only be determined once the positive domain has been understood, and this requires a common property. However serious the charge of circularity may appear, its weight mostly relies on presuppositions pertaining to the apoha theory, such as considering exclusion to be an operation of negation and assuming the preliminary establishment of the negative domain.

The Tu Quoque Argumentation

The strategy behind Dharmakīrti's reply deserves to be considered here, for his answer involving the "judgment of sameness" could have been brought forward in the first place, with no need to embark on the somewhat complicated counterargument discussed earlier. This is what some later authors such as Śāntarakṣita have done.[29] Tu quoque argumentation is used in disputation as a type of ad hominem argument whose purpose is to contest or deny the opponent's right to make an objection on the grounds that the opponent himself is guilty of the fault he is ascribing. It is usually considered a type of informal fallacy insofar as by replying in this way the debater does not address the issue itself. This type of answer to an objection can be contemplated in particular if the debater does not want to acknowledge that he is guilty of the charge ascribed to him—the tu quoque thus works as a kind of "escape strategy." It can be contemplated as well if, on the other hand, the debater acknowledges the charge but argues that it does not qualify as a fault; he thereby seeks to settle for a "two wrongs make a right" or a "partner in crime" resolution of the dispute.

This is obviously not what Dharmakīrti has in mind when resorting to a tu quoque argument, for he is neither acknowledging the circularity of the apoha theory nor is he trying to avoid the issue. And, I would add, he is also not aiming at proving that the realist's view actually involves a problematic circularity. The chosen strategy can be seen to serve here a double purpose: (1) By charging the opponent with the fault of circularity, Dharmakīrti prompts an answer on the opponent's part that he then adapts to his own case. Rhetorically speaking, it is a good move because the opponent cannot object to Dharmakīrti's final response insofar as it would amount to criticizing his own answer. What the opponent can do, however, as we see in the discussion in the PV, is object to the applicability of this answer to the other's (here Dharmakīrti's) case. Additionally, such use of tu quoque argumentation is especially convenient if the debater does not have a ready-made answer to the objection addressed to him, but this is also obviously not Dharmakīrti's problem. The interest he can have in using this sort of argumentation lies rather in the fact that bringing to the fore the apohavādin's answer as an adaptation of the realist's answer instead of stating it directly reveals how much the two views actually have in common, mostly in terms of the requirements involved in setting conventions and of the phenomenological description of the process, without however making concessions concerning the ontological grounding. (2) Second, here the tu quoque argument is a means to "defuse" the objection in an indirect way. Indeed, the attribution of the circularity objection to the realist allows the apohavādin to allude to the fact that the charge actually has to be nuanced. The realist's first rejoinder against the charge of circularity deriving from the "strong version" of the restriction thesis is a hint to this effect. As one sees at the outcome of the discussion, the apohavādin does not subscribe to the idea that knowledge of the negative domain is grounds for a negation enabling a positive ascertainment any more than the realist does: the positive domain is not understood merely as complementary to its complementary domain but through the ascertainment of a judgment of sameness.

In the works of Dharmakīrti's successors, the answers to the circularity objection pass over a number of the steps representative of Dharmakīrti's strategy. For instance, the Tibetan scholar Sakya Pandita (sa skya paṇḍita), in his epistemological work, the *Treasure of Reasoning* (RT), mentions the tu quoque argument but does not account for the prose passage following PV I.118 in which the opponent actually comes up with an answer and switches the burden of proof back to the Buddhist by claiming that this answer is not fit for the latter (2d and 2e). The Buddhist response thus reads like a refutation of the realist opponent, followed by the apohavādin's own answer to the initial objection.[30] The shortening of the argument is even more significant in Śāntarakṣita's discussion of apoha in the TS, to which I referred earlier. The response (*uttarapakṣa*) corresponding to Kumārila's circularity argument (*pūrvapakṣa*) is reduced to an equivalent of PV I.119 in TS 1063 (cited earlier), without any mention of the tu quoque argument. In both cases, the Buddhist's answer is presented independently of the realist's. Maybe these authors considered it problematic that the apohavādin's answer would be adapted from the view of the opponent or even display any similarity to it.

DHARMAKĪRTI'S ANSWER—A SATISFACTORY REPLY?

As presented earlier, Dharmkīrti's answer involves the adaptation of the realist's answer in the form of an appeal to a "judgment of sameness," as well as a brief restatement of the earlier explanation concerning the possibility and the conditions for the arising of this judgment. Additionally, Dharmakīrti inflicts a final blow on his opponent by rejecting the possibility of a universal that could be the basis for the differentiation of positive and negative domains through perception and nonperception.[31] The discussion on circularity closes on this last argument. Once the ontological basis of the realist's scenario is refuted, Dharmakīrti's scenario remains here the only option. As the entire response to the charge of circularity relies on an appeal to the evidence of judgment, it seems that what has been achieved in the course of this discussion is mostly to move the problem back one step, out of the linguistic domain and into the conceptual, to the question of the ability to distinguish positive and negative domains that is to precede the setting of conventions. Dharmakīrti's answer thus stands out as a suitable option only on the condition that this ability itself is satisfactorily grounded. And this brings us back to the more fundamental question of how the apohavādin can account for our acquisition of the basic ability to grasp as similar things that are in reality different.

Notes

My thanks to Ms. Peck-Kubaczek for correcting my English.

1. Karṇakagomin (ed. Sāṃkṛtyāyana) cites both Uddyotakara's NV and Kumārila's ŚV when commenting on PV I.113c–114 (PVSVṬ 233, 15–17 and 18–23), but Śākyabuddhi and Śaṅkaranandana only mention as the originator "Kumārila, etc." (PVṬ$_{\text{Śāk}}$ D.130b4; PVṬ$_{\text{Śaṅk}}$ D.279a4, "gZhon nu ma len la sogs pa"). Śāntarakṣita cites exclusively Kumārila on this issue (see note 7).
2. PV I.113c–121 (ed. Gnoli).
3. Uddyotakara's arguments against apoha are found in NV 679, 5–689, 10 (ed. Taranatha Nyaya-Tarkatirtha et al.) in the context of his commentary ad *Nyāyasūtra* II.66 (*vyaktyākṛtijātayas tu padārthaḥ*). They follow the statement of the apohavādin's arguments (namely, Dignāga's) against the Naiyāyika view (*pūrvapakṣa*) and Uddyotakara's defense (*uttarapakṣa*). Uddyotakara's critique of apoha involves three points: (1) the critique of Dignāga's statement "na jātiśabdo bhedānāṃ vācako" (PS V.2; NV 679, 5–680, 18); (2) the discussion of coreference (*sāmānādhikaraṇya*) (NV 680, 19–686, 6); (3) nine arguments against apoha as such (NV 686, 7–689, 10). Some of them are cited in TS 981–999 (ed. Shastri), although not the first of these nine, the one discussed in the present paper. For a synopsis, see Much 1994.

   The first of the nine arguments against apoha as such, that is, the contention that the meaning of a word is "the exclusion of what is other," reads as follows: "yat punar etat—anyaśabdārthāpohaḥ śabdārtha iti, tad apy ayuktam; vidhānaśabdārthasambhave saty ādyā pratipattiḥ, yadi vidhānaśabdārtho bhavati tato vidhīyamānaśabdārthapratipattau satyāṃ tasyānyatra pratiṣedha ity upapannaḥ pratiṣedhaḥ, yasya punar vidhīyamānaḥ padārtho nāsti tasyādyāṃ pratipattim antareṇa kathaṃ pratiṣedhaḥ, yāvac cetaraṃ na pratipadyate tāvad itaraṃ na pratiṣedhatīti. yathā gaur iti padasyārtho 'gaur na bhavatīti, yāvac ca gāṃ na pratipadyate tāvad gavi pratipattir na yuktā, yāvac ca gāṃ na pratipadyate tāvad agavīty ubhayapratipattyabhāvaḥ" (NV 686, 7–12).

   The partial citation of Uddyotakara's argument in PVSVṬ 233, 15–17 contains problematic readings: "sa yāvac *cāgāṃ* na pratipadyate tāvad *agavi* pratipattir na yuktā. yāvāc ca gāṃ na pratipadyate tāvad *gavīty* ubhayapratipattyabhāva iti."
4. Uddyotakara's scenario is in keeping with the Naiyāyika's view about absences, according to which there is no absence without an existing counterpositive. It follows that an absence cannot be apprehended unless its counterpositive has been apprehended previously.
5. In Uddyotakara's argument cited in note 3, Jhā 1984 reads: "gaur iti padasyārtho 'gaur na bhavatīti" with "pratipattir" in "gavi pratipattir na yuktā" (Jhā [1984,

1055] translates this as "he can form no idea of the cow, in the form that 'the word cow denotes what is not the noncow'"). I prefer to read "the meaning of the word 'cow' is [that which] is not a noncow" as an illustration of "yasya vidhīyamānaḥ padārtho nāsti" ("he for whom there is no object of the word that is being taken positively"), syntactically independent of the two "yāvat . . . tāvat" clauses.

6. NVTṬ (ed. Taranatha Nyaya-Tarkatirtha et al.) 686, 18–19: "agovyāvṛttirūpaṃ cet na tad asiddhaṃ gavi śakyaṃ grahītum. agauś ca goniṣedhātmeti gosiddhim apekṣata iti duruttaram itaretarāśrayatvam." Vācaspatimiśra clearly interprets Uddyotakara's "insofar as [what is a] cow is not understood" (yāvac ca gām na pratipadyate) as "insofar as the exclusion of noncow is not established."
7. ŚV V (Apohavāda) 1–176, pp. 400–435 (ed. Shastri). A number of Kumārila's objections are cited in TS 914–980 and 1000–1001.
8. See, respectively, ŚV V (Apohavāda) 81–82 and 83–85b. I follow here Śrī Pārthasārathi Miśra's *Nyāyaratnākara* (NR), where the first argument is introduced by "kiṃ ca, ko 'yam agauḥ, yasyāpoho dṛśyate. yo gośabdasyānabhidheyaḥ so 'gauḥ syāt, kasya tadanibhidheyatvam iti na jñāyata ity āha" (415, 19–20), and the second by "atha yo gor anyaḥ so 'gauḥ. tatrāha dvayena" (415, 27).
9. ŚV V (Apohavāda) 81–82 (cited in TS, ed. Shastri, 940–941): "agośabdābhidheyatvaṃ gamyatāṃ ca kathaṃ punaḥ / na dṛṣṭo yatra gośabdaḥ sambandhānubhavakṣaṇe // ekasmāt tarhi gopiṇḍād yad anyat sarvam eva tat / bhaved apohyam ity evaṃ (TS etan) na sāmānyasya vācyatā."
10. See NV cited in note 3: "yāvac cetaraṃ na pratipadyate tāvad itaraṃ na pratiṣedhatīti."
11. ŚV V (Apohavāda) 83–85b (cited in TS 942–944b): "siddhaś cāgaur apohyeta goniṣedhātmakaś ca saḥ / tatra gaur eva vaktavyo nañā yaḥ pratiṣidhyate // sa ced agonivṛttyātmā bhaved anyonyasaṃśrayaḥ / siddhaś ced gaur apohyārthaṃ (TS apohārthaṃ) vṛthāpohaprakalpanā (TS °prakalpanam) // gavy asiddhe tv agaur nāsti tadabhāve ca (TS tu) gauḥ kutaḥ."
12. See Kamalaśīla's introduction to each argument in TSP 370, 1 ad TS 940–941 (= ŚV V [Apohavāda] 81–82): "punar apy apohe saṅketāsambhavaṃ pratipādayann āha"; and TSP 370, 9 ad TS 942–943b (= ŚV V [Apohavāda] 83–85b): "itaś cetaretarāśrayadoṣaprasaṅgād apohe saṅketo 'śakyakriya iti darśayann āha."
13. This consequence is the same as that in Uddyotakara's argument, namely, *ubhayapratipattyabhāvaḥ*.
14. See the conclusion of the argument in NR 416, 5–6: "iti prathamaṃ vidhirūpagaur abhidhātavyaḥ."
15. Dharmakīrti deals with this objection in PV I.113c to PV I.121 in the course of the section on universals (*sāmānyacintā*).
16. PV I.113c–114 (ed. Gnoli): "avṛkṣavyatirekeṇa vṛkṣārthagrahaṇe dvayam // anyonyāśrayam ity ekagrahābhāve dvayāgrahaḥ / saṅketāsambhavas tasmād iti kecit pracakṣate." PVSV 58, 22–25: "yady avṛkṣebhyo bhedo vṛkṣas tasyāvṛkṣagrahaṇam antareṇa tathā grahītum aśakyatvāt, avijñātavṛkṣeṇāvṛkṣasyāpi tadvyavacchedarūpasyāparijñānāt, buddhāv anārūḍhe 'rthe na saṅketaḥ śakyata ity eke."
17. PV I.116: "anirākaraṇe teṣāṃ saṅkete vyavahāriṇām / na syāt tatparihāreṇa pravṛttir vṛkṣabhedavat." PVSV 59, 13–14: "na hi saṅkete parāvyavacchedena niveśitāc chabdād vyavahāre tatparihāreṇa pravṛttir yuktā, śiṃśapādibhedavat."

18. "This," from the realist's point of view, is the universal "treeness" (*vṛkṣatva*).
19. The technical term given to this type of negation implicit in a positive statement is "elimination of the link with other" (*anyayogavyavaccheda*). This technical term is not used in this context by Dharmakīrti, who only provides the formulation *ayam eva* (see PV I.118 and PVSV 60, 4), but it is introduced for instance by Sakya Pandita (sa skya paṇḍita) when commenting on this passage of the PVSV (see RTRG 107 ad RT IV.40, ed. Nordrang Ogyen). On the particle *eva*, see Gillon 1999; Ganeri 1999; Kajiyama 1973.
20. Cf. PVSV 59, 7–8 ad I.115: "na hi tadā pratipattā vṛkṣaṃ vetti nāvṛkṣaṃ tajjñānāyaiva tadarthitayopagamāt." The reason given here by Dharmakīrti may be understood as mirroring the last part of Kumārila's argument (not cited in the PV's rephrasing), namely, that if the apohavādin concedes a prior understanding of what is a cow, he has no need to posit that "cow" is to be understood via the exclusion of what is not cow. Parallel to this, here, if the realist were to admit that one knows what is a cow (and what is not), there is no need to establish a convention with that aim.
21. PVSV 60, 5–10: "na doṣaḥ, dṛṣṭaviparītasya sujñānatvāt. ekaṃ hi kiñcit paśyato 'nyatra tadākāravivekinīṃ buddhim anubhavatas tato 'nyad iti yathānubhavaṃ tadvivecano vaidharmyaniścaya utpadyate. sa hy ayam eva vṛkṣa iti pradarśya vyutpāditaḥ. yatraiva taṃ na paśyati tam evāvṛkṣaṃ svayam eva pratipadyate."
22. PVSV 60, 10–13: "nedaṃ vyavacchedavādinaḥ sambhavati. ekatra dṛṣṭasya rūpasya kvacid ananvayād darśanena pratipattau vyaktyantare 'pi na syāt tathā pratītiḥ."
23. He introduces PV I.119 with the words "evaṃ tarhi tatrāpi tulyam etat, yasmāt."
24. PV I.119: "ekapratyavamarśākhye jñāna ekatra hi sthitaḥ / prapattā tadataddhetūn arthān vibhajate svayam."
25. See for instance PVSVṬ 238, 26–239, 1: "ekaśākhādimadākāraparāmarśahetūn tadviparītāṃś ca pṛthakkaroti svayam eva saṅketāt prāg api"; PVṬ$_{\text{Śāk}}$ D.134a1: "brDa'i mdun rol du yang bdag nyid rnam par 'byed par byed." Sakya Pandita also in RTRG 107 ad IV.41 (ed. Nordrang Ogyen): "kho bo cag brda byed pa na brda'i snga rol du yal ga dang ldan mi ldan mtshan nyid rnam par phye nas." For Śāntarakṣita, see TS 1063 in note 26.
26. TS 1063: "gāvo 'gāvaś ca saṃsiddhā bhinnapratyavamarśataḥ / śabdas tu kevalo 'siddho yatheṣṭaṃ samprayujyate." Śāntarakṣita thus accounts for the contrast between different judgments (*bhinnapratyavamarśa*). In the preceding verses, the contrast is between a judgment and its absence (TS 1059: "tādṛkpratyavamarśaś ca vidyate yatra vastuni / tatrābhāve 'pi gojāter ago 'pohaḥ pravartate"; TS 1062: "tādṛkpratyavamarśaś ca yatra naivāsti vastuni / agośabdābhidheyatvaṃ vispaṣṭaṃ tatra gamyate").
27. PV I.121b: "uktir bhede niyujyate."
28. The term *anyonyāśraya* (or *°saṃśraya*) and its synonym *itaretarāśraya*, which can be literally translated as "interdependence," already occurs in grammatical literature in Patañjali's commentary to Pāṇini's Astdh sūtra 4.1.3. The earlier explanation of the consequences to be drawn from interdependence that make it unacceptable (namely, as seen in Uddyotakara's argument, the impossibility to establish either of the two, and hence the nonestablishment of both) is found in Pakṣilasvāmin's *Nyāyabhāṣya* (NBh). See Oberhammer et al. 1991, 66–67, 125–

126. The fault thus labeled corresponds to the English "circularity." The Sanskrit term *cakrakam* on the other hand is better translated as "circular regress," as it constitutes less a case of "circularity" than one of "regressus in infinitum" (*anavasthā, avyavasthā*). In grammatical literature, it indicates a situation where the application of a transformational rule will induce the application of a second rule, which, in turn, will induce that of the first rule again, and so on. See Oberhammer et al. 1996, 85.

29. See TS 1063 cited above, and note 26.

30. Sakya Pandita concludes his argumentation against the realist by stating that if the latter admits the weak version of the restriction thesis, although conventions can be set and applied, the realist's universal is actually established as an exclusion of what is other. This, on the one hand, means that the opponent cannot escape the problem of circularity he raised in the first place against the apohavādin. Additionally, as several commentators point out, this also stands as an ironic consequence, namely, that the realist actually subscribes to apohavāda! See RTRG 107, 15–20 ad IV.40.

    Let us note in addition that while presenting Dharmakīrti's answer, Sakya Pandita fails to cite the first part of PV I.119, namely, the part in which Dharmakīrti introduces the notion of "judgment of sameness" and attempts to ground the apohavādin's answer in the process of mutual (and simultaneous) differentiation of particulars without relying on a "judgment of sameness" as an initial positive starting point. See RTRG 107, 26–108, 16 ad IV.41.

    The section of the RTRG dealing with the question of circularity is translated in Hugon 2008, 479–485 and is discussed on 205–210.

31. See PVSV 61, 3–8: "na punar ekaṃ vastu tatra dṛśyam asti yasya darśanādarśanābhyāṃ (Gnoli darśānādarśaṇābhyāṃ) bhinnadarśane 'py eṣa vṛkṣāvṛkṣavibhāgaṃ kurvīta. tasya śākhādipratibhāsavibhāgena daṇḍavad daṇḍiny agrahaṇāt. agṛhītasya cāparapravibhāgenānupalakṣaṇāt. ākṛter apy ekatra dṛṣṭāyā anyatra draṣṭum aśakyatvāt. tadatadvator vṛkṣāvṛkṣatve vyaktir ekaiva vṛkṣaḥ syāt."

5

Apoha Theory as an Approach to Understanding Human Cognition

Shōryū Katsura

Śākyabuddhi's (ca. 660–720) analysis of the three meanings of the expression *anyāpoha* had a considerable influence upon the subsequent development of apoha theory in India and Tibet (Funayama 2000; Sakurai 2000; Dunne 2004). As Hisataka Ishida makes clear, Śāntarakṣita (ca. 725–788) developed Śākyabuddhi's classification of *anyāpoha* by interpreting it in terms of the two kinds of negation, namely, "implicative negation" (*paryudāsa*) and "absolute negation" (*prasajyapratiṣedha*), and Śāntarakṣita came to be criticized by Jñānaśrīmitra (ca. 980–1040) and Ratnakīrti (ca. 990–1050).[1] The three meanings of *anyāpoha* are as follows:

1. The excluded particular (*vyāvṛtta-svalakṣana*) that is not the object of a linguistic item (*śabdavācya*) but is rather the basis for practical activity (*vyavahāra*) based on a linguistic item (*śabda*) or an inferential mark (*liṅga*).
2. The mere exclusion of others (*anyavyavacchedamātra*) or mere negation (*pratiṣedhamātra*) that is common to all the particular instances.
3. The representation/image of an object in a conceptual cognition (*vikalpabuddhi-pratibhāsa*) that is the object of a linguistic item.[2]

Since Śākyabuddhi quotes the *Pramāṇavārtika* 1.40cd[3] in his exposition of *the first meaning of anyāpoha*, it is undoubtedly Dharmakīrti who identified the particular (*svalakṣaṇa*) with *anyāpoha* (exclusion/differentiation of others), or more precisely, *anyāpoḍha* (excluded/differentiated from others).

The particular for Dharmakīrti is the sole reality and is truly unique because it is "differentiated" from things of the same kind/class as well as things of other kinds/classes (*svabhāvaparabhāvavyāvṛttibhāgin*).[4]

According to Dharmakīrti, the particular, which consists of an aggregate of atoms in the case of material objects, possesses causal powers (*arthakriyāśakti*) such as "containing water" and "projecting its image into a resulting perception (*pratyakṣa*)." Unless there is a cause for misjudgment (*bhrānti*) or doubt (*saṃśaya*), a perception of the particular can produce a perceptual judgment (*sāṃvṛta*, *smārta*, *ekapratyavamarśa*, etc.) of, for example, "pot." The correct judgment leads to the practical activity (*vyavahāra*) of either obtaining or avoiding the particular that triggered this process of perception and judgment.

A perception cognizes the particular as it really is (*yathābhūtam*) with all its particular characteristics, that is, "differentiations" (*vyāvṛtti*, *vyavaccheda*, *bheda*). A judgment or a conceptual cognition (*vikalpa*), which occurs right after the perception, cognizes only one of the many "differentiations," that is, the universals (*sāmānyalakṣaṇa*), that characterize the particular, by excluding other "differentiations." The particular and perception are beyond our linguistic description but the universal and judgment are closely connected with our linguistic conventions. We say, "this is a pot, nothing else" (*ayaṃ ghaṭa eva, nānyaḥ*). In any case *anyāpoha* is the essential nature and function of a conceptual cognition. And this is *the second meaning* of *anyāpoha*.

The conceptual cognition may take (or, in fact, mistake) the particular as an individual pot "differentiated" from all other pots (*sajātīya-vyāvṛtta*). Or it may take/mistake the particular as a pot in general "differentiated" from things other than pots (*vijātīya-vyāvṛtta*) or as potness, that is, "differentiation" of a pot from things other than pots (*vijātīya-vyāvṛtti*). The image of an individual pot, a pot in general, or potness in the conceptual cognition is *the third meaning* of *anyāpoha*. In this connection it is to be noted that for Dharmakīrti the distinct cognition of a pot (*tat-pariccheda*) is at the same time the exclusion of nonpots (*atad-vyavaccheda*; HB 25–27). In other words, the conceptual cognition has a double nature of affirmation and negation.

Both an inferential mark and a linguistic item can enter into the process of producing a conceptual cognition/judgment. Upon seeing an inferential mark (*liṅga*) such as a bit of smoke rising on a hill, there occurs a conceptual cognition of "smoke" in general. With the help of the recollection of a special relation between smoke and fire (i.e., wherever there is smoke, there is a fire), this cognition produces an inferential judgment that there is a fire on the hill. Similarly, upon hearing a linguistic item (*śabda*) such as "pot" uttered by the speaker, there arises in the hearer a judgment of a linguistic item, "pot." Having awakened the residual impression (*vāsanā*) of the meaning of the linguistic item "pot," it produces in the hearer an inferential judgment that the speaker meant a certain vessel by uttering the linguistic item "pot."

Thus both an inferential mark such as smoke and a linguistic item such as "pot" can produce via perception a conceptual cognition of smoke and of the linguistic item "pot," respectively, which further lead to the conceptual cognition of a fire and of a pot. The distinct cognition of a fire or a pot (*tat-pariccheda*), which is at the same time characterized by the exclusion of nonfires or nonpots (*atad-vyavaccheda*), will lead to a practical activity. Since the practical activity originates from the perception of the particular (p_1) and deals with the particular (p_i) belonging to the same continuum (*santāna*) consisting of $p_1, p_2, p_3, \ldots p_i$, the particular "differentiated" from those of the same kind as well as those of the other kinds is said by Śākyabuddhi to be the basis (*āśraya*) for the practical activity.

Although Śākyabuddhi does not admit the excluded particular (i.e., the first meaning of apoha) to be the object of a linguistic item, Dharmottara and others would admit that it is the indirect object of a linguistic item, the direct object being the universal (*sāmānya, jāti*). For Dharmottara introduced the theory of two kinds of objects, namely, direct (*grāhya*) and indirect (*adhyavaseya*). Thus, the perception can be said to take the excluded particular (i.e., *anyāpoḍha*) as the direct object and the universal, that is, "exclusion of others" (*anyāpoha*) as the indirect object. On the other hand, the conceptual cognition, both verbal and inferential, takes the universal as the direct object and the particular as the indirect object. The perception takes the universal as its indirect object because its direct object, that is, the particular, is characterized by many differentiations (i.e., universals), and the conceptual cognition takes the particular as its indirect object because it is indirectly caused by the particular. Therefore, the excluded particular can be regarded as the indirect object of a linguistic item, to be more precise, that of a conceptual cognition of a linguistic item. The direct object of the conceptual cognition of a linguistic item is the universal or *anyāpoha*, which is *the third meaning of apoha*.

So far I have tried to elaborate upon Śākyabuddhi's account of the three meanings of *anyāpoha*, mainly relying upon Dharmakīrti's apoha theory. If my elaboration is correct, the apoha theory can be regarded as an approach to understanding human cognition. For it explains the whole process of human cognition, beginning from the object of perception and ending with the human action initiated by that perception: Namely, (1) the object of perception is the particular that may be defined as "excluded from others" (*anyāpoḍha*) because it is characterized by various kinds of exclusions/differentiations from others. (2) The particular object gives rise to its perception, which in turn produces a perceptual judgment/conceptual cognition of that particular object. The conceptual cognition, including verbal and inferential cognition, takes as its object one of the exclusions/differentiations (*anyāpoha*) of the particular object that appears in that conceptual cognition with the help of the residual impression. Such an exclusion/differentiation, though only a mental construction, can play the role of a universal (*sāmānya, jāti*) as understood by Indian realists. (3) The nature and the function of conceptual cognition is the exclusion/differentiation of others (*anyāpoha*) that takes the form of a definitive judgment, "this is A and nothing else"; hence, it can give rise to human activity toward the successor moment of the initially perceived particular object.

Now a question arises. Why did Buddhist logicians adopt the apoha theory to explain the process of human understanding? From my reading of Dignāga and Dharmakīrti, I have not yet come across any ready-made answer. What follows is my conjecture.

As George Cardona (1981, 79) has pointed out, Indian grammarians and philosophers applied the principle of induction consisting of *anvaya* (copresence) and *vyatireka* (coabsence) to establish a relationship between two items. When a linguistic item *X* is heard, the meaning *M* is understood, and when a linguistic item *X* is not heard, the meaning *M* is not understood. From the observations of such copresence and coabsence of *X* and *M*, we can inductively establish the semantic relationship that the linguistic item *X* means *M*. Similarly, where there is a fire, there arises smoke, and where there is no fire, there arises no smoke. From the observations of copresence and coabsence of fire and smoke, we can inductively establish the causal relationship between a fire and smoke.

Indian logicians formulated the theory of *trairūpya* (the three characteristics of a valid inferential mark) by applying the principle of copresence and coabsence. Namely, when an item is present (*anvaya*) in instances similar (*sapakṣa*) to what is to be inferred (*anumeya*) and is absent (*vyatireka*) from instances dissimilar (*vipakṣa*) to what is to be inferred, it is a valid inferential mark, provided that it is present in the object of inference (*pakṣa*).

Since smoke is present where there is a fire and absent where there is no fire, we can establish a special relationship between smoke and fire. Smoke is a valid inferential mark that causes us to know the presence of an unperceived fire, that is, what is to be inferred.

Like the grammarians, Dignāga discovered the same principle between a linguistic item and its referent. Namely, a linguistic item, say "cow," is present in (*vṛtti*, i.e., applicable to) a group of similar (*tulya*) four-legged animals with horns and other characteristics, and it is absent from (*avṛtti*, i.e., inapplicable to) a group of the dissimilar (*atulya*) animals, in short noncows. In this connection, it is to be noted that he was well aware of the problem of induction. According to him it is impossible to check all the positive cases of application of a linguistic item because there are innumerable referents of "cow." Thus, it is ultimately impossible to establish the positive relation (*anvaya*) between a linguistic item and its referent. He also recognizes that it is impossible to check all the negative cases for the same reason. Nonetheless, he proposes that it is possible to establish the negative relation (*vyatireka*) between a linguistic item and its referent on the basis of the "mere nonobservation" (*adarśanamātra*) of it being applied to the objects other than the targeted referents. In other words, as long as we do not experience the linguistic item "cow" being applied to animals other than cows, it is possible to say that it refers to a cow. (Needless to say, we are putting aside metaphorical expressions such as "*bāhulikas* are cows.") Dignāga adds that a linguistic item's denotation of its referent is an inference based on "exclusion" (*vyavacchedānumāna*; see Katsura 1991). This argument indicates the supremacy of the negative concomitance (*vyatireka*) over the positive concomitance (*anvaya*) in the linguistic as well as inferential cognition. That, I believe, is the main reason behind the origin and development of apoha theory in the Buddhist epistemological tradition.

As a matter of fact, Dignāga declares at the very beginning of his apoha analysis that both a linguistic item and an inferential mark function by "exclusion of others" (*anyāpoha*).[5] The linguistic item "cow" can refer to a cow as long as it is *excluded* from animals *other* than cows. An inferential mark, smoke, can indicate the existence of a hidden fire as long as it is *excluded* from things *other* than fires. If we rephrase the passive sentence "X is excluded from Y" to the active one "X excludes Y," we can attribute the function of "exclusion of others" (*anyāpoha*) to a linguistic item and an inferential mark and further to conceptual knowledge in general. Following Indian rhetoricians' practice of "metaphorical application" (*upacāra*), both the object and the result of the function, that is, the exclusion of others, can be called "exclusion of others," too. In this way we may obtain *the three meanings* of *anyāpoha* attributed to Śākyabuddhi.

We still face a serious problem in the theory of apoha, which is the problem of universals. The above description of apoha theory tacitly presupposes our ability to group together certain similar objects. How did Dignāga and Dharmakīrti differentiate a cow from noncows and a fire from nonfires? Unlike Indian realists such as the Naiyāyikas and Mīmāṃsakas, they could not resort to universals (*sāmānya, jāti*) to justify their practice of grouping similar things together.

From the outset, Indian Buddhist thinkers tended to take a reductionist position. For example, they analyzed a person into five components (*skandha*), a material body, sensation, perception, volition, and cognition/mind, and considered a person to be a mere concept (*prajñapti*) superimposed upon those five components. Vasubandhu clearly states that when a certain object is mentally analyzed or physically divided up, if it is not recognized anymore, it cannot be regarded as a real entity (*paramārthasat*) and it is a mere conceptual existent (*saṃvṛtisat*).[6] Furthermore, Buddhists take the position that everything is causally produced (*pratītyasamutpanna*) and momentary (*kṣaṇika*). Thus, an individual pot is nothing but a concept superimposed upon a continuum (*santāna*) of momentary phases of an aggregate of atoms that we conventionally call "pot." The universal *potness* also is a concept superimposed as a common property upon a number of individuals that we call "pot." As a matter of fact, Dignāga declares that his *anyāpoha* possesses the essential characteristics of universals, namely, "singularity," "eternity," and "pervading all the individual members."[7] For him, that is what is conveyed by our linguistic activity; in other words, the meaning of a linguistic item is a kind of universal defined as *anyāpoha*. In this sense, Dignāga may be regarded as a kind of *Jātivādin* in Indian semantics.

Now, how could Dignāga group similar things together? He seems to have presupposed a hierarchy of "universals" that must have been commonly accepted by Indian intellectuals of his time. The highest universal/category is "what is to be known" (*jñeya*), which is further divided into "existent" (*sat*) and "nonexistent" (*asat*). What is existent is divided into three categories, namely, "substance" (*dravya*), "property" (*guṇa*) and "action" (*karman*). Substance is divided into "what is made of the earth elements" (*pārthiva*), "what is made of the water elements" (*āpa*), etc. What is made of the earth elements is divided into "tree" (*vṛkṣa*), "pot" (*ghaṭa*), etc. Trees are divided into "Śiṃśapā," "Palāśā," etc. Śiṃśapās are divided into "flowered" (*puṣpita*), "fruited" (*phalita*), etc.

As I once described in detail (Katsura 1979), Dignāga sets up several rules to explain how one universal excludes another. For example, the universal *A* excludes the other universals *B*, *C*, and so forth that belong to the same order of the hierarchy of universals. It does not exclude the universals of the higher orders or those of lower orders. In short, he grouped together similar things on the basis of the hierarchy of universals that might have been accepted by his rivals. Unlike his rival Indian realists, Dignāga has no metaphysical justification for reconstructing this hierarchy. His final resource was our verbal convention (*lokavyavahāra*). He says that we employ a certain linguistic item for a certain group of similar objects only because we follow the general linguistic convention adopted by people. We call a certain object "pot" because people say so. In other words, things are grouped together on the basis of our cultural background. This may remind us of the Sapir-Whorf hypothesis.

Tillemans (in this volume) reports that Dharmakīrti took a different position on this matter. Namely, Dharmakīrti employs the causal relation as the ground for grouping similar things together. Things that are produced by a similar set of causes can be grouped together, and things that produce a similar effect can be grouped together. However, since he presupposes the similarities among causes and effects, he will be further questioned how he can justify those similarities, which will lead to an infinite regress. Besides, as Masahiro Inami demonstrated, Dharmakīrti regards the theory of causation not as the ultimate truth but as the conventional truth (2000). Therefore, the causal relation cannot give the final answer to our initial question. I think Dharmakīrti, like Dignāga, has to rely on verbal convention as the final recourse to explain our act of grouping similar things together.

In a broader sense the Buddhist theory of apoha is definitely an approach to understanding human cognition. It explains every phase of human cognition, beginning with the object of cognition and ending with the judgment that prompts human action or nonaction (*pravṛtti, nivṛtti*). In a narrower sense it is a Buddhist theory of the universal as well as a semantic theory. According to it, a linguistic item (*śabda*) directly refers to the universal, such as a pot in general, that is defined as *anyāpoha* (exclusion of others belonging to the same or different classes). It indirectly refers to a particular object, such as a pot, that is also characterized by *anyāpoha* (exclusion of others belonging to the same as well as different classes). Buddhist logicians do not postulate any real universal to justify our habit of grouping similar things together; instead, they resort in the end to the verbal conventions or linguistic customs that reflect our cultural background. Such a Buddhist attitude reminds me of the theory of pattern recognition of Satoshi Watanabe (1910–1993). He proposed the "Theorem of the Ugly Duckling" and said:

> From the formal point of view there exists no such thing as a class of similar objects in the world, insofar as all predicates (of the same dimension) have the same importance. Conversely, if we acknowledge the empirical existence of classes of similar objects, it means that we are attaching non-uniform importance to various predicates, and that this weighting has an extralogical origin.
>
> When we employ a concept, we usually understand that there is a group of objects corresponding to this concept that any two members of the group resemble each other more than a member and a nonmember. Two sparrows are very much alike, while a sparrow and a rose are not alike. It is natural to translate the term "to resemble" as "to share many predicates in common." But this interpretation can be shown to lead to a denial of the existence of a class of similar objects by the following theorem, which I have dubbed the theorem of the ugly duckling. The reader will soon understand the reason for referring to the story of Hans Christian Anderson, because this theorem, combined with the foregoing interpretation, would lead to the conclusion that an ugly duckling and a swan are just as similar to each other as are two swans. (Watanabe 1969, 376)

Watanabe mathematically proved that any given pair of objects shares exactly the same number of predicates. Therefore, it is meaningless to talk about the resemblance between two objects that may help us to group things together. He thinks that unless we superimpose some cultural bias, it is impossible to differentiate a swan from ducks. In other words, to recognize different patterns in our cognition and to identify a certain object, we must first weigh a number of predicates with our cultural background and determine which predicates are more important or relevant than others. I think Indian Buddhist logicians would have gladly accepted Watanabe's interpretation of pattern recognition.

Notes

1. See Ishida forthcoming. I would like to thank Mr. Ishida for providing me with his preprint.
2. For the critical edition of the relevant passage, please see Ishida forthcoming.

3. PV I.40 (ed. Gnoli): "sarve bhāvāḥ svabhāvena svasvabhāvavyavasthiteḥ / svabhāvaprabhāvābhyāṃ yasmād vyāvṛttibhāginaḥ."
4. Mokṣākaragupta defines the particular (*svalakṣaṇa*) as *sajātīyavijātīyavyāvṛtta* in the *Tarkabhāṣā* (TBh), ed. Rangaswami Iyengar, 21.
5. S V.1: "na pramāṇāntaraṃ śābdam anumānāt tathā hi tat / kṛtakatvādivat svārtham anyāpohena bhāṣate."
6. AK (ed. Pradhan) VI.4: "yatra bhinne na tadbuddhiranyāpohe dhiyā ca tat / ghaṭāmbuvat saṃvṛtisat paramārthasad anyathā."
7. PSV ad V.36d: "jātidharmavyavasthitiḥ."

6

The Apoha Theory as Referred to in the *Nyāyamañjarī*

Masaaki Hattori

The *Nyāyamañjarī* (NMJ) was written by Jayanta, a Naiyāyika active in Kashmir during the second half of the ninth century. It deals with the sixteen topics enumerated at the start of the *Nyāyasūtra* (1.1.1), beginning with instruments of knowledge (*pramāṇa*) and objects of knowledge (*prameya*). But while it ostensibly adopts the format of a commentary on the first and fifth books of the *Nyāyasūtra*, about two-thirds of the entire work is devoted to an examination of instruments of knowledge, and it differs in its approach from other commentaries in that its arguments are developed freely without being overly concerned with the interpretation of the wording of individual *sūtras*. Jayanta was a scholar of great learning who was well versed in the doctrines of many other schools, starting with the Mīmāṃsā school in which his father specialized, and he weaves into his detailed examination of various topics dealt with in the Nyāya school descriptions and critiques of the doctrines of other schools and references to the views of earlier scholars not found in any other works. The *Nyāyamañjarī* is thus a valuable source of material on doctrinal differences within the Nyāya school and on the intellectual climate as a whole during the period when he was active.

In the first section of the fifth *āhnika* of the *Nyāyamañjarī*, Jayanta undertakes an examination of the referent of the word (*śabda*). After having first presented the standard Nyāya view that the word refers to an individual (*vyakti*) qualified by a particular genus or universal (*jāti*),[1] he touches on the counterarguments of the Buddhist school, which reject the reality of the genus or universal (*sāmānya*). He then goes on to deal in some detail with the Buddhist response to the question of why there should arise the unitary concept of *cow* with respect to many mutually different individual cows, such as spotted cows, black cows, and so on, if one does not recognize the existence of the universal. This Buddhist response takes the form of the apoha theory, namely, the view that the referential or denotative function of the word consists of nothing other than the "negation [apoha] of others" (*anyāpoha, anyavyāvṛtti*).

The section dealing with the apoha theory is divided into the following three parts:

1. An overview of the apoha theory (NMJ 10.7–14.13; 275.23–277.17)
2. Kumārila's criticism of the apoha theory (NMJ 14.15–21.15; 277.19–278.25)
3. The Buddhists' rejoinder (NMJ 21.18–29.4; 278.27–282.16)

Following on from 3, rather than presenting his own critique of the apoha theory, Jayanta lays the basis for the widely accepted theory that the individual qualified by a particular genus or universal (*jātiviśiṣṭavyakti*) represents the referent of the word by demonstrating the reality of the genus or universal.[2]

I

The apoha theory was established by Dignāga (ca. A.D. 480–540). At the start of the fifth chapter of the *Pramāṇasamuccaya*, his main work, he writes: "cognition based on the word (*śābda*) is not different from inference, for just as [an inferential mark (*liṅga*) such as] the quality of having been produced (*kṛtakatva*) [proves the probandum (*sādhya*) through the exclusion of others (*anyāpoha*)], it [i.e., the word] expresses its own object through the exclusion of others."[3] He then goes on to show that the individual (*bheda*), the genus (*jāti*), the relation between the two (*sambandha*), and the locus of the genus (*jātimat*) (i.e., an individual qualified by a genus, although this last term is interpreted in various ways) cannot act as the referent of a word (Hattori 2000, 137–146).

That an inferential mark in the case of inference proves that which is to be proved by the "negation of others" is discussed by Dignāga in the second chapter of the *Pramāṇasamuccaya* (Svārthānumāna). Take for instance the case in which one infers fire on a mountain from smoke, its inferential mark. In this case the "fire" is not a real, blazing fire possessing various attributes such as flames, heat, and so on, but fire in general, common to all individual fires. But fire in general does not exist as an independent entity and is nothing but a concept constructed through the "exclusion of others," or the negation of everything that is not fire. Dignāga's view regarding inferential marks was that "an object has various properties, but [they] are not cognized in their totality by means of an inferential mark. It [i.e., an inferential mark] produces the cognition of those [properties] to which it is connected through the exclusion of others."[4]

The word functions in exactly the same way to denote its object. "Objects have various aspects, but [the diversity of the object] is not understood in its totality by the word. It [i.e., the word] produces the effect of the exclusion of others in accordance with its relation [with the object]."[5] Let us take the example of a particular kind of tree. People may use the word *khadira* to refer to it, but they can also refer to it by terms such as "tree" (*vṛkṣa*), "substance" (*dravya*), "earthy" (*pārthiva*), or "existent" (*sat*). These words all refer to the same object. In other words, they merely refer to one aspect among the various properties of the object, such as its *khadira*- ness, treeness, and substanceness. If words referred to the object itself, this would mean either that the various words applied to the same object were all synonyms or that the object denoted by these various words was one and yet had many distinct realities. Therefore, the function of the word is deemed to consist solely in differentiating the referent from other things. The word "tree" (*vṛkṣa*) functions solely to differentiate the object from "nontrees" (*avṛkṣa*) and there does not exist any real entity corresponding to this word. Thus, the gist of Dignāga's apoha theory is that the word does no more than denote the object through the "exclusion of others" (*anyāpoha, anyavyavaccheda, anyavyāvṛtti*) or denote that portion (*aṃśa, bhāga*) of the object that is differentiated from other things.

While Dharmakīrti basically accepted these views of Dignāga, in the first chapter of the *Pramāṇavārttika* he developed his own apoha theory, which has several original features. Since there have already been published some outstanding studies of Dharmakīrti's apoha theory (Frauwallner 1932–33, 1935; Akamatsu 1980; Katsura 1989, 1991), here I wish to mention only one distinctive feature of his theory, namely, the fact that he interprets "others" in the "negation of others" as "that which does not have the [same] effect" (*atatkārya*). For example, when several different kinds of medicinal herbs all have the effect of alleviating fever, they are understood in terms of the general concept of "febrifuge" and distinguished from everything else that does not produce the same effect. In this case, there is no single entity corresponding to the concept of "febrifuge" that exists in individual kind of herb, and their shared quality is no more than a mental construct produced by conceptualization.

In the first part of his discussion of the apoha theory, Jayanta provides an overview of the Buddhists' theory of apoha, and he does so on the basis of Dharmakīrti's views. According to Dharmakīrti, our perception differs from one individual object to another. The perceptual image of a spotted cow is not the same as the perceptual image of a black cow. But immediately after having perceived a spotted cow there arises the judgment (*pratyavamarśa*) "[this is] a cow," and likewise immediately after having perceived a black cow there also arises the judgment "[this is] a cow." Insofar as they both have the effect of producing the same judgment, the two perceptions can be said to be the same.[6] Moreover, two individual objects producing the same perception are not differentiated. Jayanta quotes a verse from Dharmakīrti that alludes to this: "Since the knowledge [perceptual image] [of individual objects] is the cause of the same judgment [*ekapratyavamarśa*] [about them], it is not differentiated. And on account of the fact that individual objects are the cause of the same knowledge [perceptual image], they are not differentiated from each other."[7]

II

Stating that "[Kumārila] Bhaṭṭa let loose a heavy shower of criticism on the apoha theory,"[8] Jayanta goes on to summarize Kumārila's criticism of the apoha theory. In the "Apohavāda" chapter of his *Mīmāṃsāślokavārttika*, Kumārila criticizes Dignāga's apoha theory from various angles, and Jayanta bases his criticism on these arguments. In the following, I wish to take up some points raised by Kumārila in his criticism of the apoha theory.

According to Kumārila, apoha ("exclusion," "negation") is nothing other than *abhāva* ("nonexistence," "absence"), and *abhāva* is not complete nothingness but signifies a certain kind of existence.[9] But whereas jars (*ghaṭa*) and so on exist independently, *abhāva* has some other thing as its locus (*adhikaraṇa*, *āśraya*) and exists in dependence on it. For example, when one says, "there is no jar on the ground," the nonexistence of the jar has the ground as its locus, and when one says that "a cow is not a horse," the nonexistence of the cow has the horse as its locus. The nonexistence of the jar or the cow is ascertained by the perception of its locus, namely, the ground or the horse, and by the recollection of its counterpositive (*pratiyogin*) in the form of the jar or the cow whose existence is being negated.[10] Kumārila first questions the locus of the "negation of noncows" (*ago-nivṛtti*, *-vyāvṛtti*, *-apoha*), that is, the locus of the "nonexistence of noncows," denoted by the word "cow." Its locus cannot be a unique particular (*svalakṣaṇa*) not common to anything else, nor can it be a genus such as the black cow or the spotted cow, since the word "cow" is applied to all cows. Again, if it had as its locus the totality (*samudāya*) of all cows, the "negation of noncows" would not be comprehended as long as not all cows were known. On the basis of this inquiry, Kumārila concludes that the "negation of noncows" is nothing but another way of referring to the universal "cowness" (*gotva*).[11]

Next, it is pointed out that since, according to the apoha theory, every word denotes apoha, all words end up being synonyms (*paryāya*). The thesis that words denote real objects (*vastu*) does not entail this difficulty, for real objects each have their own essence and are mutually distinguished.[12] Of course the apoha theorist will reply that words are not all synonymous on their theory either. Dignāga, who established the theory of apoha, recognizes distinctions in apoha according to differences in what is negated (*apohya*). He makes this clear when explaining how in the apoha theory the two words *nīla* ("blue") and *utpala* ("lotus") in a term such as *nīlotpala* ("blue lotus") can have colocusness (*sāmānādhikaraṇya*) and how a qualifier-qualified relation (*viśeṣaṇaviśeṣyabhāva*) can obtain between them.

If one accepts the thesis that a word denotes a real object, *nīla* refers directly to an attribute (*guṇa*) residing in a particular locus and *utpala* refers directly to the genus (*jāti*) to which that locus belongs, which means that the locus in which both the attribute and the genus inhere is denoted by both words, and so colocusness is established between *nīla* and *utpala*. Furthermore, the genus referred to by *utpala* encompasses all particular varieties of lotus such as red (*rakta-*), blue (*nīla-*), and so on, but because *nīla* restricts it to only blue lotuses, the qualifier-qualified relation (*viśeṣaṇaviśeṣyabhāva*) is also established between the two words. But according to the apoha theory, *nīla* refers to the "negation of nonblue" (*anīlāpoha*), while *utpala* refers to the "negation of nonlotuses" (*anutpalāpoha*). Since neither are real existents, they neither inhere in the same locus nor do they stand in a qualifier-qualified relation to each other. This means that neither colocusness nor the qualifier-qualified relation is established between the two words *nīla* and *utpala*. Anticipating this sort of counterargument against the apoha theory, Dignāga writes as follows:

> Depending on differences between what is negated (*apohya*), [the two words *nīla* and *utpala*] have different objects, but they are unclear regarding the understanding of the particular belonging to their own object.

> [But] because [both words] do not operate separately toward the same thing (i.e., that which is neither nonblue nor a nonlotus), they stand in a qualifier-qualified relation.[13]

In Dignāga's view, a real object is completely unique insofar as it constitutes a combination of infinite aspects, and it cannot be directly denoted by a word. The function of the word consists solely in differentiating it from other objects. The word *nīla* differentiates a real object from nonblue objects, while the word *utpala* differentiates it from nonlotuses. A different word is applied to the same real object, depending on what the object is to be differentiated from. Therefore, the "negation of others" (or "differentiation from others") expressed by the word may be defined as differentiation based on distinctions among what is to be negated. Thus, it goes without saying that since, according to Dignāga's thinking, the two words *nīla* and *utpala*, which have as their function the "negation of nonblue" and the "negation of nonlotuses" respectively, refer to the same thing, differentiated from both the nonblue and nonlotuses, there is established between them a relationship of colocusness. Moreover, the word *utpala* differentiates the object from nonlotuses, but it cannot indicate whether it is red, blue, or some other color. The word *nīla*, on the other hand, differentiates the object from nonblue things. The limitation by *nīla* of the range of denotation of the word *utpala*, which extends to many individual lotuses of any color, corresponds to the qualifier-qualified relation between *nīla* and *utpala*, and therefore the qualifier-qualified relation (*viśeṣaṇaviśeṣyabhāva*) also holds true in the apoha theory.

Leaving the question of colocusness and the qualifier-qualified relation in the apoha theory for a later discussion, Jayanta focuses first on Kumārila's criticism of the anticipated response from proponents of the apoha theory, who argue that because "there are distinctions in apoha depending on differences in what is negated (*apohya*)," words are not all synonymous. According to Kumārila, the universal (*sāmānya*) stands in an intrinsic relation (*svābhāvikaḥ sambandhaḥ*) to the individual (*vyakti*), and it exists only where there is an individual, or, even assuming that it is omnipresent, manifests only in the individual.[14] The individual is the substratum (*ādhāra*) of the universal. In the same way, in the apoha theory too the "apoha of noncows" denoted by the word "cow" has individual cows as its substratum. Moreover, the "apoha of noncows" is not differentiated in accordance with differences in the substratum. It is impossible for such an apoha to be differentiated with "negatees" (*apohya*) that are only distantly related to it. Even supposing that it were "differentiated," this would not be differentiation in any true sense of the word, but differentiation in a secondary sense (*bhākta*).[15]

Kumārila's criticism of the thesis that apohas are differentiated according to differences in the objects of negation (*apohya*) moves on to repudiate all distinctions whatsoever in the objects of negation. The noncows that represent the object of negation in the "negation of noncows" correspond to horses, elephants, lions, tigers, and so on, and the nonhorses that represent the object of negation in the "negation of nonhorses" correspond to cows, elephants, lions, tigers, and so on. Noncows and nonhorses each encompass an infinite number of species, only one of which differs (horse or cow), and they share all other species (elephant, lion, tiger, etc.). Since one species among an infinite number of species is equivalent to null, there can be no distinction between noncows and nonhorses. Therefore, one is still left with the irrational conclusion that the "cow" and the "horse" that denote the "negation of noncows" and the "negation of nonhorses" respectively are synonyms.[16]

Jayanta also alludes to another point made by Kumārila, namely, that since the concept of noncow is premised on an understanding of cows, the thesis that "cow" denotes the "negation of noncows" results in circular reasoning. This point is also discussed by the Naiyāyika Uddyotakara when criticizing Dignāga's apoha theory.[17] Kumārila's criticism that neither colocusness nor the qualifier-qualified relation holds true in apoha theory is based on his equating of apoha (= *abhāva*) with a certain kind of existent. The gist of Kumārila's argument, succinctly summarized by Jayanta, is that since the "negation of nonblue" and the "negation of nonlotuses" are separate existents, where one of them exists the other cannot exist, which means therefore that the qualifier-qualified relation does not obtain between their respective denoters *nīla* and *utpala*, and since there is no locus on which the two apohas both depend, there is no colocusness between the two words either.[18]

Again, according to the apoha theory, the referents of the words *prameya* and *jñeya* ("object of cognition") are the negations *aprameya* and *ajñeya* ("that which is not an object of cognition"). But because the object of negation has to be cognized, *aprameya* and *ajñeya* end up being the same as *prameya* and *jñeya*, and the "negation of others" cannot hold true in this case.[19]

In the fifth chapter of the *Pramāṇasamuccaya*, after having examined various theories about the meaning or referent of a word and established the theory of apoha, Dignāga touches on the question of sentence meaning (*vākyārtha*). Influenced by the grammarian Bhartṛhari, he recognizes that words are extracted (*apoddhṛta*) from a unitary indivisible sentence and that sentence meaning is a flash of intuition (*pratibhā*).[20] On hearing the words "a tiger is coming," one feels fear, and on hearing a love poem, feelings of passion are aroused. Even when the objects denoted by individual words do not exist in the outer world, a sentence will still produce in the listener an intuition possessing positive content, and in Dignāga's view it is this intuition that corresponds to the meaning of the sentence. He then goes on to argue that the intuition representing the meaning of a sentence arises on the basis of the understanding of individual words, which have as their essence the "negation of others," and that sentence meaning is also characterized by the "negation of others." Kumārila criticizes this thesis of Dignāga's, that words in the form of sentences give rise to intuition, as being inconsistent with the apoha theory, and his arguments are also summarized by Jayanta.[21] Dignāga's apoha theory was also criticized by Uddyotakara of the Nyāya school, but his criticism is not mentioned by Jayanta, nor does Jayanta make any attempt to present his own rebuttal of Dignāga's apoha theory.

III

At the start of the third part of his discussion of the apoha theory, Jayanta presents the Buddhists' rejoinder to Kumārila's criticism. Kumārila bases his arguments on the understanding that apoha is the same as absence (*abhāva*), which is of the nature of the real (*bhāvātmaka*), but according to the Buddhists this is unreasonable: "Apoha is neither something inner [knowledge] nor something outer [external object] and is different from knowledge and objects."[22] "Because it is neither inside nor outside, it is said to be false (*mithyā*) and mentally constructed (*kālpanika*)." "It is superimposed [upon the real object] (*āropita*), is mere appearance/form (*ākāramātra*), and dyes (or affects) an idea (*vikalpoparañjaka*)."[23]

It is clear from the terminology used here that it is the views found in Dharmottara's *Apohaprakaraṇa* that are being used as representative of the arguments of Buddhists. In its commentary on this section, the *Nyāyamañjarī-granthibhaṅga* cites the following verses, which it attributes to Dharmottara,[24] and they correspond in fact to the verses of salutation at the start of the *Apohaprakaraṇa*: "Paying reverence here with my head to the teacher who is the conqueror of all foes of defects and who, stating that the form (*rūpa*) drawn by the fabricating mind as being distinct (*vivikta*) from others is neither the cognition [within] nor the external [object] and that it is unreal (*nistattva*) and superimposed [upon the real] (*āropita*), taught people the truth, I shall definitely explain in detail [the doctrine of] apoha."[25]

Some distinctive features can be seen in Dharmottara's apoha theory as developed in the *Apohaprakaraṇa*. Stating first that the word denotes the object (*viṣaya*) of a concept (*vikalpa*) (APD, Narthang ed., fol. 254a.5), he discusses the objects of concepts. Perception arises with things in the outer world as its objects, but because concepts arise even when there are no external objects, things in the outer world are not the objects of concepts. However, concepts do not arise spontaneously with no relation to things in the outer world; they are formed on the basis of the perception of external objects. There then arises on the basis of perception an image that has as its object something in the outer world, and the intrinsic nature of the concept lies in ascertaining (*ṅes pa* = *niścaya*) it to be a "jar," "cloth," and so on (APD, N., fol. 254b.1–2.). But this ascertainment is produced by mental construction and does not correspond to any external object per se. These are the basic tenets of Buddhist epistemology going back to Dignāga, but Dharmottara stresses the fact that the objects of concepts, which may be characterized in the above fashion, are neither things in the outer world nor inner knowledge, have no real existence, are false or unreal (*alīka*), and are subjective constructs.

Dharmottara goes on to base the "negation of others" on this nonreality of the objects of concepts. There is no need to ask whether or not something that has been perceived exists, but the existence of the objects of concepts must be questioned. For example, when water is perceived, the existence of water is self-evident. But in the case of the concept *tree*, whether or not it exists becomes clear only through its union with the separate concept of "existence" (*bhāva*) or "nonexistence" (*abhāva*). That which is ascertained or determined by a concept is an "object that is common to both affirmation and negation (*vidhiniṣedhasādhāraṇa*) and can be combined with either" (APD, N., fol. 258a.2–3). In other words, the object of a concept superimposed (*āropita*) on a real entity does not exist as something possessing any positive qualities. A tree representing an object of perception possesses the positive quality of the tree's existence and the negative quality of not being some other thing (a nontree), but since the existence of the object of a concept cannot be affirmed, it has as its intrinsic nature merely "differentiation from others (nontrees)," that is, the "negation (apoha) of others." Dharmottara's originality in the history of the development of the apoha theory can be recognized in the way in which he thus based the "negation of others" expounded by Dignāga on the nonreality of the referent of a word, that is, the object of a concept.

Jayanta's summary of the Buddhists' rejoinder to Kumārila's criticism of the apoha theory is extremely succinct, and it is often difficult to grasp the gist of the argument. But if one interprets it by taking into account Dharmottara's ideas found in the *Apohaprakaraṇa*, it becomes possible to decipher its meaning. The following passage summarizes the arguments set forth by Jayanta in part 3 (NMJ 25.4–15; 281.3–12):

> An entity residing on the ground of a concept (*vikalpabhūmir arthaḥ*) [i.e., the object of a concept] is understood (1) to require [supplementation by union with the concept of] existence (*bhāva*) and nonexistence (*abhāva*) due to another thought, (2) to have a determinate (*niyata*) quality, and (3) to resemble something in the outer world. These three qualities do not hold true for real entities in the outer world. This is because (1) union with the concept of "existence" due to another thought is not established in a real entity in the outer world understood in its own form [since it is self-evident that it "exists"], for it would be meaningless. Union with "nonexistence" is also not established, for it would be contradictory. (2) Determinateness inheres in the object of a concept. Since [determinateness], which is understood in the form "this is indeed a cow and not a horse," is not possible without the negation of something else, [it] will necessarily have the exclusion or negation (*vyavaccheda*) of others as its object (*viṣaya*). Otherwise, a determinate decision (*niyata-pariccheda*) would be impossible. It is not the case that something doubtful [in which it is unclear whether it is a cow or a horse] is understood [by a concept].
>
> Since the fact that concepts have things in the outer world as their objects has thus been negated, . . . (3) it [i.e., a superimposed form (*āropitaṃ rūpam*) which is not something in the outer world] manifests as if it were something in the outer world. And since the resemblance between something in the outer world and something that has been superimposed does not exist apart from the "exclusion or negation (*vyāvṛtti*) [of others]," this means that the concept does indeed have the "exclusion [of others]" as its object (*viṣaya*).

Jayanta describes the difference between Dharmottara's and Dharmakīrti's views on the objects of concepts in the following manner.[26] According to Dharmottara, the object of a concept "is neither something inner [knowledge] nor something outer [external object] and is a superimposed form (*āropitākāra*) different [from knowledge or an external object]," and "because it is tinged with exclusion or negation (*vyāvṛtti*) [of others], it is said to be the referent of the word 'apoha.'" According to Dharmakīrti, on the other hand, the object of a concept "is a constructed or conceptual image (*vikalpa-pratibimba*), is a mere form of knowledge (*jñānākāramātra*), has distinctions of form due to distinctions in various residual impressions (*vāsanā*) [of past experiences], manifests as if it were something in the outer world, and sustains worldly activity," and it is called "apoha" because it has a tinge of "exclusion or negation (*vyāvṛtti*) of others." Jayanta characterizes the former as the theory that the unreal is cognized (*asatkhyāti-vāda*) and the latter as the theory that the self [i.e., one's own mental image] is cognized (*ātmakhyāti-vāda*).

Dharmakīrti's above ideas as pointed out by Jayanta can be found, for instance, in the following passages:[27]

> The objects of words are dependent on concepts born of residual impressions (*vāsanā*) [of experiences] since beginningless time.
>
> (PV I, k. 205)

> The image of the object appears in knowledge on account of the word as if it were something [in the outer world] removed [from knowledge], but it is not the thing itself [in the outer world]. The error [that identifies it as something in the outer world] arises on account of residual impressions (*vāsanā*) [of past experiences].
>
> (PV III, k. 165).

In contrast, Dharmottara, stating that the form which appears in knowledge is not the object of a concept since it does not have "ascertainment" as its intrinsic nature,[28] maintained that the objects of concepts are neither forms appearing in knowledge nor things in the outer world.

Why do people engage in activities directed at things in the outer world on the basis of the ascertainment of an object with concepts in spite of the fact that the objects of concepts are not things in the outer world? Jayanta gives the Buddhists' response to this in the following terms. The activities of people result from the unification (*ekīkaraṇa*) of perceptual objects (*dṛśya*) and conceptual objects (*vikalpya*). When a concept arises immediately after an object has been perceived, one is misled by the quick succession of events and, assuming that the perceived object has itself been grasped by the concept, identifies the two (NMJ, 26.16–20; 281.22–25).

Dharmottara also provides an explanation of the unification (*ekīkaraṇa*) of perceptual objects and conceptual objects (APD, N., fol. 258a.7–258b.3), but in view of the fact that Jayanta clearly indicates that this unification arises through conceptual error and that people's activities are based on this, it is to be surmised that Dharmottara has based himself on Dharmakīrti's arguments. When describing how conceptual cognition perceives many individual entities, yet conceals (*saṃ- vṛ-*) their individual forms with its own form, and manifests in a unitary form, Dharmakīrti writes as follows:

> Cognitive knowledge grasps the image (*pratibimba*) itself [of the object] residing in knowledge, but because of an intrinsic error (*prakṛtivibhramāt*) on the part of the concept it manifests as if it were grasping an [external] entity. . . . [Question:] Is it not the case that things in the outer world are individuals? And is it not the case that concepts do not arise in response to them? [Answer:] The commentators (*vyākhyātāras*) do indeed differentiate [external individual entities and conceptual objects] in this manner, but people engaged in everyday activities (*vyavahartāras*) do not. Thinking that [the conceptual image representing] their object of cognition (*ālambana*) has itself the potential for efficacy (*arthakriyāyogya*) [as does an external entity], they act (*pravartante*), having united (*ekīkṛtya*) the perceptual object (*dṛśya*) and the conceptual object (*vikalpya*).[29]

Dharmottara was an original thinker, and while he basically followed Dharmakīrti's views, he made some amendments. Jayanta exemplified the Buddhists' rejoinder to Kumārila's criticism of apoha theory chiefly with the arguments of Dharmottara, but he also referred to the views of Dharmakīrti, on which they were based.

Dharmakīrti provides answers to several of the points raised in Kumārila's criticism. Although not mentioned by Jayanta, the criticism made by Kumārila by applying the relationship obtaining between the individual (*vyakti*) and the universal (*sāmānya, jāti*) that exists as its qualifier (*viśeṣaṇa*) to the relationship between perceptual objects and apoha was rebutted by Dharmakīrti with a lucid theory, also adopted by Śāntarakṣita (see Hattori 2000, 387–94). Jayanta touches on the criticism that the apoha theory lapses into circular reasoning and that the words *prameya* and *jñeya* have no objects of negation, and Dharmakīrti gives clear-cut responses to these points too (PV, I, k. 113cdff., kk. 122–123). A full-scale rebuttal of Kumārila's criticism of the apoha theory was essayed by Śāntarakṣita, who, while basing himself on Dharmakīrti's theories, propounded an apoha theory that stressed the positive functions of words, according to which mental images are produced by words. In the "Śabdārtha" chapter of his *Tattvasaṃgraha*, he first presents Kumārila's views and then sets forth detailed counterarguments. The gist of his arguments is also clearly explained by his disciple Kamalaśīla in his commentary, the *Tattvasaṃgraha-pañjikā* (TS & TSP, 1021–1183). But Jayanta makes no mention of the responses to and countercriticism of Kumārila's criticism by Dharmakīrti, Śāntarakṣita, and Kamalaśīla.[30] It is to be surmised that this was perhaps because in late ninth-century Kashmir, where he lived, the main current of Buddhist philosophy was represented by the thought of Dharmottara, who had also been active in Kashmir, but for the moment I wish to refrain from being any more explicit in this regard.

Notes

1. NMJ 5.4–5; 271.12–14: "tatra gavādijātiśabdānāṃ gavādijātyavacchinnaṃ vyaktimātram arthaḥ, yas tadvān iti naiyāyikagṛhe gīyate." Cf. ibid., 59.5–6, 10; 295.24–25, 30: "Anyeṣu tu prayogeṣu gāṃ dehy ity evamādiṣu / tadvato 'rthakriyāyogāt tasyaivāhuḥ padārthatām . . . tasmāt tadvān eva padārthaḥ." In *Nyāyasūtra* 2.2.66, *vyakti* (individual), *ākṛti* (configuration, characteristic of a species), and *jāti* (species, universal) are mentioned as the object of words (*vyaktyākṛtijātayas tu padārthaḥ*). Jayanta states: "Though it is established that a *tadvat* [= *jātimat*, a *vyakti* qualified by a *jāti*] is the meaning of a word, in some cases of application *jāti* is the primary and *vyakti* is subordinate, as for example, 'cow should not be touched by foot.' . . . in some [other] cases *vyakti*/*ākṛti* is the primary and *jāti*/*vyakti* is subordinate" (NMJ 63.18–64.6; 297.25–31; "sthite 'pi tadvato vācyatve kvacit prayoge jāteḥ prādhānyaṃ vyakter aṅgabhāvaḥ, yathā 'gaur na padā spraṣṭavyā' iti . . . kvacid vyakteḥ/ākṛteḥ prādhānyaṃ jāter/vyakter aṅgabhāvaḥ."

   We have consulted the following editions of the *Nyāyamañjarī* of Jayantabhaṭṭa: (K) ed. Surya Narayana Sukla, Kashi Sanskrit Series, (M) ed. Vidwan K. S. Varadacharya. Part 2. (Mysore: Oriental Research Institute, University of Mysore, 1983). Page and line numbers of (M) are given first, and then follow those of (K).
2. The main portions of Jayanta's arguments to prove the reality of *jāti* are translated into German in Frauwallner 1992, 156–170.
3. PS V k. 1 (cf. TSP, ed. Shastri, 539): "na pramāṇāntaraṃ śābdam anumānāt tathā hi tat / kṛtakatvādivat svārtham anyāpohena bhāṣate."
4. PS II k. 13 and *Vṛtti*. Cf. Hayes 1980, 256–257. Cf. also Frauwallner 1959, 102.
5. PS V k. 12 (cf. ŚVṬ, ed. Kunhan Raja, 46.7–8): "bahudhāpy abhidheyasya na śabdāt sarvathā gatiḥ / svasaṃbandhānurūpyena vyavacchedārthakāry asau."
6. NMJ 11.14–16; 276.5–6: "yathaiva śābaleyādipiṇḍadarśane sati gaur ity anantaram avamarśaḥ, tathaiva bāhuleyapiṇḍadarśane 'pi gaur ity evāvamarśa iti tad ekatvam ucyate."

7. NMJ 12.1–2: 276.7–8 (= PV I k. 109, ed. Gnoli): "ekapratyavamarśasya hetutvād dhīr abhedinī / ekadhīhetubhāvena vyaktīnām apy abhinnatā." In APD fol. 262b.5ff. (ed. Narthang), Dharmottara defends Dharmakīrti's thought expressed in this verse against some hostile criticisms; cf. Frauwallner 1937, 284–285.
8. NMJ 14.15; 277.19: "apohavādaviṣaye mahatīṃ dūṣaṇavṛṣṭim utsasarja bhaṭṭaḥ."
9. In ŚV (ed. Shastri), Abhāva, k. 2cdff., Kumārila explains that *abhāva* is divided into four kinds, namely, *prāg-abhāva, pradhvaṃsā°, anyonyā°*, and *atyantā°*, and states: "There could not be these divisions for the unreal; therefore it (= abhāva) is real" ("na cāvastuna ete syur bhedās tenāsya vastutā"; k. 8ab). Cf. also ŚV, *Nirālambanavāda*, k. 118c: "bhāvāntaram abhāvaḥ."
10. ŚV (ed. Shastri), Abhāva, k. 27: "gṛhītvā vastusadbhāvaṃ smṛtvā ca pratiyoginam / mānasaṃ nāstitājñānaṃ jāyate 'kṣānapekṣanāt."
11. NMJ 14.15–16.4; 277.19–31. ŚV, Apohavāda, kk. 1–10.
12. NMJ 16.14–17.5; 278.4–9. ŚV, Apohavāda, kk. 42–46.
13. PS V k. 14 (cf. TSP, ed. Shastri, p. 379.5–6, 365.10): "apohyabhedād bhinnārthāḥ svārthabheda-gatau jaḍāḥ / ekatrābhinnakāryatvād viśeṣaṇaviśeṣyatā."
14. Cf. ŚV, Ākṛtivāda, kk. 25–26, 31cd.
15. NMJ 17.11–17; 278.12–15. ŚV, Apohavāda, kk. 47–52.
16. NMJ 18.1–7; 278.16–21. ŚV, Apohavāda, kk. 53–57.
17. NMJ 19.8–10; 288.30–32. ŚV, Apohavāda, kk. 83–85ab. Cf. NV pp. 328.17–329.3.
18. NMJ 20.5–9; 279.6–10. ŚV, Apohavāda, kk. 115–119.
19. NMJ 20.11–15; 279.10–13. ŚV, Apohavāda, k. 144.
20. PS V k. 47 (cf. TSP p. 363.15–16): "apoddhāre padasyāyaṃ vākyād artho vivecitaḥ / vākyārtho pratibhākhyo 'yaṃ tenādāv upajāyate." Cf. Hattori 1980, 61–73.
21. NMJ 21.12–13; 279.22–23. ŚV, Apohavāda, k. 40.
22. NMJ 22.6–7; 279.32: "nāyam antaro na bāhyo 'pohaḥ. kiṃ tu jñānārthābhyām anya eva."
23. NMJ 22.13–15; 280.5–6: "yata eva tan nāntar bahir asti tata eva mithyeti kālpanikam iti ca gīyate. . . . āropitaṃ kiṃcid ākāramātraṃ vikalpoparañjakam."
24. NMJG (ed. Shah) p. 132. A German translation from the Tibetan version (the Sanskrit text was not available) is in Frauwallner 1937, 255.
25. "buddhyā kalpikayā viviktam aparair yad rūpam ullikhyate / buddhir no na bahir yad eva ca vadan nistattvam āropitam / yas tattvaṃ jagato jagāda vijayī niḥśeṣadoṣadviṣām / vaktāraṃ tam iha praṇamya śirasā 'pohaḥ sma vistāryate."
26. NMJ 26.1–8; 281.15–19: "so 'yaṃ nāntaro na bāhyo 'nya eva kaścid āropita ākāro vyāvṛtticchāyāyogād apohaśabdārtha ucyate, itīyam asatkhyātivādagarbhā saraṇiḥ. atha vā vikalpapratibimbakaṃ jñānākāramātrakam eva tad abāhyam api vicitravāsanābhedopāhitarūpabhedaṃ bāhyavad avabhāsamānaṃ lokayātrāṃ bibharti, vyāvṛtticchāyāyogāc ca tad apoha iti vyavahriyate, seyam ātmakhyātigarbhā saraṇiḥ." Cf. Frauwallner 1937, 280n1.
27. PV I (ed. Gnoli) k. 205a–c: "anādivāsanodbhūtavikalpapariniṣṭhitaḥ / śabdārthas." PV III (ed. Tosaki) k.165: "vyatirekīva yaj jñāne bhāty arthapratibimbakam / śabdāt tad api nārthātmā bhrāntiḥ sā vāsanodbhavā."
28. Cf. APD fol. 255a.6–255b.1 (ed. Narthang).
29. PVSV (ed. Gnoli) 39.2–8: "tam (= buddhipratibhāsam) eva gṛhṇatī sā (= buddhiḥ) prakṛtivibhramād vikalpānāṃ vastugrāhiṇīva pratibhāti. . . . nanu bāhyā vive-

kino na ca teṣu vikalpapravṛttir iti kathaṃ teṣu bhavati. 'vyākhyātāraḥ khalv evaṃ vivecayanti na vyavahartāraḥ. te tu svālambanam evārthakriyāyogaṃ manyamānā dṛśyavikalpyāv arthāv ekīkṛtya pravartante.'" is quoted in NMJ 25.17–26.1; 281.14–15.

30. Jayanta criticizes Dharmakīrti's thought as expressed in the verse: "ekapratyavamarśasya" (PV I k. 109, cited in note 9), asserting that the judgments which follow the perceptions of individual objects are different from each other, just like perceptual images, thus denying the formation of *ekapratyavamarśa* (the same judgment). He also criticizes Dharmottara, stating that our everyday activities could not be carried on if the objects of conceptual cognition were merely imaginary and unreal. Cf. NMJ 40.11ff., 43.13ff.; 287.11ff., 288.18ff. However, he does not enter into a detailed, critical examination of the apoha theory.

7

Constructing the Content of Awareness Events

Parimal G. Patil

EXCLUSION AT VIKRAMAŚĪLA

In the hands of the exclusion theorists Jñānaśrīmitra (ca. 975) and Ratnakīrti (ca. 1000)[1]—perhaps the last two innovative exclusion theorists in India—the scope of the theory of exclusion was extended far beyond the semantic context in which it was originally developed.[2] While Dignāga (ca. 480–540) himself recognized the very close relationship between semantics and inferential reasoning,[3] and Jñānaśrīmitra and Ratnakīrti's predecessors—Dharmakīrti (ca. 600–660), Dharmottara (ca. 740–800), and Prajñākaragupta (ca. 800)—recognized, with ever increasing clarity, the extent to which the theory of exclusion also applied to perception,[4] it was in the hands of Jñānaśrīmitra that the distinction between perception and inference was, for practically all philosophical (but not exegetical or commentarial) purposes, finally obliterated,[5] and the theory of exclusion became (at least for members of the Buddhist epistemological text tradition) *the* Buddhist theory of mental content—that is, *the* Buddhist theory of what our experiences and thoughts are about.[6] Jñānaśrīmitra's version of the theory of exclusion was thus invoked to explain not only the content of awareness events produced though testimony or inferential reasoning but also the content of perceptual awareness events (and related issues about the phenomenology of perceptual experiences).[7] Like all exclusion theorists, Jñānaśrīmitra and Ratnakīrti shared the (by then pan-Indian) Buddhist commitments to the theory of "momentariness" and "nominalism" about universals.[8] The challenge that they set for themselves was to provide a satisfying account of mental content—and especially conceptual content—given these two constraints. What I want to do in this paper is try to develop a generic interpretation of the theory of exclusion that is based on the work of Jñānaśrīmitra and Ratnakīrti, and then use it to raise questions that a contemporary exclusion theorist should, but may not, have very good answers for. I have partially defended my reconstruction of the theory by referring, in the notes, to passages from the work of both Jñānaśrīmitra and Ratnakīrti.[9]

A GENERIC VERSION OF THE THEORY OF EXCLUSION

The Story of Exclusion

The theory of exclusion is a back story—a story from hindsight—about how what we take the conceptual content of our awareness episodes to be is constructed from what we take to be directly, and nonconceptually, present to awareness.[10] Its overall structure is a kind of inference to the best explanation, which makes use of a variety of important subarguments along the way.[11] What follows is a generic version of this story, which, like many good stories, has a beginning, a middle, and an end.

> *The Beginning*: We start with some object *p* with "identity" conditions *I*. *I* is the set of conditions that individuate *p*.

Exactly why we start with this *p* and related questions about its ontology and the precise nature of its identity conditions are, for the moment, set aside. However these questions are answered, let us refer to this object as a "particular" (*svalakṣaṇa*)—a momentary entity that has neither spatial nor temporal extension.[12]

> *The Middle*: Let *S*—a subset of *I*—define a set of "selection" conditions. This set of selection conditions is the basis for the construction of a dissimilarity class, non-*P*—the set of objects that do not satisfy *S*, that is, the set of objects that exclusion theorists refer to as "non-*p*'s."[13]

Again, questions about why *this S* are to be temporarily set aside, as are questions about exactly how this "construction" is supposed to take place.

The End: Finally, construct the similarity class like-*P* by the exclusion of the dissimilarity class non-*P*. The similarity class, like-*P*, consists of objects that satisfy *S*, that is, it consists of all "*p*'s." Here, the construction process is described in terms of an "exclusion," which, at least for the time being, can be understood as the process of constructing the complement of non-*P*.

On the basis of *p*, a set of selection conditions *S*, and two analytically separable, but not necessarily separate, processes—one of which is an exclusion—exclusion theorists suppose that a similarity class, like-*P*, can be constructed.[14] They further suppose that it is this similarity class that best accounts for and characterizes conceptual content and is, more specifically, what we take semantic value, the objects of inferential awareness, and the phenomenal objects of perceptual awareness to be. Nonconceptual content is that on the basis of which conceptual content is constructed. Since the idea of nonconceptual content is contrastive, in providing an account of conceptual content, the theory of exclusion also provides, at least partially, an account of nonconceptual content.[15] Thus, in my view, the theory is best understood as the Buddhist theory of mental content—whether conceptual or nonconceptual.[16]

Filling in the Story

This generic version of the theory is, of course, incomplete and raises a number of questions that a successful exclusion theorist will have to answer and that, in fact, specific exclusion theorists in the past have answered in very different ways.

Questioning The Beginning

Perhaps the most obvious questions about the beginning of the exclusion story have to do with what sort thing *p* is supposed to be, and why we start with this *p* instead of some other *p* or *q*. Often, *p* is supposed to be a mental image (*ākāra*) whose specific form (*rūpa*) or set of identity conditions (*I*) is due to (F1) purely personal factors (such as those encoded in one's *vāsanā*'s, or latent karmic dispositions);[17] (F2) contextual factors (such as one's location, current interests/concerns, the conversational or inferential context in which one is involved, etc.;[18] (F3) well-established social conventions or norms;[19] or (F4) for some, but not all exclusion theorists, properties of the object/s that in some sense produced the image.[20] It is also important that for many exclusion theorists (e.g., Dignāga, Dharmakīrti, Prajñākaragupta, Jñānaśrīmitra, and Ratnakīrti) we do not have full "access" to *p*—it is itself phenomenally unavailable to us—that is, it is not like anything for us to be aware of *p*.[21] It is the nonconceptual or, more accurately, "manifest" content of awareness.

Questioning The Middle

This is the heart of the story and raises some of the most difficult (and interesting) questions. In my version, four sets of questions seem to be particularly pressing. One set of questions has to do with how *S* is constructed or selected: What accounts for this *S* instead of some other subset of *I*? The answer usually given is that *S* is formed on the basis of factors such as (F1) through (F4), with special emphasis given either to personal interests, tastes, dispositions, and past experiences, etc., or to social conventions and practices, depending on the context. Some but not all factors that result in *S* are transmitted from those that result in *p*. Let us refer to the nontransmitted factors, if and when there are any, as (F5). A second set of questions has to do with exactly what it means to say that a dissimilarity class non-*P* is *constructed* on the basis of *S*. The answer usually given is that the dissimilarity class non-*P* is defined by objects that do not satisfy *S*—objects that exclusion theorists call "non-*p*'s." The kind of process through which non-*p*'s are collected or grouped together is not usually specified by traditional exclusion theorists—neither are any further details about what sort of an object the dissimilarity "class" non-*P* really is. What Jñānaśrīmitra and Ratnakīrti's view seems to require is that this dissimilarity class not be an object or mental image at all. Instead, it seems as though there is a subprocess in the overall process of exclusion whereby non-*P* is constructed without it itself figuring into the account given of that process.[22] A third set of questions has to do with what it means to say that non-*p*'s are objects that *do not satisfy S*. The basic idea seems to be that non-*p*'s are those objects that are, according to us, not able to function within the expectation parameters encoded in *S*. A fourth set of questions has to do with why there needs to be a middle of the story at all: Why is it necessary, for example, to construct a dissimilarity class only to exclude it? Is it not sufficient to argue simply and directly that like-*P* consists of all of those objects that we take to satisfy *S*? The intuition of exclusion theorists seems to be that without the middle of the story, especially the construction of the dissimilarity class non-*P*, the story would be incomplete, since it would not be able to address issues having to do with "indeterminacy," such as the "indeterminacy of pointing." In this context, such indeterminacies may be more accurately referred to as the "indeterminacy of conceptualizing," the "indeterminacy of conceptual content," or, as we will see, the indeterminacy of "actionable objects," including those that we point to.[23] Exclusion theorists reason that conceptual content is structured, in the sense that the similarity class like-*P* is defined to include *p*'s and only *p*'s, that is, it includes *p*'s but also excludes all non-*p*'s. Without the middle of the story, there would be no way, they reason, to account for this structure in a way that avoids existential commitments to entities such as universals.

Questioning The End

The most pressing questions about the end of the exclusion story have to do with the kind of thing that the similarity class like-*P* is; with whether it is generic enough to account for the semantic value of any expression, the content of any inferential awareness event, and the phenomenal content of perception; and with what it means to say that it can be formed by the second subprocess of exclusion—usually referred to by "exclusion" itself. These questions are primarily about the product of the entire exclusion process. But questions must also be raised about exactly how to understand the process whereby the similarity class like-*P* is constructed from the dissimilarity class non-*P*. For example, if the dissimilarity class non-*P* is not an object of any sort, it is not at all clear what it would mean to say that like-*P* is the "complement" of non-*P*. According to Jñānaśrīmitra and Ratnakīrti, it is clear that like-*P* is a mental image (*ākāra*). Ratnakīrti most often describes it as either a positive object characterized by its exclusion of others (*anyāpohaviśiṣṭavidhi*), a thing-in-general separated out from things that are nonthat (*atadrūpaparāvṛttavastumātra*), or a generalized image. This object is understood to be constituted by two components, which are only analytically or conceptually separable from one another—the so-called positive component, the mental image, and the so-called negative component, exclusion.[24] It is claimed that this object accounts for conceptual content, since it is what we end up taking to be the object (or patient) of any act—whether linguistic, mental, or physical. It is, in other words, the only sort of "actionable" object.[25] Regardless of a particular exclusion theorist's ontology, it is never the case that what we experience/notice/think there to be is, in fact, what there really is. There is always a "mismatch" between what is *available* (*p*) and what is *actionable* (like-*P*). According to Jñānasrīmitra and Ratnakīrti, the gap between what is available (i.e., what is directly present in/manifest to awareness) and what is actionable (i.e., conceptual content) is bridged by determination (*adhyavasāya*), which, for them, just is exclusion.[26]

WHAT ELSE IS EXCLUSION SUPPOSED TO EXPLAIN?

According to traditional exclusion theorists like Jñānasrīmitra and Ratnakīrti, the theory of exclusion is supposed to provide an account of a variety of actionable objects—including semantic value and what we may call "inferential" and "perceptual" value—and relations.[27] As an example, here is how Jñānasrīmitra and Ratnakīrti's version of the theory could be used to account for semantic value.

Exclusion at Work: Semantic Value

What the theory of exclusion is most often invoked to explain is what it is that I am aware of when I understand a word—for example, the word "cow" in the sentence "bring a cow." Ratnakīrti argues that hearing the term "cow" in such a context will (for a competent speaker of English) trigger the memory of a previously observed "cow" (RNA [AS, 63.20–63.21]). Triggering one's memory in this way is said to result in the manifestation of a paradigmatic cow image. In this context, this image is the object *p* with which the story of exclusion begins. Insofar as the exclusion story requires that this image be manifest in awareness, it has to be a particular; but, insofar as we retrospectively think of it as a manifest "*cow* image," it has to have the informational structure of a universal—either a vertical universal, as when it is taken to be like the picture of a specific cow, or a horizontal universal, as when it is more genericized.[28] It should be clear that a great deal must already be in place for such an image to become manifest in awareness: for example, the person must already be able to understand language; she must have been able to individuate objects well enough to learn which object was being labeled "cow" by the person who first taught her the conventions regarding the use of the word; and she must recognize that the unique set of sounds *c-o-w* that she now hears is similar enough to the sounds that she heard when the convention was established. In addition to providing an account of memory, a successful exclusion theorist must also be able to explain everything that goes into establishing the "conventions"—whether verbal or perceptual—that the exclusion story relies upon.

Now suppose that *I* is the set of the recalled cow image *p*'s identity conditions. On the basis of a subset of these identity conditions *S*, a dissimilarity class non-*P* is supposed to be constructed.[29] Suppose *S* is defined by what I take to be the dewlap, horns, tail, and shape (but not the color) of the cow in the recalled cow image.[30] Given this, non-*P* would be the class of all things that I do not take to satisfy *S*—that is, those things that do not have what I take to be a dewlap, horns, tail, and cow shape. The construction of non-*P* is clearly due to whatever accounts for what I take to be a dewlap, horns, tail, and cow shape—that is, the expectation parameters of *S*. According to exclusion theorists, what accounts for this is a combination of (F1) through (F5), with a great deal of emphasis being placed on (F3), well-established social/linguistic conventions.

By excluding the dissimilarity class, non-*P*, exclusion theorists suggest that I can construct the similarity class like-*P*, which is supposed to contain all of the things that I take to be cows.[31] Like-*P* is the complement of non-*P*. What I understand when I hear the word "cow" in the sentence "bring a cow" is such a similarity class/generalized image/thing-in-general/etc. Again, what ensures that "what I take to be cows" today and "what I take to be cows" tomorrow or what you take to be cows are, for all practical purposes, "the same" is said to be well-established social/linguistic conventions (F3). These conventions are supposed to provide the norms that inform what we all take cows to be.[32]

Exclusion theorists thus suppose that what I understand from hearing the word "cow" in a sentence such as "bring a cow" is the complex object like-*P*, which is constructed by excluding the dissimilarity class non-*P* from *p*. It is this "object" that exclusion theorists take semantic value to be. It is important to keep in mind that on the basis of the manifest cow image *p* described earlier, any number of similarity classes could be constructed, for example, "cow," "brown cow" (by including the "color" component), "draught animal," etc. Moreover, according to the story of exclusion, all of this can be accounted for without recourse to universals or shared properties of any kind.

QUESTIONING EXCLUSION

There are, in my view, (at least) five problem areas that a successful exclusion theorist will have to address. In considering these "problem areas," it is important to keep in mind, however, that traditional exclusion theorists develop and argue in support of their theory of exclusion only once they have argued against what they take to be all of the other accounts of semantic value (and conceptual content), such as universals, individuals, or some combination of the two.[33] It is only given the conclusion that none of the available alternatives are acceptable that the story of exclusion is told.

Problem 1: Agreement

(F1) through (F4) are the factors that an exclusion theorist can appeal to in order to explain why we start with the *p* that we start with and, with the addition of (F5), why this *p* provokes or triggers the construction of the image that it does.[34] It is also by appealing to these factors that selection set *S* is accounted for, and the exclusion process is said not to be arbitrary. Suppose that (F1) through (F5) define a context *C*. In my view, exclusion theorists' appeal to (F1) through (F5) cannot adequately explain either intrasubjective agreement or consistency in what seem to be similar contexts or intersubjective agreement in what seems to be a single context. Let us first consider cases of intrasubjective agreement—where an exclusion theorist needs to explain why, for example, in a number of different contexts, the word "aspirin" in a sentence such as "please bring me some aspirin" (or a personal desire for aspirin) routinely triggers the memory of a particular *p*, which then routinely leads to the construction of a generalized aspirin image. Why is it that in each of these different—but from a first-person perspective relevantly similar—contexts, a generalized aspirin image is constructed? Why does "this" context and not many others trigger the appropriate exclusion? Despite claims to the contrary, in my view, the best that an exclusion theorist can do is appeal to brute facts and simply say, "because it does." Similarly, it is only by appealing to brute facts that an exclusion theorist seems to be able to account for intersubjective agreement in what seems to be a single context *C*. Why do competent speakers of English construct a generalized cow image on the basis of which we all generally say "cow" when asked to identify a specific animal in front of us? Given that for each of us what is "available" is different from the unique "actionable" object that we each individually construct through exclusion, it seems to be a startling coincidence. How can an exclusion theorist explain this? Again, in my view, the best that an exclusion theorist can do is say, "because it does."

An exclusion theorist should protest, however, by arguing that (F1) through (F5)— especially (F3), the well-established social and linguistic conventions that we are all subject to—can in fact explain what seem to be amazing coincidences.[35] These conventions are supposed to provide the norms in accordance with which we all construct conceptual content. While the theory may be able to use (F3) to explain why the selection set *S* and the expectation parameters it defines are "shared," it is very difficult to see how it can explain why the other relevant factors—(F1), (F2), (F4), (F5)—are similar enough (i.e., why we construct a similar enough dissimilarity class) and why the exclusions that are constructed on the basis of *p* are, in so many cases, "functionally" equivalent. Appeals to latent karmic dispositions or their contemporary stand-in, natural selection, to account for this only seem to push back the problem. If expectation parameters, the conditions of their satisfaction, and functional similarity more generally are to be explained by natural selection, the problem remains: why does natural selection account for this—just because it does? It is not at all clear to me that an exclusion theorist has any chance of explaining this, or explaining the basis for how and why natural selection works, without violating the traditional exclusion theorist's commitments, for example, to momentariness—and the closely allied issues of object individuation, relations, and pragmatic efficacy (*arthakrīyā*)—and uncompromising nominalism.

Problem 2: Tolerance

How fine grained is like-*P*? Suppose that we agree that we all act in accordance with a remarkably wide-ranging set of social conventions such that (F1), (F3), and (F4) are fixed. Let us also suppose that there is some naturalistic explanation for these similarities, based on similar human dispositions, the ways in which we are enculturated, facts about our environment, etc. Still, as exclusion theorists recognize, there are other, more personal factors such as our personal interests, experiences, abilities, etc., that affect the exclusion process and account for the construction of slightly different similarity classes—even when (F1), (F3), and (F4) are assumed to be fixed. In addition to these other factors, there are also the two construction processes—the construction of the dissimilarity class non-*P* and the construction of the similarity class like-*P*. These processes—especially the first—seem to be epistemic and, at least to some extent, to be indexed to us and our abilities to discriminate and "know" what we think will satisfy *S*. The issue is whether the theory can explain the degree of tolerance in our constructed images—which are both fine grained and coarse grained enough to account for intra- and intersubjective agreement on a wide variety of things. The more general issue here is whether the epistemology of exclusion can in fact adequately explain the "levels of grain" in conceptual content. Specifying precisely what constrains and governs the exclusion process and determines the level of grain in what is constructed is, in my view, a central task for a contemporary exclusion theorist.

Problem 3: Capacities, Abilities, and Powers

There are a number of (seemingly mysterious) capacities, powers, and abilities that the story of exclusion seems to rely upon. If we assume, for example, that all of our minds (or the mind) are preconditioned by an unending series of latent karmic impressions, we need an account of exactly what kinds of things these karmic impressions are. Under what circumstances do they "ripen" and contribute to the manifestation of *p*? Exactly what is manifestation? Parallel questions can be raised about minds conditioned through natural selection. Exactly what accounts for the kind of mental dispositions that the theory of exclusion seems to require? In my view, questions also need to be raised about exactly what sort of ability allows us to construct dissimilarity classes. Identifying the members of this class as non-*p*'s and pointing out that non-*p* can be understood in terms of an implicative form of negation doesn't seem to be sufficient. Neither does it seem sufficient to say that exclusion is the ability to construct a complement. The mental capacities, abilities, and powers in terms of which the epistemology of dissimilarity and similarity class construction is explained seem rather mysterious.

Problem 4: Concerns About Content

My version of the exclusion story suggests that the theory of exclusion is best understood as a theory of mental content—that is, as a theory of the kinds of objects that are directly available to us and, as I have emphasized, the kinds of objects upon which we act. I have referred to these "actionable" objects as the *conceptual content* of awareness events. In my version of the exclusion story, I have suggested that one way to describe conceptual content is in terms of similarity classes. It is not at all clear, however, whether this is appropriate. Exactly what kind of content is this? As I have described it, the content of awareness sometimes seems to be like the contents of a bucket—the material that fills awareness (whether directly or indirectly).[36] It also seems, however, like the content of a newspaper, with a kind of "informational" content that can be described, perhaps, in terms of accuracy conditions.[37] It also seems as though this content could be an ability, for example, the ability to discriminate cows from noncows and make inferences.[38] Traditional theories of exclusion provide resources for each of these characterizations.

From what Jñānasrīmitra and Ratnakīrti have to say, it is clear that states of awareness have two kinds of content—nonconceptual/manifest and conceptual/determined.[39] It is also clear that conceptual content can be factored into its positive and negative features. What is not clear, however, is the extent to which, by insisting that all conceptual content can be described in terms of positive entities characterized by their exclusion from others, traditional exclusion theorists really think that the conceptual content of perceptual and inferential/verbal awareness is "the same." It is also not clear whether a contemporary exclusion theorist could agree that there is generic description of mental content that works for both perceptual and inferential/verbal awareness events.

Another important (and interesting) feature of Jñānasrīmitra and Ratnakīrti's view of conceptual content is that there doesn't seem to be any room for so-called propositional attitudes—and not just because they don't accept propositions (or any abstract objects). For them, the "attitude" part is also in question, since the attitude component is built into mental images. The desiring-aspirin image is different from the fearing-aspirin image, the believing-in-aspirin image, or the visually-appearing-aspirin image. There is no generic aspirin image that can be factored out from the attitude in which it is encoded.[40] According to most exclusion theorists, intentional mental states are not understood as relations to mental representations that can themselves be explained in terms of the semantic properties of the representations. It is not clear to me whether a contemporary exclusion theorist is willing to accept the consequences of this view.

Related to this, is the issue of phenomenality—and the distinction between the conceptual content of perceptual and inferential/verbal awareness events. Earlier, I referred to the conceptual content of perception—the "determined object" of perception—as the content of an experience (usually brought about by one of the sense modalities) and the conceptual content of inferential/verbal awareness—the "determined object" of inferential/verbal awareness events—as the content of a thought. According to exclusion theorists like Jñānasrīmitra (and less explicitly) Ratnakīrti, what distinguishes the two types of awareness events is just their phenomenal character:[41] What it seems like to us when we perceive something—for example, fire on a mountain—is different than when we infer its presence on a mountain or learn of its presence on a mountain on the basis of testimony. For simplicity, let us just consider the phenomenal character of experiences. According to Jñānasrīmitra, the phenomenal character of an experience is not exhausted by its so-called representational content. Like its representational content, its phenomenal character is also built into the mental image—and, in this sense, mental images are like phenomenal concepts.[42] To put it somewhat differently, phenomenal character is constitutive of constructed mental images and must, somehow, be accounted for by (F1) through (F5). What is not at all clear to me is how the phenomenal character of mental images should be dealt with by contemporary exclusion theorists who are simply interested in semantic value.

Problem 5: Presuppositions

In addition to the issues outlined earlier, it is clear that the theory of exclusion assumes (1) that arguments in support of real universals fail—and that there are equally good arguments to support the view that there are neither real world-given differences between things nor relations to connect them; (2) a sophisticated account of memory—which is rarely, if ever, provided; and (3) the notion of pragmatic efficacy. I take it that a contemporary exclusion theorist will either have to agree with Buddhist arguments for nominalism or provide arguments of her own, provide an account of memory that is consistent with her nominalism (and her naturalism), and provide an account of pragmatic efficacy. One question that this raises for me is to what extent the theory of exclusion as a semantic theory can be insulated from what seem to me very challenging issues in the philosophy of mind and metaphysics. By restricting its scope, it may be possible to avoid having to confront all of its weaknesses. The question is whether in restricting it in this way it is actually able to provide a meaningful account of semantic value.

Fundamental Trade-Off

Where does this leave us and my version of the story of exclusion? I think the generic version of the exclusion story is philosophically rich and full of possibilities. As one fills in the details, however, the cost of accepting it goes up, especially as one tries to specify (F1) through (F5) and describe *p*, non-*P*, and like-*P*, and explain how we are able to construct similarity classes. In exchange for its seemingly minimal metaphysical and ontological commitments, and its appeal to a small set of mental capacities, the theory requires us to accept brute facts—including unexplained powers and coincidences—and just stop there. Exactly which brute facts it appeals to will depend on exactly how a specific exclusion theorist understands its scope and fills in the details of the story. There is, then, a philosophical cost-benefit analysis that the theory leaves us with: The least problematic version of the story is not detailed enough to be of any real use, but as we fill in the details and the explanatory power of the theory increases, we are required to accept brute facts and underdescribed objects and capacities. Appealing to some form of naturalism to explain these facts is, it seems to me, no solution at all, since for a true heir to traditional exclusion theorists, it is the theory of exclusion that is supposed to explain naturalism. The task for a contemporary exclusion theorist is to clearly state what costs—ontological, metaphysical, epistemological, semantic—she is willing to assume for what presumed benefits. My sense is that a contemporary exclusion theorist is going to have to abandon the idea that there are only differences between things (to provide a story about our shared physiology or environment), and thereby accept an ontology that most traditional exclusion theorists would reject.[43] If this is correct, there is no real middle ground that traditional and contemporary exclusion theorists can share. This should not be a problem for a contemporary exclusion theorist, as long as she is not committed to defending traditional versions of the theory.

CONCLUSION: JÑĀNAŚRĪMITRA'S "SOLUTION"

In conclusion, I want to very briefly present Jñānaśrīmitra's not nonconclusion to his own "Monograph on Exclusion" (*Apohaprakaraṇa*). Jñānaśrīmitra argues that the two-component model of conceptual content—according to which conceptual content is best described as a positive entity characterized by its exclusion of others—is the very best account that can be given for the conceptual content of experiences and thoughts, including, of course, semantic value. According to him, this account is phenomenally adequate; it is in accordance with our perceptual experiences and our inferential and linguistic practices—and therefore meets the demands that they place on representational content; and it is less vulnerable to philosophical critique than any of its rivals. Despite this, however, Jñānaśrīmitra also thinks that it is mistaken. According to him, this account of conceptual content is, at the end of day, not a philosophically adequate account of what it is that our experiences and thoughts are really about. While Jñānaśrīmitra thinks that his complex, two-component model does provide an accurate account of what I have called the representational content and phenomenal character of mental content, it does not, he thinks, do justice to its "manifest" content—that is, *p*, the object that is grasped by awareness. The manifest content of awareness is undistorted by any mental processes that may follow upon its appearing in awareness—such as exclusion, conceptualization, and determination. In this sense, when compared with determined/conceptual content, it alone is epistemologically adequate.[44]

According to Jñānaśrīmitra, for something to meet all of the desiderata for being the content of an awareness event, it must be both manifest in awareness and determined by it. But, there is nothing that meets this criterion. The objects that are manifest in awareness are not determined and those that are determined are not manifest. While manifest objects are not distorted, they do not meet the criterion for phenomenal or representational adequacy. While determined objects do meet this criterion, they are not epistemologically adequate. There is, therefore, no single object that satisfies Jñānaśrīmitra's criteria for mental content.[45] A multiple-content view of mental content—such as the view that he inherits, with modification, from Dharmottara—is, he thinks, useful, but only as a convenient framework for helping us to understand the words of his predecessors in the Buddhist epistemological tradition, and for helping us to see clearly why there can be no philosophically adequate account of mental content. It helps us to understand the words of his predecessors by explaining why classic statements such as "perception is free from conceptual construction" and "exclusion is the meaning of a word" are nothing but pedagogically useful "white lies."[46] It helps us to understand why there can be no philosophically adequate account of mental content by showing us that there is no single object that can satisfy the criteria of epistemological, representational, and phenomenal adequacy.

Notes

1. Vikramaśīla is the name of the Buddhist monastic and educational complex where Jñānaśrīmitra and his student Ratnakīrti are said to have lived and worked. Both are said to have been "gate keepers" of this complex. For an introduction to their work, and what little we know of their lives, see Thakur 1975 and Thakur 1987. Vikramaśīla, which is supposed to have been founded by the Pāla king Dharmapāla (ca. 775–820 C.E.), was located in the Bhagalpur district of modern day Bihar.
2. For the "prehistory" of the theory of exclusion, see Bronkhorst 1999; Hattori 1977; and Raja 1986. For the philosophical context in which this theory was developed, see Dravid 1972; Dunne 2004; Hayes 1988; and Scharf 1996. For Dignāga's view, see Hayes 1986, 1988; Hattori 1968, 1980, 2000; Katsura 1979, 1991. For post-Ratnakīrti Buddhist discussions of exclusion in Sanskrit, see Mokṣākaragupta's *Tarkabhāṣā* (trans. in Kajiyama 1998), Vidyākraśānti's *Tarkasopana*, and the *Vādarahasya* and *Tarkarahasya*.
3. That Dignāga recognizes such a connection is clear from PS 5.1 in Hattori 2000, 139n3. For a translation of PSV ad PS V.1–PS V.12, see Hattori 2000, 137–146. For an account of how the theory of exclusion developed in Dignāga's work, see Frauwallner 1959; Katsura 1983; and Katsura 1991, 139.

4. For Dharmakīrti's view, see Pind 1999; Dunne 2004; and McCrea and Patil 2006, and the references contained therein. For Dharmottara's view, see Frauwallner 1937 and Hattori in this volume. For Bhāviveka's (ca. 490–570) extremely important but neglected account, see Hoornaert 2001; Saito 2004; and Eckel 2008. The Buddhist philosophers identified as "Jñānaśrīmitra and Ratnakīrti's predecessors" are only those whom Jñānaśrīmitra and Ratnakīrti seem to have been influenced by, especially with respect to their thoughts on exclusion. Other important exclusion theorists not included in this group include Śāntarakṣita and his commentator Kamalaśīla—for a discussion of their version of the exclusion theory, see Hugon in this volume and Siderits 1986—and Śaṅkaranandana, whose work on exclusion is being studied by Helmut Krasser and Vincent Eltschinger in Vienna. See Krasser 2002 for more on Śaṅkaranandana.
5. For Jñānaśrīmitra's view, see Katsura 1986 and McCrea and Patil 2006. For Ratnakīrti's view, see Patil 2003.
6. By "experiences" I generally mean awareness events produced by any of the six sense modalities (the standard five plus "mind"—for a discussion of mental perception, see Hattori 1968; Nagatomi 1968; and Kajiyama 1998, 44–47. To this list should also be added reflexive awareness, *svasaṃvedanā*). By "thoughts" I generally mean awareness events produced through testimony or inferential reasoning. In claiming that the theory of exclusion became *the* theory of mental content, I mean that while this theory was, most explicitly, used to explain the *conceptual* content of awareness events (i.e., "determined content" [*adhyavaseya*]), in so doing it also provided an account of the *nonconceptual* content of awareness events (i.e., "manifest content" [*grāhya*]).
7. According to Buddhist epistemologists—that is, the intellectual heirs to the Dignāgan and Dharmakīrtian philosophical traditions—there are only two valid modes of awareness, perception and inferential/verbal. As a result, it is the content of these awareness events that is of greatest interest and importance for them. For an excellent introduction to this philosophical tradition, see Kajiyama 1998; Hattori 1968; Dreyfus 1997; and Dunne 2004. For a detailed discussion of this specific issue, see McCrea and Patil 2006. The first clear statement of the "multiple-contents" view is given by Dharmottara: "The *object* of this fourfold perception—that is, the thing that is cognized—is a particular. A *particular* (*svalakṣaṇa*) is a property *lakṣaṇa*—that is, a unique character that is its own (*sva*). For a thing has both a unique character and a general character. And of these, that which is unique is what is *grasped* by perception. For the object of valid awareness is twofold: a grasped object whose image is produced and an attainable object that one determines. For the grasped object is one thing and the determined is something else, since, for perception, what is grasped is a single moment, but what is determined—through a judgment that arises by the force of perception—can only be a continuum. And only a continuum can be the attainable object of perception because a moment cannot be attained" (NBṬ ad NB I.12, ed. Malvania, 1971).

   It is worth highlighting just how radical Dharmottara's position is here. Never before has anyone connected with the Buddhist epistemological tradition even suggested that perception has more than one object. What Dharmakīrti himself says is simply that the object of perception is a particular. By importing the term "grasped" (*grāhya*) into his gloss on Dharmakīrti's text, without any clear basis in either the *Nyāyabindu* or any of Dharmakīrti's other works, Dharmottara has introduced into his account of perception precisely what Dharmakīrti sought to avoid—a bifurcation between two different kinds of objects which creates a gap between them that needs to be bridged by determination. Both of the extant Sanskrit commentaries on Dharmottara's text try to minimize his break with Dharmakīrti by suggesting that he does not mean to literally claim that perception itself has both a grasped and determined object. The author of the anonymous Ṭippana NBṬṬ, 3 (ed., Shastri 1984) comments as follows: "[Objector:] But how is the continuum an object of perception since it is [in fact] the object of conceptualization? [Reply:] We say that it is due to figurative usage. Because it is made into an object in such a way that it is determined by that conceptualization which is the functional output of perception, it is called the 'object of perception' on the basis of figurative usage—thus there is no problem." And Durveka Miśra (DhPr, 71.21), in his commentary, says: "Since the judgment that follows perception functions only with respect to what was grasped in perception, adding nothing to it, therefore, what is determined by that [judgment] is [said to be] 'determined by perception itself'—this is the idea." This paragraph is based on a parallel passage in McCrea and Patil 2006.

8. The standard Buddhist argument for momentariness is based on a particular understanding of causality. Roughly: Experience tells us that, after some time, a seed that has been planted and properly cared for will produce a sprout. Buddhists argue that the seed, at the moment of producing a sprout, has to be different, in some way, from the seed in previous moments, since the seed at just that moment produces a sprout while the seed in previous moments did not. But if this is the case, one must also admit that the seed that existed just prior to and therefore produced the sprout-producing seed is itself different, in some way, from the seeds in each of the moments that preceded it: it produced the sprout-producing seed and they did not. A similar argument can be made about the seed at the moment prior to that, that is, the seed that produced the seed that produced the sprout-producing seed, and the moment before that, etc. Thus, the single observed event—the production of the sprout from the seed—requires that we accept that each moment in the history of the seed is different from any other. If the seed were the same at each and every moment, then it would produce its effect, the sprout, in each and every moment of its existence. Thus, the continuity of the seed over time is not based on the persistence of a single entity. The "continuity" is only apparent. And it is this appearance of continuity over time that Buddhists designated by the term "continuum" (*santāna*). By analogy, all pragmatically effective objects must be momentary in this way. For more on this, see Stcherbatsky 1984, 79–118; Steinkellner 1969; Von Rospatt 1995. For a discussion of the early history of this idea, see Yoshimizu 1999 and Oetke 1993. For a discussion of Dharmakīrti's famous *sattvānumāna*—the inferential proof of momentariness from "existence"—see Dunne 2004, 91–97; see Frauwallner 1935 for an edition and German translation from the Tibetan of Dharmottara's "Proof of Momentariness" (*Kṣaṇabhaṅgasiddhi*); see Tani 1997 for an analysis of this text;

and see Mimaki 1976; Woo 1999; and Tani 1999 for a discussion of this theory in the work of later Buddhist epistemologists. This paragraph is based on a parallel passage in McCrea and Patil 2006.

9. A translation of Ratnakīrti's "Demonstration of Exclusion" (*Apohasiddhi*, AS) may be found at the companion website for this volume, www.cup.columbia.edu/apoha-translation. A translation of Jñānaśrīmitra's "Monograph on Exclusion" (*Apohaprakaraṇa*, AP) will be published in McCrea and Patil (2010). For an example of what I am calling "contemporary exclusion theory," see Siderits 2006.
10. One way of thinking about the theory is that it provides an account of how conceptual content—determined content (*adhyavaseya-viṣaya*)—is constructed from nonconceptual content—manifest content (*grāhya-viṣaya*). It is a "backstory" in the sense that, according to traditional exclusion theorists, independently of determined content, we do not have access to manifest content. A "bottom-up" approach is only possible in hindsight. See the contributions of Tillemans and Ganeri in this volume for a discussion of "top-down" and "bottom-up" strategies for thinking about the theory of exclusion.
11. Many of the subarguments have to do with showing that alternative accounts of conceptual content—especially semantic value—are indefensible. For a brief discussion of this, see Patil 2003; Patil (2009); and especially Sen's contribution to this volume. Also see Dravid 1972 and Siderits 1991. A helpful way to consider the overall structure of the argument is to consider RNA 66.08ff., where Ratnakīrti presents his argument in the form of an inference and then defends it.
12. In his *Tarkabhāṣā* (TBh), Mokṣākaragupta describes particulars in a way that is entirely consistent with how Jñānaśrīmitra and Ratnakīrti describe them. He says that particulars are "excluded from those that belong to the same class and those that belong to a different class" (*sajātīya-vijātīya-vyāvṛtta*).
13. See, for example, the simple example at JNĀ AP, 221.26–222.02. For a discussion of selectivity in conceptualization, see Dunne 2004; Kellner 2004; and Patil 2003.
14. For other accounts of the theory of exclusion, see Ganeri 2001; Herzberger 1975; Patil 2003; Siderits 1991, 1999, 2006.
15. For a discussion of relativization, see McCrea and Patil 2006. It is also worth noting that "nonconceptual" refers to the way in which content is presented to awareness and does not mean "never conceptual." In other words, an image that (factoring out momentariness) was previously constructed conceptually can be nonconceptually presented to awareness. This is, for example, often the case with recalled image *p*.
16. The term "nonconceptual" refers to the content of an awareness event in the sense that the image that is the content of that awareness event is directly presented to awareness (*pratibhāsa*). In the history of the continuum that best describes that image, there may have been some form of conceptual construction. But, in the awareness episodes in question, the image itself is presented directly. For example, consider what Jñānaśrīmitra has to say about habitual behavior: "And even if, in the case of completely habitual behavior, things are seen to be objects of activity merely by appearance [and without any accompanying determination], nevertheless, that very habituation could not exist without determination. Thus, this [fact that when we act habitually we do so on the basis of

appearance alone] is itself [due to] the power of [previous] determination. Therefore, this qualification [to the criterion given above] is required, "whatever is not determined by it when there is no habituation [is not the object of an episode of awareness]" (JNĀ [AP, 230.27–231.02]).

17. See RNA (AP, 65.28–66.05)
18. See Kellner 2004 for a discussion of what she calls "situational factors."
19. See JNĀ (AP, 204.08–204.16): "If you say this, then in the same way, if the form of the conventional association were stated with words of this sort, 'the word "cow" refers to what is excluded from noncows,' then there would be this problem. But what possible problem could there be if:

    (a) The language learner has, with respect to the individuals intended by the speaker, a reflexive awareness containing a single image, and
    (b) On the basis of context, he is caused to form a determinate awareness of them, and then
    (c) The speaker makes the conventional association "this is a cow"?

    For that language learner understands that all of the individuals that fall within the scope of his own conceptual awareness—which are themselves excluded without relying on words such as 'excluded from noncows' from all individuals that do not belong to that class—are expressed by the word 'cow.' Therefore, by the word 'cow' he refers only to those individuals in which he apprehends the exclusion of what does not belong to that class. And in virtue of this, the statement that 'the word "cow" refers to what is excluded from noncows' is just a gloss on the empirically established fact and is not a form of the convention itself. This is because once this semantic function of the word 'cow' has been accepted, then everything else can be denoted by the word 'noncow.' But whether that shared image, which is excluded from what does not belong to that class, is the appearance, in conceptual awareness, of a universal or is the real nature of the individual will be determined later. But it is established that there is no circularity." Also see McCrea and Patil 2006 and Hugon's contribution to this volume for a discussion of how "conventions" are formed.
20. For Ratnakīrti's view of this see, RNA (CAPV, 129.19–129.21). For an extremely interesting discussion, see Sen's references to "the form of four and a half" in his contribution to this volume.
21. This is, but only in part, a way of parsing the claim that particulars are "inexpressible." For references to the inexpressibility of particulars, see Hattori 1968; Dunne 2004; McCrea and Patil 2006, where JNĀ is discussed; and Patil 2003, where RNA is discussed.
22. I owe this way of formulating the point to Mark Siderits.
23. See Patil 2003, 237–240 (and the RNA references contained therein), RNA (AS, 59.16–59.20), and JNĀ (AP, 204.08–204.16). My point here is to suggest that both Ratnakīrti and Jñānaśrīmitra are sensitive to the idea that fixing the referent of a term through ostension presupposes some form of conceptualization and that a successful account of reference fixing should be able to explain why a term that has a specific semantic value has only that semantic value.
24. For a more detailed and textually grounded discussion of Ratnakīrti's account of this complex-positive object, see Patil (2009).

25. JNĀ (AP, 226–227): "And just as the determination 'there is fire here' produces bodily activity, in the same way it produces verbal [activity] as well: 'Fire has been apprehended by me.' It also produces mental activity, that is, a reflective awareness having the same form [as the verbal statement]."
26. For more on determination, see McCrea and Patil 2006.
27. For a discussion of this, see Patil (2009).
28. See Patil 2003 for a discussion of these concepts in RNA.
29. The elements from which there is exclusion are also described by Ratnakīrti in a number of different ways. He says, for example, that these entities are "other than" (*anya*), "different from" (*para*), "nonthat" (*atad*), those that "belong to a class which is nonthat" (*atajjātīya*), those that "belong to a different class" (*vijātīya*), "other objects" (*viṣayāntara*), or those that "have the form of nonthat" (*atadrūpa*). What these expressions help clarify is the sense in which the elements from which there is exclusion are different or other; they are "different" or "other" in the sense that they have a different form or belong to a different class.
30. See JNĀ (AP, 220.02–220.4): "From the word 'cow' in the sentence 'there are cows grazing on the far bank of the river,' dewlap, horn, tail, etc., appear—accompanied by the form of the letters [which make up the word 'cow']—in effect, 'lumped together' because of inattention to differences between things belonging to the same class. But, that [conglomeration of dewlap, horn, etc.,] is not itself a universal" RNA (AS, 63.10–63.16).
31. Ratnakīrti uses five different terms to denote the process of excluding: "exclusion" (*apoha*), "taking away" (*parihāra*), "separating out" (*vyāvṛtti*), "covering up" ([*para*]-*āvṛtta*), and "absence" (*abhāva*). These terms are used synonymously. What these expressions suggest is that exclusion is the capacity of differentiating or selecting between elements in what could be multi-entitied awareness events.

    Ratnakīrti describes this collection as "an object that is characterized by its exclusion from others and excluded from those which belong to a different class" (*anyābhāva-viśiṣṭo vijātīyavyāvṛtto 'rthaḥ*), as "a thing-in-general that is determined and excluded from those which belong to a different class" (*adhyavasita-vijātīyavyāvṛtta-vastumātra*), as "a thing in general that is excluded from the entire collection of things which do not have its form" (*sakalātadrūpaparāvṛttavastumāt ra*), as "that which has the nature of a generic particular that is excluded from those that do not have its form" (*atadrūpaparāvṛttasvalakṣaṇamātrātmaka*), and as "a determined external object" (*adhyavasita-bāhya-viśayatvam*). See RNA (AS, 66.05–66.06); RNA (AS, 66.13); RNA (KSA, 73.21–73.23); RNA (VNi, 109.17); and RNA (AS, 66.20).
32. Exclusion theorists like Ratnakīrti usually appeal to causal factors to explain why exclusion is not arbitrary. See, for example, RNA (KSA, 74.07–74.15) and RNA (KSA, 73.12–73.17), which are translated in Woo 1999. Also see RNA (AS, 65.25–66.05).
33. For an excellent discussion of these arguments, see Sen's contribution to this volume. For Ratnakīrti's account, see, for example, RNA (AS, 60.20–63.09) for his arguments against sensible particulars and RNA (AS, 63.10–65.14) for his arguments against universals.
34. See Tillemans's and Dreyfus's contributions to this volume, where this issue is also discussed.
35. See note 34.

36. This seems to be an apt description of "manifest content." See Siegel 2005, which is my source for the use of the phrase "content of a bucket." Also see Lawrence and Margolis 1999 and Horgan and Woodward 1985.
37. This seems to be an apt description of "determined content." See Siegel 2005.
38. This seems to capture the idea of exclusion processes. For an "abilities view" of concepts see Brandom 1994; Dummett 1993b; and Millikan 2003.
39. There are, given this terminology, two states of awareness—perceptual and inferential/verbal—each of which is constituted by two awareness events/episodes, a "nonconceptual" awareness event and a "conceptual" awareness event. This is the structure of the famous four-object model (two states multiplied by two awareness events) developed by Dharmottara and accepted by Jñānaśrīmitra and Ratnakīrti. Jñānaśrīmitra and Ratnakīrti describe this typology of objects in a number of different places in their work.
40. It has been suggested that I may be overstating the point here since, perhaps, "fearing *x*" could be explained by the presence of an image *x* in awareness along with a distinct "fear" impression (*saṃskāra*). One problem with this suggestion, however, is that neither Jñānaśrīmitra nor Ratnakīrti suggest that there are, in fact, such "impressions" or that anything like them (e.g., a fear image) is needed to account for conceptual content. More generally, it seems to me that Buddhist philosophers do not accept the idea that propositional content and propositional attitude or even content and force are independent, in the sense that they do not accept the idea that conceptual content can be factored into, for example, a content component and a force component. This does not mean, of course, that they do not recognize the distinction.
41. See, for example, JNĀ (AP, 231.10–231.16), which is discussed in McCrea and Patil 2006. I think that Jñānasrīmitra, like Dharmottara, would agree that both awareness events have a phenomenal character. The point that Jñānaśrimitra is making in the context of the passage referred to earlier is that while ordinary people may confuse the content of perception with that of inference and vice versa, careful attention to one's mental states can easily correct this since the phenomenal character of the two types of awareness is, in fact, different.
42. For a discussion of "phenomenal concepts," see Block 2003 and Chalmers 2003.
43. It is, perhaps, worth noting that neither Jñānaśrīmitra nor Ratnakīrti accept that there is a *real* difference (*bheda*) between things. They argue that all differences are only apparent and are constructed by us. For more on this, see Patil (2009), where Ratnakīrti's doctrine of multifaceted nonduality (*citrādvaita*) is discussed.
44. For both Jñānaśrīmitra and Ratnakīrti, the problem is that there is a "mismatch" between manifest and determined content. It is this mismatch that is the basis for claiming that the view is mistaken. This is where the so-called misplacement theory of error becomes relevant.
45. JNĀ (AP, 229.03–229.10): "There is no way of really affirming either the mental image or the external object. Conventionally [there is affirmation] only of externals, whereas even conventionally there is no [affirmation] of the mental image. For this mental image, which is indubitable and an object of reflexive awareness, cannot be what is affirmed or denied by means of words etc., since this would be useless [in the case of affirmation] and impossible [in the case of denial]. Neither

can the external object, which does not appear in conceptual awareness, [really be affirmed or denied]. Since this object is not cognized, what could be affirmed or denied. Therefore, just as, on the basis of determination, an external tree is conditionally adopted (*vyavasthāpita*) as what is denoted by the word 'tree,' in the same way, it is only on the basis of determination that one talks about affirming or denying [any] external object. Even when, due to certain circumstances, one examines a mental image, having brought it to mind by means of another conceptualization, then too there is affirmation and denial of what is external to this conceptualization."

Interestingly, an opponent explicitly suggests that Jñānaśrīmitra ought to accept that there are two kinds of mental content—manifest content and determined content—and that such a multiple-content view is a proper one. For Jñānaśrīmitra's response, see JNĀ (AP, 231.07–231.10), which is discussed in McCrea and Patil 2006.

46. See Patil 2007, and McCrea and Patil 2006 for a discussion of the philosophical and pedagogical significance of such "white lies." The central idea discussed in both of these papers is the idea of a "conditionally adopted position" (*vyavasthā*), which is a philosophical position that is based on a little bit of the truth and is to be used for a specific purpose. "In response to this, we say: by relying on a little bit of the truth (*tattvaleśa*), a certain conditionally adopted position is, for a specific purpose, constructed [by us], in one way, even though the actual state of affairs is different" (JNĀ [AP, 204.26–205.03]). "Here too, [the idea that] linguistic expression takes a positive entity as its object is just the same [in that it too is a conditionally adopted position]. Here, we conditionally adopt the position that exclusion, even though it is [really just] a necessarily attendant awareness, is the object of conceptual [including inferential/verbal] awareness, to set aside any suspicion that we accept [the position] pushed by [our] opponents that it is only the positive entity that is really expressed. And therefore, we don't just talk in terms of the positive entity [when describing the semantic value of a word]. But, when [someone] pushes the position that 'exclusion alone is the primary meaning of a word,' then we put forth the positive entity as well. As stated, 'first of all, it is the [external] object that is primarily expressed by words.' But, in perception, because there is no disagreement [of this sort], it is proper that one should not conditionally adopt this position" (JNĀ [AP, 205.03–205.09]).

8

The Apoha Theory of Meaning

A CRITICAL ACCOUNT

Prabal Kumar Sen

The Buddhist view regarding the relation between thought, language, and reality has been expressed in the *Laṅkāvatāra Sūtra* in the following manner: "The nature of things cannot be ascertained when they are subjected to critical examination. Consequently, they have been declared [by Buddha] to be beyond the ken of language, and also devoid of ultimate nature."[1] This doctrine was stated and expounded further by Nāgārjuna in his *Madhyamakaśāstra*; and his followers such as Āryadeva, Candrakīrti, and Śāntideva followed suit. Their common refrain was that thought and language fail to capture the nature of ultimate reality.[2] For Nāgārjuna and his followers, all concepts are infested with some sort of contradiction, and hence, they are devoid of ultimate nature, though they are useful tools in our mundane life. Hence, they are only "empirically real" (*vyavahārikasat/prajñaptisat*) and not ultimately real (*paramārthasat*). They are also called *samvṛtisat*, since they cover or conceal the real nature of things.[3] Nāgārjuna also maintains that one cannot speak of the ultimate truth without taking recourse to the empirical truth, and that unless one comprehends the nature of the ultimate truth, one cannot attain *nirvāṇa* (PP ad MMK XXIV.10). Concepts are thus useful but unreal tools for the discourse about ultimate truth, and the same holds good of thought and language, which are totally dependent on the employment of concepts.

Dignāga, the original propounder of the apoha theory of meaning also subscribed to the Buddhist thesis that ultimate reality cannot be revealed by language. But while philosophers like Nāgārjuna and his followers simply said that thought and language can deal with only what is "empirically real" or "conventionally real," they did not spell out in detail how language can nevertheless function as the medium of communication, and how it assumes such an important role in our behavior, which is so heavily dependent on language. The apoha theory of meaning propounded by Dignāga is intended as a solution to this vital problem. Just as Nāgārjuna and his followers drew a distinction between "ultimately real" and "empirically real," similarly Dignāga also neatly bifurcates entities into two groups: (i) unique particulars (*svalakṣaṇas*), which are ultimately real (*paramārthasat*); and (ii) "universals" (*sāmānyalaksaṇas*) or "imaginary constructions" (*vikalpas*), which are only "empirically real" or "conventionally real" (*samvṛtisat*). The unique particulars, which are endowed with causal efficacy (*arthakriyāsāmarthya*), are objects of perception (*pratyakṣa*), and they produce impressions that are vivid (*sphuṭābha*). "Universals" or "imaginary constructs," which lack causal efficacy, are the objects of inference (*anumāna*), and the awarenesses that they produce are faint (*asphuṭābha*).[4] The unique particulars, which are devoid of all common properties, cannot be brought under any concept; and hence, they cannot be objects of discursive thought, nor can they be expressed by words in a straightforward manner. Like inference, language can deal only with "universals"; and on this ground, Dignāga has subsumed cognitions generated by linguistic expressions under inference. In support of this view, Dignāga has claimed that since both inference and language function through what is known as "exclusion of the other" (*anyavyāṛtti/anyāpoha/apoha*), there is no basic difference between them.[5] This claim needs to be explained further. In the words of B. K. Matilal,

> [A] word expresses a concept, and a concept, being a fiction, cannot POSITIVELY qualify or characterize the particular . . . but it can NEGATIVELY disqualify the particular from being claimed by other fictions or concepts. Thus, construction and verbalization are to be understood as "exclusion" of all rival claims. If we associate the name *cow*, or *the cow*, with a particular, it *means* that it is not what we cannot call a cow.[6]

One advantage that this view may yield is that we need not assume the existence of real universals like *cowness* as the basis of the application of general terms like "cow." We would not also require real universals for identifying a particular entity. When we look at an animal and identify it as a cow, we do not do so due to the fact that the animal is characterized by the universal *cowness* (*gotva*), which is one and eternal. What happens is that when we look at its features like the dewlap, the hump on its back, the horns on its head, and its cloven hooves, we can determine that it is not a goat or a camel or a tiger and so on, none of which is a cow. Thus, we are aware that the animal is not a noncow. Thus, the positive feature, *cowness* (*gotva*), can be dispensed with, because its purpose can be served very well by "difference from noncows" (*a-go-vyāvṛtti*). The meaning of words consists in differentiation, and hence, it is negative in character.

While the arguments in favor of the apoha theory were originally given by Dignāga, some modifications in the apoha theory and some new arguments in its favor were gradually introduced by subsequent Buddhist philosophers like Dharmakīrti, Dharmottara, Śāntarakṣita, Jñānaśrīmitra, and Ratnakīrti. The tenability of a philosophical theory (say, T_1) depends primarily on the soundness of the arguments that are given in favor of it and of the arguments that are given for rejecting rival theories (say, T_2, T_3, and so on). One certainly has also to consider how far such a theory can be defended against the criticisms that are leveled against it by its critics, but one can do so only when some convincing arguments have been adduced for admitting that theory. One can certainly come up with an entirely new set of arguments in favor of T_1, and thus make the examination of the arguments given so far in favor of T_1 unnecessary. Still, it seems to me that a short account of the arguments given by Dignāga and his successors in favor of the apoha theory and an examination of those arguments may not be entirely out of place.

Before we start discussing the arguments given by Dignāga in favor of the apoha theory, some remarks about the views of his predecessors about the classification of meaningful words may be in order. Words were divided by authors like Yāska and Vātsyāyana into four groups: (i) noun (*nāma*), (ii) verb (*ākhyāta*), (iii) prefix or preverb (*upasarga*), and (iv) particle (*nipāta*) (NU 1/1/8; NBh 5/2/7). Another classification of words mentioned by Patañjali in his *Mahābhāṣya* divides words into four groups: (i) "universal word" (*jātiśabda*), (ii) "quality word" (*guṇaśabda*), (iii) "action word" (*kriyāśabda*), and (iv) "arbitrary word" (*yadṛcchāśabda*) (MB 90). This seems to be a subdivision of nouns (*nāma*), the principle of division being the types of "ground of application" (*pravṛttinimitta*) of the word or the absence of such a ground. Thus, a "universal word" (*jātiśabda*) is a word that can be applied to an object that is characterized by a specific universal (e.g., the word "cow" can be applied for describing an animal that possesses the universal *cowness*), and thus, the ground of application here is a universal (*jāti*). In the cases of a "quality word" (*guṇaśabda*) and an "action word" (*kriyāśabda*), the grounds of application are some quality or some activity respectively, their examples being "white" (which is applicable to a thing that has the quality known as white color) and "cook" (which is applicable to someone who performs the action of cooking). In the case of an "arbitrary word" (*yadṛcchāśabda*), however, there is no ground of application whatsoever, that is, the employment of such words is entirely arbitrary; and examples of such words are proper names like "*Caitra*," "*Maitra*," "Tom," "Dick," etc.

There has been, however, no unanimity among Indian thinkers regarding this fourfold classification of words. Patañjali has mentioned in his *Mahābhāṣya* the views of Vājapyāyana and Vyāḍi, two ancient grammarians. For Vājapyāyana, words stand for universals, while for Vyādi, words express individuals.[7] Bhartṛhari maintains in his *Vākyapadīya* that "reality" or "existence" (*sattā*), the highest universal, is the referent of all words (VP 3.1.33–4). *Nyāyasūtras* 2:2.59–2.2.65 records the dispute between some rival views regarding the meaning of words. According to some thinkers, the word "cow" means the universal (*jāti*) *cowness*; according to some others, it means the individual (*vyakti*) cow(s); and according to some others, it means a particular configuration (*ākṛti*) that is present in individual cows. *Nyāyasūtra* 2.2.66 maintains that all these three (namely, *jāti*, *vyakti*, and *ākṛti*) can be the meaning of words, depending on the context in which the word has been employed.

Dignāga, however, maintains that words express "something that is different from others" (*anyāpoha*). The arguments given in favor of this view are found in verses 1–12 of chapter 5 of his celebrated work *Pramāṇasamuccaya*.[8] In these verses, Dignāga seeks to prove that kind terms (*jātiśabda*) cannot in fact mean either (i) a particular entity (*bheda*) or (ii) a universal (*jāti*) or (iii) the relation (*sambandha*) that is supposed to obtain between a universal and the particular (where this universal is located) or (iv) a particular that is characterized by a universal (*jātimat*). According to Dignāga, once these four possibilities concerning the possible meaning of a kind term are ruled out, it would follow that words can express only "exclusion from others."[9]

Let us now see how Dignāga proceeds to establish that a kind term like *sat* (existent), which is supposed to designate the universal *sattā* (existence), cannot mean any of the four alternatives mentioned above.

Dignāga first points out that for two reasons, the word "existent" (*sat*) cannot designate individual things, that is, particulars. These two reasons are (i) innumerability (*ānantya*) of existent individuals and (ii) deviation or lack or invariance (*vyabhicāra*). We can claim that a certain word designates a certain object only when the "designative relation" (*vācyavācakasambandha*) between them can be established. Since the individual entities that are existent are also infinite in number, it is humanly impossible to establish (or know) the "designative relation" between such entities and the word "existent." Unless such a relation between a word and some object(s) is established (or is known), when we listen to that word, we will be aware of the word alone and not of the object(s) supposed to be designated by it. Thus, the word "existent" cannot designate *all* existent individuals. Nor can it be claimed that for our practical purposes, we need to know that the word "existent" designates only those existent things that are known to us, because then our use of the term "existent" would have to be restricted to those existent things alone; but those who admit "kind terms" (*jātiśabda*s) like *sat* (existent) maintain that it can designate any existent entity whatsoever. In the latter case, if the designated entity is not known to us, then it would have to be admitted that the employment of a word for designating a particular entity is justified even when the designative relation between that word and that entity has not been established (or is not known). This is inconsistent with the claim that the establishment (or knowledge) of the designative relation is a precondition of the proper employment of that word, and is thus tantamount to the admission of a deviation (*vyabhicāra*) from the claim made earlier. Moreover, the universal *sattā* (existence) that is admitted by the Nyāya-Vaiśeṣika school is supposed to inhere in substance (*dravya*), quality (*guṇa*), or action (*kriyā*). Accordingly, when we listen only to the word "existent," we will not be in a position to know with certainty whether this word has been employed for designating a substance or a quality or an action. Unless the utterance of a particular word leads to a definite cognition about some specific object, one should not claim that the word concerned designates that object. Accordingly, Dignāga maintains that kind terms (i.e., the words that are supposed to express some universal or other) cannot designate individuals.

Words like "existent" also cannot designate a universal, because there are legitimate linguistic expressions like "the existent substance" (*sad dra-*

vyam), where the word that is supposed to designate the universal is in apposition with the word that designates some individual, since the same "case inflexion" (*vibhakti*) has been associated with both these words. As per the rules of Sanskrit grammar, this happens in the case of two words only when they are coreferential, that is, when they have the same "locus of reference" (*sāmānādhikaraṇya*; in other words, when they designate the same thing). Unless we admit this rule, we would have to admit that even when two words designate different things, they can nevertheless be in apposition. But such expressions (e.g., "the pot cloth" [*ghaṭo paṭaḥ*]) are not regarded as meaningful or legitimate. Thus, it cannot be claimed that the word "existent" designates simply the universal called "existence" (*sattā*).

One might suggest here that the coreferentiality of the two terms "existent" (*sat*) and "substance" (*dravya*) in the expression "the existent substance" (*sad dravyam*) can be explained by the fact that the word "existent" designates existence, which, in its turn, is located in the substance. But this defense is not tenable, because in that case these two words (one of which designates a certain property and the other of which designates the bearer of that property) should have different case endings, as we have in the expression *dravyasya sattā* (the existence of the substance), where the first word has the genitive case ending and the second word has the nominative case ending; and consequently, they cannot be in apposition.

The word "existent" also cannot designate here the relation (*sambandha*) that obtains between existence (which is the supposed meaning of the word "existent") and substance, because like properties (which belong to the bearer of the property), relations also belong to (or obtain between) their respective relata; accordingly, two terms, one of which expresses a relation and the other a relatum of that relation, cannot be in apposition.

Dignāga now proceeds to reject the fourth alternative, namely, that the word "existent" designates something that is characterized by the universal. He rejects this possibility on several grounds. The first of them is "lack of independence" (*asvātantrya*). From the word "existent," we do not usually understand specific existent things like pot, cloth, etc. The term "taste," for example, brings to our mind specific tastes like sweet, sour, bitter, and so on; and the term "color" brings to our mind specific colors like red, blue, green, and so on. In such cases, we can have apposition, that is, agreement in the case endings of the two terms, for example, "the blue color" or "the sweet taste." But we cannot treat expressions like "the sour color" as meaningful.[10] The two words "existent" and "pot" thus should not be in apposition, but we nevertheless use expressions like "the existent pot." This is possible if the word "existent" can designate the pot not by itself but only when it is accompanied by the word "pot"; and this is what Dignāga calls "lack of independence."

One may try to avoid the difficulty that has been pointed out here by claiming that in expressions like "the existent pot," the word "existent" primarily designates existence; and then, it is metaphorically used to mean something that has existence. If something *X* is related to the primary meaning *Y* of a word *W*, then *X* may be regarded as the secondary meaning (*lakṣyārtha* or *aupacārika artha)* of *W*. Here also, what is characterized by existence is certainly related to existence, which happens to be the primary meaning (*mukhyārtha* or *śakyārtha*), and as such, the word "existent" may be metaphorically used to mean some such entity. In this way, the coreferentiality of the two terms "existent" and "pot" can take place, and hence, they can also be in apposition. If an influential royal employee begins to behave like the king, then one can say in a metaphorical sense that the employee is the king (*rājā bhṛtyaḥ*); but even in such cases, our awarenesses of the king and his employee are not similar, even though there are some similarities in this case between the king and the employee. But in the case under consideration, there is no such similarity of properties between a universal and what is characterized by that universal. Even if there is metaphorical transference due to some relation or other, what is characterized by existence can never be the primary or proper meaning of the term "existent." Moreover, the expression "the existent pot" (*san ghaṭaḥ*) is supposed to be a *genuine* description of the pot, which cannot be the case if the word "existent" is used here in a metaphorical sense.

Besides, in standard usage where the similarity of two or more things is stated, there is no apposition between the term that expresses the common property and the terms that designate the things that are similar (e.g., "kundasya kumudasya śukteśca śauklyam," i.e., the whiteness of the *kunda* flower, water lily, and mother of pearl), where the terms applied for designating similar things have the genitive case ending, while the term used for designating the common property, namely, the white color, has the nominative case ending). Moreover, the similar things are expressed here by different words that are uttered in sequence. But the word "existent" is claimed to express existence and that which has existence at the same time through a single utterance. Hence, the usage "the existent pot" is not a case of metaphorical transference on the basis of some similarity.

One may say here that that we should treat the expression "existent pot" like the expression "the red crystal," which describes a piece of crystal that has been dyed red with lac dye (*lākṣā*). Even though the crystal is not red, the word "red" is applied to it due to the presence of an adventitious condition (*upādhi*), namely, the red dyeing stuff that is applied to it, and which makes us ascribe its property (namely, redness) to the crystal. In like manner, due to the presence of the universal known as existence in the pot, the word "existent," which designates this universal, is applied to the pot, which is the locus of that universal. But this defense is not tenable, because when we are aware of the "red crystal," we need not have a separate and previous cognition of the adventitious condition, namely, the red stuff that has been applied to the crystal. In like manner, after hearing the word "existent," one should be aware of the existent pot, even though there is no prior cognition of the universal known as existence, since the mere presence of this universal is supposed to make us apply the epithet "existent" to the pot. But this is not the case here, because if the primary meaning of the word "existent" is the universal known as existence, then the latter should at first be presented before our mind by the word "existent."

There are some further difficulties that arise if we treat the expression "the existent pot" as a metaphorical one. When we listen to such an expression, we do not become aware of the literal meaning and metaphorical meaning of the term "existent" in succession, because a word is incapable of expressing one meaning through one significatory function, and then expressing another meaning through another significatory function (*śabda-buddhi-karmaṇām viramya vyāpārābhāvaḥ*). Nor can we be simultaneously aware of both these meanings, because a single utterance of a word can generate the awareness of only a single meaning (*sakṛduccaritaḥ śabdaḥ sakṛd arthaṃ gamayati*). Moreover, if the expression "the existent pot" is like the expressions "the employee is the king" or "the crystal is red," then the cognition generated by it will not be veridical, because it would not present things as they are in reality. Thus, the suggestion that the term "existent" justifiably designates only individuals that are characterized by the universal "existence" does not stand to reason.

One may still say that even though the pot is not the primary meaning of the word "existent," the word "existent" may nevertheless make us aware of the pot by implication (*arthākṣepa*). When we hear the sentence "this person, who is in good health, does not eat during the day," we understand by implication that the person eats during the night (since otherwise he would have become emaciated), even though this is not the literal meaning of that sentence. If in this manner one can have an awareness of things like pot from the awareness of the word "existent," then the apposition between the words "existent" and "pot" can take place, since they may be related as substantive and adjective. But such a position is not tenable, because in the example of meaning grasped by implication that has been given earlier, one can be sure about what the statement implies, since there is no other possibility that may be entertained. But from the mere awareness of the word "existent," one cannot be sure whether the thing meant by this word is a pot or not. Hence, the question of being aware of something through implication does not arise in this case.

The individuals supposed to be characterized by the universal known as existence are also infinite in number, and they may also be of different types, namely, substance, quality, and action; accordingly, it cannot be known with certainty what exactly is designated by the word "existent" on a certain occasion. Thus, the defects of infinity (*ānantya*) and deviation (*vyabhicāra*) that vitiate the thesis that bare individuals *alone* are the designata of the word "existent" also vitiate the view that individuals characterized by existence are designated by the word "existent."

As a last resort, one may say that words like "existent" designate some particular entity that is characterized by some uncommon property (*asādhāraṇaviśeṣaṇa*), that is, some property that belongs to this entity alone. But this solution, too, is not acceptable, because in that case, this word will not be applicable to any other entity. There is no other alternative that can be considered here. From this, Dignāga concludes that words can express something only by excluding entities other than that thing.[11]

These arguments of Dignāga have not remained unchallenged. Uddyotakara, Kumārila Bhaṭṭa, Vācaspati Miśra, Jayanta Bhaṭṭa, Bhaṭṭaputra Jayamiśra, and Pārthasārathi Miśra have criticized these arguments from their respective standpoints. Jaina thinkers like Mallavādin have also criticized these arguments in a thoroughgoing manner. It will not be possible for us to deal with all the counterarguments in this paper, though some indication of the directions in which these counterarguments proceed can be given. Before that, however, we would like to point out some other problems that trouble us.

Let us grant, for the sake of argument, that Dignāga has succeeded in showing that kind terms (*jātiśabdas*) like "existent" are not capable of designating either (i) universals, or (ii) bare particulars, or (iii) the relation between universals and particulars or (iv) particulars that are characterized by universals. Let us also grant the obvious claim that such words are not capable of designating unique particulars. Would it then follow necessarily that *all* words designate apoha alone? Among the four types of words that have been admitted by Patañjali and others, only the possible designata of "universal word" (*jātiśabda*) have been discussed by Dignāga in the verses quoted earlier. Is it evident that the arguments given about kind terms (*jātiśabda*) hold *equally* in the cases of "quality word" (*guṇaśabda*), "action word" (*kriyāśabda*), and "arbitrary word" (*yadṛcchaśabda*)? Moreover, has it been shown by Dignāga that *all* the possible alternatives about the designatum of a kind term (*jātiśabda*) have been rejected by him after due consideration? Unless one can answer both these questions in the affirmative, it cannot be claimed that the arguments given by Dignāga that we have considered so far can conclusively establish the thesis that all words express apoha.

So far as the first question is concerned, it may be said by a follower of Dignāga that the arguments given in connection with kind terms (*jātiśabda*) may, with some necessary modifications, be applied to "quality word" (*guṇaśabda*) and "action word" (*kriyāśabda*). Like universals, qualities and actions are also supposed to be located in their respective substrata through the relation known as "inherence" (*samavāya*), though they differ in some important respects (e.g., while universals are supposed to be eternal and located in many substrata, actions are always noneternal, and no action can be located in more than one substratum, and qualities may be either eternal or noneternal, with no eternal quality located in more than one substratum). It seems that so far as "quality words" (*guṇaśabda*s) are concerned, one may at least ask whether they designate only qualities, or things characterized by those qualities, or the relation that obtains between a quality and the thing qualified by it, then proceed to show that since "quality words" (*guṇaśabda*s) can also be in apposition with the words that are supposed to designate the qualified thing (e.g., in the expression *śuklaḥ paṭah* [white cloth]), it cannot designate merely a quality or the relation that obtains between a quality and what is qualified by it. It can then be argued that the reasons that have been given for rejecting the view that a kind term (*jātiśabda*) can designate a particular entity or individual characterized by the universal can be applied, mutatis mutandis, against the view that a "quality word" (*guṇaśabda*) designates a thing characterized by a quality. This *may* perhaps be true of "action words" (*kriyāśabda*s) as well. But I fail to see how any of these arguments can be employed in the case of "arbitrary words" (*yadṛcchāśabdas*); and in case they are not so applicable, it is hard to see how the arguments of Dignāga can establish the thesis that *all* types of words can mean only *anyāpoha*.[12]

Let us now consider whether Dignāga's arguments in favor of the view that kind terms (*jātiśabdas*) like "existent" (*sat*) cannot designate bare individuals or the relation between universals and individuals or individuals characterized by universals are tenable. These arguments seem to be effective against the view of the Pūrva-Mīmāṃsakas, who maintain that the primary meaning (*śakyārtha*) of words like "cow" are universals like *cowness*, though in the case of expressions like "this cow has been born today," the word "cow" is used in its secondary meaning (*lakṣyārtha*), which is an individual cow. But as Uddyotakara has tried to point out, these arguments are *not* applicable to the Nyāya view that the primary meaning of the word "cow" is an individual characterized by the universal known as *cowness*. We will briefly indicate why Uddyotakara thinks that the Nyāya view on this matter is immune to the criticisms of Dignāga.

Uddyotakara maintains that according to the Nyāya view, universal (*jāti*), configuration (*ākṛti*), and individual (*vyakti*) can all be designated by words like "cow," "horse," etc., and there is no hard or fast rule regarding the primary and secondary meanings of those words. When the word "cow" is applied to a real cow individual, the word designates an individual that is characterized by *cowness*. But when we speak of a toy cow made of clay (*mṛdgavaka*), the individual designated by the word "cow" has merely the configuration of a cow, that is, the arrangement of its constituent parts resembles the arrangement of the limbs of a genuine cow. Thus, when Dignāga seeks to prove that the only designata of words like "existent" cannot be either bare particulars (i.e., substances, qualities, and actions that are not characterized by existence) or the universal known as existence or the relation between this universal and the particulars or individuals to which existence belongs, his arguments may succeed in rejecting the views of those who maintain that *any one* of the four alternatives rejected by Dignāga is the *sole* designatum of the word "existent," but the Nyāya view is not affected in any way, since the extremist views rejected here were never admitted by the Naiyāyikas in the first place.[13]

But even if the Nyāya view remains unaffected when the first three of the four possible designata of the word "existent" are thus ruled out, is it not adversely affected when Dignāga adduces a number of arguments for proving that the word "existent" cannot designate individuals that possess existence? Uddyotakara, however, is unfazed by such arguments. We have seen that according to Dignāga, the word "existent" cannot designate things like substance, etc., (or pot, etc.) due to "lack of independence"; and this argument is based on the premises that (i) when the word "existent" is uttered, things like substance, etc., are not presented to us immediately, and that (ii) since entities like substances are in no way implied by the word "existent," there cannot be any coreferentiality of the two terms "existent" and "substance," just as there cannot be coreferentiality of the two terms "color" and "sour." Uddyotakara, however, turns this argument on its head and claims that since we use expressions like "the existent substance" in a significant manner, these two terms (namely, "existent" and "substance") should be treated as coreferential, because they have the same suffix, that is, the same case ending, as can be seen in the expression *sad dravyam*. The relation between "existent" and "substance" is not one of incompatibility, as is the case with the words "color" and "sour," because entities like substance, quality, and action *are* characterized by existence, and hence they can be regarded as "existent." Thus, existence and substancehood can share the same locus in some cases. On the other hand, colorhood and sourness can never be present in the same locus, and that is why there cannot be any apposition between the terms "color" and "sour," that is, the absence of apposition here is not due *merely* to the fact that the words "color" and "sour" designate different things. Moreover, since the word "existent" designates both the universal called "existence" and its instantiator through a *single* function, the objection of Dignāga that the word "existent" cannot designate these *two* designata through *two* functions that occur either successively or simultaneously is simply off the mark. In this case, there is also no question of metaphorical transference that is either due to some relation (as happens in the case of expressions like "the platforms are shouting" [*mañcāḥ krośanti*]) or due to some similarity (as happens in the case of expressions like "the servant is the master" [*svāmī bhṛtyaḥ*]) or due to the presence of some adventitious condition (as happens in the case of usages like "the red crystal" (*raktaḥ sphaṭikaḥ*), when the crystal is smeared with some red stuff). Accordingly, one cannot also say that the expression "the existent substance" will invariably produce an erroneous cognition.

Finally, the claim of Dignāga that, like the first alternative regarding the designata of the word "existent," the last alternative is also vitiated by the defects of innumerability (*ānantya*) and deviation (*vyabhicāra*) does not stand to reason. This objection would have been tenable if the word "existent" had designated substances, qualities, or actions bereft of existence. This, however, is not the case. Since existence is a common property of substances, etc., we understand from the word "existent" things that are characterized by existence, irrespective of their mutual differences. Thus, the fact that the designata of the word "existent" are innumerable in number does not pose an insuperable problem.[14] Since there is no uncertainty about the meaning of the word "existent," the charge of "deviation" also does not hold. There would be deviation in this instance if the word "existent" designated nonexistent things along with existent ones. But this certainly is not the case.[15]

After rejecting the arguments of Dignāga in favor of the apoha doctrine, Uddyotakara has proceeded to show that this doctrine, too, is not free from difficulties. Some of his criticisms may smack of sophistry and, hence, need not be discussed here. But some of his objections against the apoha doctrine seem to me to be quite substantial. I shall *not* discuss the familiar objection that this doctrine leads to some sort of mutual dependence or circularity, since this issue has been already discussed by Pascale Hugon (in this volume). Instead, I shall concentrate on a few other objections.

(a) When Dignāga says that the word "cow" designates "not noncow," what sort of entity does he have in mind? Is it a positive entity or a negative entity? If it is a positive entity, then is it different from the cow or not? If it becomes identical with the cow, then there is no reason for disputing it. If it is a positive entity that is also different from the cow, then we have a queer view, namely, the word "cow" designates something different from the cow (e.g., a horse or a donkey). If, however, it is claimed to be a negative entity, then that would go against our experience, because nobody grasps a purely negative entity while listening to the word "cow" (NV 314).

(b) Even if this doctrine may seem to work in the case of words like "cow," it cannot do so in the case of some other words. There are entities that are called "cow," and there are also entities that are not so called. In this case, when one negates the cow, one may be aware of something that is not a cow and vice versa. But such a dichotomy (*dvairāśya*) does not hold in the case of every word. Consider the word "all" (*sarva*). There is no entity called "not all" (*asarva*), which may be obtained by negating the meaning of the word "all." Nor can it be suggested that the expression "not all" is applicable to unity, duality, etc., which only form a part of the totality that is supposed to be designated by the word "all," because the word "all" stands for a collection, the components or constituents of which may be things that are one, two, and so on in number; and once all the constituents of this collection are negated, the word "all" will become an empty expression, because a collection (*samudāya*) is nothing over and above its constituents (*aṅgas*). The same objection will apply to other words expressing collections (e.g., "two," "three," and so on) (NV 314).

(c) Is the exclusion of cows something different from the cow, or is it identical with it? If it is different, is it located in the cow or not? If it is located in the cow, then it would become a property (*guṇa*) of the cow, and in that case, then expressions like "this cow exists" would not indicate the existence of the cow. If it is not located in the cow, what would be the meaning of the genitive or possessive suffix (i.e., sixth case ending) in the expression "exclusion *of* cow" (*gorapohaḥ*)? If it is not different from the cow, what purpose would it be for Dignāga? (NV 315)

(d) Does the exclusion of noncows differ from cow to cow, or is it the same for all cows? If it is identical and is also related to all the cows—past, present, and future—then it is hardly distinguishable from the universal called *cowness*, which is also one and related to all the cows. If it differs from individual to individual, then, like the individual cows, the exclusions of noncows would also become infinite in number. In that case, the objection of Dignāga that kind terms (*jātiśabdas*) cannot designate individuals, since the latter are infinite in number, would lose its force (NV 315).

Uddyotakara has raised other objections against the apoha doctrine, but it is not feasible to discuss them in this paper, which has already become quite lengthy. Instead, let us at least take note of the fact that some critics have tried to show that all the objections that Dignāga has raised against the possible designata of kind terms are equally applicable to the possible designata of "exclusion words" (*apohaśabda*). The locus classicus of such counterreplies is the *Ślokavārttika* of Kumārila Bhaṭṭa. This has been explained clearly by Pārthasārathi Miśra in his *Nyāyaratnākara*, which is a commentary on *Ślokavārttika*. Thus, for example, verses 115–116 of the *Apohavāda* section of *Ślokavārttika* constitute a rejoinder to verse 2 of the *Pramāṇasamuccaya* of Dignāga, where Dignāga claimed that kind terms (*jātiśabdas*) cannot designate individuals due to the innumerability (ānantya) of the latter and also due to "deviation" (*vyabhicāra*). Kumārila Bhaṭṭa points out that the doctrine of apoha fares no better in this respect, because it is humanly impossible to know (or establish) the designative relation between an "exclusion word" (*apohaśabda*) and *all* individuals to which it is supposed to be applicable; and if we still employ that particular exclusion word to all such individuals in spite of the fact that our knowledge of this designative relation is restricted to only some of these individuals, then this would amount to a "deviation" from the alleged rule that the establishment of the designative relation between a word and the thing designated by it is a prerequisite for the employment of that word. Dignāga had also claimed that a kind term (*jātiśabda*) cannot designate a universal (*jāti*) or the relation (*yoga*) that obtains between a universal and its instantiating individuals, since such a possibility is inconsistent with the fact that there are expressions in which the alleged kind term is found to be in apposition with words that designate individuals. Kumārila Bhaṭṭa points out that with suitable modifications in the arguments given by Dignāga, it can be shown that this difficulty remains even if we treat words like "cow," "horse," etc., as exclusion words. Thus, nothing is actually gained by rejecting kind terms and admitting exclusion words in their place.[16] Again, in verse 4 of *Pramāṇasamuccaya*, Dignāga has given a host of arguments for rejecting the claim that kind terms (*jātiśabdas*) can designate individuals characterized by universals (*jātiviśiṣṭavyaktis*). In verse 120 of the *Apohavāda* section of *Ślokavārttika*, Kumārila Bhaṭṭa has pointed out that the difficulties envisaged by Dignāga with respect to kind terms also beset the claim of Dignāga that words like "cow," "horse," etc., are exclusion words. Elsewhere, Kumārila Bhaṭṭa has pointed out that when two rival theories suffer from the same shortcomings that are also sought to be obviated in the same manner, it is not fair for the proponents of one of these theories to find fault with the proponents of the other theory.[17]

In like manner, Pārthasārathi Miśra has indicated that verse 121–133 of the *Apohavāda* section of *Ślokavārttika* constitute a rejoinder to the other verses of Dignāga that have been mentioned in this paper (NR 423–433). Bhaṭṭaputra Jayamiśra has also tried to show in his commentary known as *Śarkarikā* that Kumārila Bhaṭṭa has answered Dignāga verse by verse.[18] Incidentally, it may be noted here that Mallavādin has also adopted a similar procedure in his *Dvādaśāranayacakra*.[19]

Kumārila Bhaṭṭa has also pointed out that the meaning of verbs, prefixes, suffixes, and particles cannot express apoha and also that nouns like *prameya* (knowable), *jñeya* (cognizable), etc., which are applicable to all entities, cannot express apoha either, because there are no unknowable or uncognizable entities that might be sought to be excluded by the use of these terms. What, again, would be the meaning of words like "not," "different," etc., which stand for absence or difference? It has, however, been said by Dignāga that words like "knowable" are applied to entities like chairs and table by differentiating them from some imaginary construction called "unknowable" (*ajñeyaṃ kalpitaṃ kṛtvā jñeyaśabdaḥ pravartate*).[20] But this solution can be accepted only if it is entertained that a negation having an imaginary negatum or countercorrelate (*pratiyogin*) can be admitted. Not many would agree with this view. Besides, if we have to solve the problem at hand with the help of imaginary constructions, then what is the harm in admitting imaginary universals in the place of apoha, since that would be at least consistent with the fact that when we listen to a meaningful word like "cow," we seem to become aware of some entity that at least *appears* to be *positive* in nature? (ŚV, ed. Shastri, 428–429)

Kumārila has also pointed out that the doctrine of apoha is not consistent with the fact that one can speak about the gender, case, number, etc., associated with a noun. Nor can it be said that the gender, case, number, etc., are applicable to the individuals and through it are associated with the apoha, because according to Dignāga, individuals cannot be expressed by words (ŚV, ed. Shastri, 427). Nor can one get rid of such difficulties by conceding that even though apoha is negative in nature, it has all the other properties of a universal (e.g., it is one, eternal, and resident in all the instances to which the word is applied), because for the Buddhists, negation is an imaginary construction, and if one can admit such properties in an imaginary entity, one may as well admit the possibility that a cloth can be made without threads (ŚV, ed. Shastri, 433).[21] In view of these and similar problems, Kumārila concludes that *anyāpoha* can be expressed by a word only when it has a negative component (e.g., "non-Brāhmin" [*abrāhmaṇa*])—in other cases, the meaning that we understand from a word is something *positive* in nature (ŚV, ed. Shastri, 433).

This is not to suggest that the objections raised by Kumārila against the view of Dignāga cannot be answered. In his *Tattvasaṅgraha*, Śāntarakṣita tried to refute the arguments of Kumārila, Bhāmaha, and others who have rejected Dignāga's version of the apoha theory. According to Śāntarakṣita, these criticisms leveled against the apoha theory are based on misunderstanding; and accordingly, he has castigated the critics of apoha theory as evil persons who are themselves misguided and who also indulge in misguiding others.[22] I would like to conclude this section by pointing out that while formulating the apoha theory, Dignāga has not raised any ontological issues in the first twelve verses of the *Apohavāda* section of *Pramāṇasamuccaya*—the arguments used by him depend primarily on rules of grammar and also on some considerations regarding the manner in which one is supposed to learn the designative function (*śakti*/*samaya*) of a word. If some conclusive arguments were given here by Dignāga for rejecting the existence of universals, then it would automatically follow that the meaning of a noun cannot be a universal, or the relation between that universal and an individual, or an individual characterized by that universal; and the task of

Dignāga would have become easier—he could have achieved by a single stroke the result that he has obtained after a series of long-winded and complicated arguments. This move, however, was made by Dharmakīrti, whose *Pramāṇavārttika* contains some serious objections against the reality of universals. Some of these objections are as follows:

(i) According to the Nyāya school, each universal is one (*eka*), eternal (*nitya*), partless (*niraṃśa*), immobile (*niṣkriya*), and inherent in many individuals (*anekasamaveta*) at the same time. On the other hand, the individuals that instantiate any one of these universals are many in number (*aneka*), many of them being noneternal (*anitya*); and among the latter, many are "wholes" (*avayavins*), that is, they are formed out of constituent parts (*sāvayava*). Let us assume for the sake of argument that pothood (*ghaṭatva*) is one such universal, which is supposed to be located (*āśrita*) in all pots through the relation known as inherence (*samavāya*). These individual pots occupy different locations. Now, how can *pothood*, which is a *single* entity, be simultaneously present in *all* the individuals that have different locations? If an entity X has many constituent parts (say, x_1, x_2, x_3, etc.), then it may be simultaneously located in different substrata (say, A, B, C, etc.), if it so happens that X is simultaneously present in A by virtue of one constituent part (say x_1), in B by virtue of another constituent part (say x_2), and so on. But universals, being devoid of constituent parts, cannot be multiply located in this manner.

(ii) According to the Nyāya school, universals are "pervasively occurrent" (*vyāpyavṛtti*), that is, no universal can coexist with its absence in any locus. In other words, no individual can be such that some part of it is characterized by a universal, while some other part of it is characterized by the absence of that universal—a universal characterizes the entire instantiator. Now, before an earthen pot is made by the potter, the universal known as *pothood* (*ghatva*) is not present in the lump of clay (*mṛtpiṇḍa*) or the pot halves (*kapāla-kapālikā*) out of which the pot is made. But once a pot is made, *pothood* is supposed to characterize the entire pot. How can a property be present in a thing in its entirety, and yet be absent in each of the constituent parts of the latter?

(iii) Moreover, how can *pothood*, which was *earlier* absent in the place occupied by the pot halves, become *subsequently* located in that place, once the pot is made? It cannot be said that *pothood* is *produced* along with this pot, because that would be incompatible with the eternality of *pothood*. Nor can it be said that *pothood comes* from some other substratum and thus becomes related to the pot, because universals are supposed to be immobile. Moreover, such an assumption would go against the fact that the pots, where *pothood* was hitherto present, continue to be substrata of *pothood*.

(iv) When an earthen pot is broken, *pothood* is no longer apprehended in the place where this pot was located earlier, nor is *pothood* present in the potsherds that are left behind when the pot is broken. How can one explain this? It cannot be said that *pothood* is destroyed along with its substratum, because *pothood* is supposed to be eternal, and eternal things cannot be destroyed. Moreover, under such an assumption, *pothood*, which is a *single* entity, would also cease to exist in other substrata; and in that case, the latter would once again cease to be pot individuals. Nor can it be said that *pothood* moves to some other location, because it is devoid of motion.

(v) Things that can have downward movement or can flow in some direction are in need of receptacles or substrata (*ādhāra*). Unless water is kept in a vessel, it would begin to flow—the vessel thus prevents or checks the movement of the water and causes it to remain at a definite place. But why should universals, which are immobile, require any substrata?[23]

Once the existence of universals is thus rejected, it becomes easier for Buddhists like Dharmakīrti to show that the only way in which some unique particulars can be grouped together is to distinguish them from other things. Dharmakīrti also adds some new dimensions to the apoha doctrine by giving a causal account of the process through which our perceptual cognitions about unique particulars result in the formation of imaginary constructions (*vikalpas*), which have an illusory form (*ākāra*) of sameness and which are produced through many cognitions of different individuals or particulars, which nevertheless have a *negative* feature in common—namely, the absence of the capacity of performing the task that can be performed by things other than these individuals.[24] We will try to explain this with the help of a few examples.

According to Dharmakīrti, a brindled cow (*śābaleya*) is totally different from a nonbrindled cow (*bāhuleya*), but both of these cows have the capacity to yield milk and to carry a load (*bāhadohādisāmarthya*). Moreover, none of them can perform functions that may be performed by different types of animals that are not regarded as cows (e.g., climbing trees, eating flesh, flying in the sky, and so on). These awarenesses of similarities, when repeatedly experienced by us, produce a uniform cognition of the form "this is a cow" and "this is also a cow." Since these perceptions result in the same sort of judgment, we mistakenly believe that there is some common property, namely, *cowness*, that is present in all the animals that are treated as cows. But the so-called common property is simply an imaginary construction (*vikalpa*) that is formed through conceptualization. The form (*ākāra*) of sameness that belongs to this *vikalpa* has no objective basis, and being the property of an imaginary construction, it is also something internal, which cannot belong to external objects that are called "cow." Yet, people habitually indulge in mundane behavior and linguistic usage by amalgamating the object of perception (i.e., the particular individual) and the object of imaginary construction (i.e., the concept). As a result, the imaginary construction, endowed with this form, *conceals* the uniqueness of the individuals and imposes upon them a uniform or common nature, which has no basis whatsoever. For this reason, it is called *samvṛti* (i.e., concealment). Our employment of words like "cow" thus does not depend on any objective common factor.

In his *Nyayavārttikatātparyaṭikā* and *Nyāyakaṇikā*, Vācaspati Miśra has examined some arguments given by Buddhists in favor of admitting a slightly different version of apoha, and in these arguments, one starts from the assumption that there are no universals (NVTṬ 440–447, VV 131–135). In the absence of universals, the existent things would be devoid of common properties, and each thing would be unique, being totally dissimilar to all other things. This uniqueness is what is meant to be expressed by the term *svalakṣaṇa*. For the Buddhists, words cannot even touch these unique particulars. When we apply the same word to different things, we do so on the basis of some awareness, which is taken by us to be the same (or similar) in all these relevant cases. Again, when we utter the word concerned on different occasions, the listener has some awareness that is taken by them to be the same (or similar) on each occasion. Such awarenesses are called *kalpanā* or *vikalpa*. The Buddhists claim that these *vikalpas*, which are such that their forms (*ākāra/pratibhāsa*) can be expressed in language, are the basis for the application of words; and the words, in turn, give rise to further *vikalpas*.[25]

These *vikalpas* alone can be the objects that can be designated by words, because these *vikalpas* are felt or "recognized" to be the same, even though the locations, times, and conditions associated with them may vary. Now, the unique particulars are not of this nature, and accordingly, they cannot be the objects revealed in such *vikalpas*. Nor is there any universal, which could possibly be the object of such a *vikalpa*. Nor can it be said that these *vikalpas* have as their objects the individual entities where universals are located, because in the absence of universals, nothing whatsoever could be characterized by any universal. Since unique particulars are not objects of *vikalpa*, neither the cognition (*jñāna*) produced by these unique particulars nor the form of the object (*grāhyākāra*) that is impressed on such a cognition can be the objects of *vikalpa*, because both of them also happen to be unique particulars. What, then, is the object that is revealed by a *vikalpa*? One possible answer is that such *vikalpas* project their own form (which is *not* actually external) *as* something external, and this is regarded as the object of *vikalpas*.[26]

Some Buddhist philosophers, however, do not admit this. They ask, what is the nature of this projection (*adhyavasāya*)? Is it (i) some sort of apprehending (*grahaṇa*), or (ii) some sort of creation or making (*karaṇa*), or (iii) some sort of joining or putting together (*yojanā*), or (iv) some sort of ascription or superimposition (*samāropa*)? The first alternative is not tenable, because the form of the *vikalpa* is not its object, and hence, it cannot be apprehended by it. An entity that is not an object of a *vikalpa* can never be made into an object of the latter. Hence, the second alternative is also inadmissible. Nor can the *vikalpa* mix up or conjoin its own form with the external, unique particular, because it is incapable of apprehending such particulars. This fact effectively rules out the third alternative. The fourth alternative, too, is not tenable, because a cognition cannot impose its own form upon what is its object, since it cannot do so unless it apprehends its own form. Now, do these two acts or functions (namely, apprehension and imposition) occur successively or simultaneously? The first alternative cannot be admitted, because the *vikalpa*, being momentary, cannot perform these two functions successively. If we admit the second alternative, then we have to choose from the following two options: either the *vikalpa* imposes its own form upon itself, or it imposes its form on some external object. Neither option is, however, available to us, because the *vikalpa*, while apprehending itself through self-revelation (*svasaṃvedana*), cannot also impose its own form upon itself. Nor can it impose its own form upon a particular entity that is external (*bāhya*) to it, because the particular external object, being a unique particular (*svalakṣaṇa*), can be revealed by perception (*pratyakṣa*) alone—it cannot be revealed or grasped by a *vikalpa*. Thus, what is revealed in a *vikalpa* is neither a cognition (*jñana*), nor the form of a cognition (*jñānākāra*), nor an external object (*bāhyārtha*); and hence, it must be totally fictional or unreal (*alīka*). This is the view of Dharmottara.[27] This form of *vikalpa*, which is actually fictional and yet appears as an external object, is the object of *vikalpa*, and it is also the designatum of words.

One may ask here, why does this form of *vikalpa* appear as something external? Dharmottara's answer, as recorded by Śrīdhara Bhaṭṭa in his *Nyāyakandalī* (NK), is as follows: When a person points to an animal and utters the sentence "that is a cow," the listener experiences a particular entity (1), which has the causal efficacy for yielding milk and carrying a load (*bāhadohādisādisāmarthya*). But in his awareness, there is also an image or representation, a *pratibhāsa* (2). Through repeated experience of such images, he has the awareness of a certain structure or form—an *ākāra* (3) that belongs to this image. Next, he imagines that the particular cow has such a repeatable form (*sāmānya*) of its own (4). These four are conflated together through nondiscrimination (*aviveka*), and then the imaginary form is projected (*āropita*) onto the external object. This is the doctrine of "the form of 'four-and-a-half'" (*ardhapañcamākāra*) that was propounded by Dharmottara.[28]

Now, this form of *vikalpa* is of the nature of "exclusion of others" (*anyavyāvṛttirūpa*), and this can be established by the following three reasons.

(i) The first reason is that only "exclusion from others" (*anyavyāvṛtti*) can be the common element that can consistently feature as qualifier in statements affirming or denying the existence of something, because it alone can be legitimately asserted of existent and nonexistent entities. For example, cognition (*jñāna*) is existent or real, while a hare's horn (*śaśaviṣāṇa*) is totally nonexistent or fictional; and yet, both of them have the common property of being different from things that have a visible shape or body (*amūrtatva*). If the object of the *vikalpa* produced by the word "cow" is something that can belong to only existent entities, then the sentence "the cow does not exist" would be a contradiction, while the sentence "the cow exists" would be repetitive. Again, if it were something that is common to nonexistent entities alone, then the statement "the cow does not exist" would be repetitive, whereas the expression "the cow exists" would involve a contradiction. Such, however, is not the case. Hence, what is revealed by the *vikalpa* that is produced by the word "cow" must be such that it is compatible with existence as well as with nonexistence (*bhavābhāvasādhāraṇa*), and as we have seen, such a thing can be of the nature of "exclusion from others" (*anyāpoha*) alone. Thus, what the imaginary construction (*vikalpa*) produced by the word "cow" has as its object is "difference from things that are not cows" (*a-go-vyāvṛtti*).

(ii) The second reason is that a number of things that are absolutely dissimilar can have only a negative property that is *common* to them, namely, the lack of identity with what is different from each of them (*a-tad-vyāvṛtti*). For example, no camel is identical with a cow, a horse, an elephant, or a lion. Even the Naiyāyikas would not say that the cow, horse, elephant, and lion are animals of the same type, or that they instantiate any common positive property that may be regarded as a universal. At the same time, it is true that none of these animals is a camel. Thus, the property of being different from a camel is a common (albeit negative) property of this heterogeneous group of animals. The cow individuals, being unique particulars, can likewise have a negative common property, namely, being different from things that are not cows (*a-go-vyāvṛtti*).

(iii) Whenever we hear a word whose meaning is known to us, we are aware of some sort of "exclusion from others" (*anyavyāvṛtti*). That is why when someone is asked to tie up a cow, they do not tie up a horse. This cannot be explained satisfactorily unless it is granted that the "imaginary construction" (*vikalpa*) produced by the word "cow" has difference from noncows as its object.[29]

Since according to this theory apoha is fictional in character and yet "colors" the *vikalpa*, it has been treated by Jayanta Bhaṭṭa, the author of *Nyāyamañjarī* (NMJ), as subscribing to the view that in illusion, something totally unreal appears to us (*asatkhyātigarbhā saraṇiḥ*) (NMJ, ed. G. Shastri, II.17). This view of Dharmottara has not been accepted by other Buddhists, and according to them, the form of *vikalpa*, though it is something internal (since it is an aspect of a conscious state), nevertheless appears as external due to the influence of our latent residual impressions (*vāsanā*). According to Jayanta Bhaṭṭa, those who maintain such a view maintain that, in illusion, what is actually internal appears as external (*ātmakhyāytigarbhā saraṇiḥ*).[30]

In the later phase of apoha theory, we come across a significant change in the attitude of the Buddhist philosophers. The meaning of words is no longer taken by all of them to be something *purely* negative, as this supposition led to many problems.[31] But unlike the Naiyāyikas, they never maintained that positive entities *alone* (like universals, qualities, etc.) can be the designata of words like "cow," "horse," "sweet," etc. In their opinion, what is presented by a word has two aspects—positive and negative. Some of the Buddhist thinkers like Karṇakagomin, Śāntarakṣita, and Kamalaśīla maintained that words convey a positive meaning in the first instance, and then difference from other things is grasped through implication. Others maintained that the negative aspect, namely, "difference from other things" (*anyāpoha*) is apprehended first, and then the positive aspect is grasped through implication.[32] Jñānaśrīmitra and his disciple Ratnakīrti, however, differed from the adherents of such views. In their opinion, the positive and negative aspects of word meaning are grasped together, even though among these two aspects, the positive aspect is the primary one and the negative aspect (namely, *anyāpoha*) features as its property. The words "*tāmarasa*," "*puṇḍarīka*," and "*indīvara*" provide good examples in favor of such a position. These three words designate a red lotus, a white lotus, and a blue lotus respectively. In such cases, we do not need two words (as is the case with the expression "red lotus") for being aware of a substantive and its adjective—since both of them are invariably grasped in one sweep. In like manner, the word "cow" presents to us an animal that is invariably characterized by difference from noncows. There is no *sequence* in the awarenesses of these two features, that is, it is not the case that once we are *initially* aware of a cow, we are *subsequently* aware of its difference from noncows; nor is it the case that in such instances, we are initially aware of the fact that it is not noncow, and then apprehend subsequently that it is a cow. Thus, neither the awareness of a positive feature nor the awareness of the negative feature has any precedence in such cases—both of these features are grasped simultaneously by the *same* awareness, even though these two aspects are *not* apprehended as unrelated—because the negative aspect (namely, the "exclusion of others") features as the *attribute* of the positive aspect, and being an attribute, it is *subsidiary* (*gauṇa*) in nature.[33]

I have attempted to give some idea of the arguments that have been given in favor of the doctrine of apoha (*Apohavāda*) and the different modifications that occurred in this doctrine in the face of hostile criticisms from the opponents. I do not think that *Apohavāda* is a very satisfactory doctrine, and the reasons thereof are (i) that the Buddhist version of *Apohavāda* is a consequence of the doctrine of momentariness (*kṣaṇabhaṅgavāda*), which, in its turn, is based on questionable assumptions; and (ii) that acceptance of this doctrine would imply that *all* our empirical experiences, behaviors, and linguistic usages are based on everlasting illusions, where unique particulars are always cognized as possessing common features, which is a bit difficult to believe. How long can we survive if all our cognitions invariably deceive us? The fact that we manage to survive in spite of such pervasive and persistent errors must be considered to be great good luck!

Here, a supporter of the apoha doctrine might urge that while the Buddhists admit momentariness as well as *Apohavāda*, the latter does not necessarily presuppose the former; and one may adopt *Apohavāda* simply on the ground that acceptance of universals leads to more problems that it can solve. Thus, one may not admit momentariness and yet admit *Apohavāda*, and hence, even if the doctrine of momentariness is rejected, the doctrine of apoha may still be tenable. It can also be argued that even though the Buddhists describe our usages and cognitions as mistaken (*bhrānta*), this term is used in a special sense. These are "mistaken" in the sense that they cannot withstand critical examination, even though they can provide an adequate basis for our mundane behavior. In some cases, even erroneous cognitions can lead to successful activity. Thus, the mere fact that some cognition is conducive to our mundane behavior is no guarantee for its correctness. It can even be pointed out that in the light of recent discoveries in neurophysiology, the explanation of the manner in which we apprehend color suggests that we are wrong in ascribing colors to the things that we perceive, although our survival depends to a large extent on the way we apprehend colors. Hence, my argument that (i) our very survival depends to a large extent on the correctness of our experience and (ii) that consequently one cannot treat all our experiences as erroneous is not sound.

I am ready to admit that not any and every possible version of *Apohavāda* can be deduced straightway from the doctrine of momentariness. But the fact that the Buddhists admit both momentariness and *Apohavāda* is not a coincidence. Once we admit that all things are momentary, it becomes evident that entities like universals (which are supposed to be eternal) will have to be rejected—but this is only half of the story. A momentary thing, because of its extremely short duration, cannot be the substratum of anything else, and it cannot also be located in any other thing. Moreover, there cannot be any genuine relation (*at least* as understood in the Indian tradition) between momentary entities, because every relation must satisfy three conditions: (i) it must reside in both its relata; (ii) it must be different from its relata; and (iii) it must be identical or *one* in respect of both the relata. But since nothing can reside in a momentary entity, the first condition cannot be satisfied here; and accordingly, it would follow that momentary entities are entirely unrelated. In that case, the momentary entities cannot also be the bearers of any property, because to be the bearer of a certain property, an entity must also be related to that property. Under such a circumstance, we will not be in a position to say that two real entities may be distinguished in terms of the different properties that belong to them. Thus, two entities will be distinguishable only if they are self-differentiating, that is, if each of them is unique. But if all entities are unique, then the only common property that such entities have in common must be a negative one—we have already mentioned the argument in favor of this thesis while discussing the views of Dharmottara. So far, we have shown the link between the doctrine of momentariness and the rejection of "horizontal" universals like *cowness* (which is supposed to be present in all cows, even though they may be at different places and times). If all sorts of universals are denied, including "vertical" universals like "*Devadatta*-hood" (which are supposed by some Naiyāyikas to be present even in a single individual like Devadatta throughout the different phases of his life) then nothing will be the same for even two consecutive moments, and this would amount to the doctrine of momentariness. Perhaps one could show in this way some hidden connection between momentariness and *Apohavāda*. A full-fledged defense of *Apohavāda* as formulated by the Buddhists will certainly require the establishment of some other important doctrines of the Yogācāra-Sautrāntika school of Buddhism—for example, (i) that every cognition has a form (*ākāra*), (ii) that every cognition is self-revealing, (iii) that every cognition, while revealing itself, also reveals its form, (iv) that some of these forms are produced by objects, and some of them are produced by residual impressions (*vāsanā*), and so on. None of these doctrines is accepted unanimously by the rival schools, and the establishment of any one of them would require a considerable amount of argument. What I am suggesting is that the Nyāya-Vaiśeṣika and Buddhist doctrines about universals are inextricably bound up with some other ontological and epistemological theses; one cannot simply admit the Buddhist version of *Apohavāda* and at the same time reject the rest of the Yogācāra-Sautrāntika ontology and epistemology.

It may be claimed by the supporters of nominalism that one may decide not to admit abstract entities like universals, simply because it is difficult to see how such eternal and abstract entities could enter into causal relation with other entities, or because it is difficult to explain how any relation between the universals and their instantiators can be established. Besides, rejection of universals is likely to give us a more parsimonious ontology. While I admit that one may choose *not* to have universals in one's ontology for a host of good reasons, what I fail to see is how admitting *Apohavāda* is going to be less problematic than admitting universals. So far, we have noted four versions of the apoha theory, and each of them seems to involve some difficulty or other. Following the footsteps of Jayanta Bhaṭṭa, some of these difficulties will be pointed out.

In the formulation of apoha doctrine as proposed by Dignāga, apoha was said to be of the nature of "difference from others" (*anyavyāvṛtti*). Now, difference per se cannot even be understood—we have to speak of the difference of some specific entity (say, a cow) from some other specific entity (say, a horse). Thus, any difference requires a locus (*āśraya*, *anuyogīn*) and also a counterpositive or countercorrelate (*pratiyogin*). (In the example given earlier, the cow is the locus (*anuyogin*) and the horse is the counterpositive (*pratiyogin*) of a specific difference.) Now, the question is, where would this difference be located? It cannot be said that some unique particular is its locus, because apoha is said to be revealed by some image (*vikalpa*), and images can never have unique particulars as their objects. Hence, on this alternative, the difference from horse that characterizes the cow can never be revealed in any *vikalpa*. Perception *alone* can reveal the unique particular; but imaginary constructions like an apoha can never be revealed in perception. Thus, on this alternative, it becomes well-nigh impossible to be aware of a unique particular that is characterized by some apoha. Nor can it be said that properties like being a brindled cow (*śābaleyatva*) could be the locus of such an apoha, because a nonbrindled cow (*bāhuleya*), while being different from a brindled cow, is nevertheless a cow; and hence, it cannot be treated as a noncow (NMJ, ed. G. Shastri, II.11). It cannot also be suggested that the collection or sum total (*samudāya*) of all the individual cows taken together would be the locus of the difference from noncows, because a collection of individuals cannot be apprehended unless the individuals constituting such a collection are apprehended. How would the act of bringing all the cow individuals into a single collection (*vargīkaraṇa*) be possible in view of the fact that the individual cows are infinite in number and that they are also located in different places and different times? Barring an omniscient person, no one can ever be aware of the collection of all possible particular cows at any given time. The only way in which we can know that all cows are different from noncows is to admit some common *positive* property that is present in all cows, which is incompatible with the property of being a noncow. But then, such a property would be nothing but the universal *cowness* (NMJ II.11). Moreover, if the word "cow" means something other than noncow, then how are we to understand the expression "not noncow"? Does it mean the difference from *some* particular individual (or individuals) that is (or are) different from cow, or does it mean the difference from *all* individuals that are not cows? Under the first alternative, even a lion, which is different from a tiger, would be meant by the word "cow," because the tiger is a noncow, and the lion is different from the tiger. This difficulty can be avoided if we adopt the second alternative, but a similar difficulty crops up here: how can one be aware of all individuals that are noncows? Had there been a genuine common property in all noncows that is also incompatible with being a cow, then all noncows could be brought under one homogeneous group by virtue of such a property. But nominalists like the Buddhists would be totally unwilling to admit any such common property (NMJ II.12).

Let us now see if the version of apoha theory as formulated by Dharmakīrti fares any better. According to this version, apoha is not an external object like the allegedly objective difference (*bheda/anyonyābhāva*) that is supposed to obtain between disparate entities; it is the form (*ākāra*) of an image (*vikalpa*), which is imposed on this image by residual impressions (*vāsanā*) that are due to beginingless ignorance (*anādi-avidyā*). This *vikalpa* is mistakenly taken to be identical in nature and also mistakenly ascribed to external objects, and as a result, objects like cows are also supposed to have some identical (i.e., common) nature. But this cannot be the case, because different perceptual cognitions of totally different objects cannot produce an identical image. Moreover, how is the sameness or identity of this *ākāra* (form) apprehended by us? It cannot be revealed to us by perception, because perception, according to the Buddhists, reveals only unique particulars. Nor can it be revealed by any *vikalpa* (image), because such images reveal either some imaginary construction, or they reveal only themselves. Hence, the form of one *vikalpa* cannot be revealed by another *vikalpa*. Accordingly, there is nothing that could apprehend all these *vikalpas*, compare them, and find any common feature in them. If it were said now that there is actually no identical form in the *vikalpa* and that they are grasped as similar or the same due to the fact that their mutual differences are not apprehended, then one may still also point out that these *vikalpas*, being cognitive states, are momentary; and hence, they are all unique and mutually different. Now, is the form (*ākāra*) that characterizes each particular *vikalpa* (image) identical with that image or different from the latter? Under the first alternative, the forms, like the images, would be different; and hence, there would be no scope for an *identical* form that could characterize all the images. Under the second alternative, some common and identical feature present in all the images would have to be admitted. But if this can be admitted, then what is the harm in admitting some common feature in the object of perceptions? (NMJ II.11)

Dharmakīrti's claim that we ascribe an identical feature to some unique particulars because the particulars have the same sort of causal efficacy is also open to question. According to him, the concept *cowness* is formed by us because all cow individuals can produce milk or carry a load. But this view cannot be admitted, because once we accept it, we would have to include also she-buffaloes, she-camels, etc., within the class of cows, since they can also produce milk and carry loads. Moreover, a cow that has become incapable of yielding milk or carrying a load would cease to be considered a cow. Such a consequence is certainly unwelcome.

The alternative version of *Apohavāda* proposed by Dharmottara fares no better. According to this view, the form that "tinges" or "colors" the images is neither internal nor external—it is simply unreal. But how can a fictional entity "tinge" or "color" any image? How, again, can such a fiction be even imposed on anything else? (NMJ II.29)

Since Jñānaśrīmitra and Ratnakīrti were much later than Jayanta Bhaṭṭa, we cannot find any criticism of their views in *Nyāyamañjarī*. Nevertheless, what Jayanta says about the theory of sentence meaning proposed by *Vyaktiśaktivādins* (i.e., the supporters of the view that words designate only individuals) such as Vyāḍi can easily be applied to the theory of word meaning that has been proposed by Jñānaśrīmitra and Ratnakīrti. According to Vyāḍi and his followers, the expression "the white cow" means something that is different from things that are noncow and nonwhite. Jayanta Bhaṭṭa maintains that in such cases, once we are aware of the positive features of the thing, we *subsequently* grasp by implication what the thing is not—the latter is *not revealed by words*. The same argument can be employed here as well (NMJ II.48).

Besides, from the standpoint of the Naiyāyikas, negative facts like mutual difference are as real as the positive ones; and in their opinion, each instance of mutual difference is eternal and capable of characterizing innumerable entities. Thus, replacing the universal (*jāti*) by "exclusion from others" (*apoha*) is not going to reduce the number of multiply located eternal entities in the world—though one could still say that by reducing all universals (*jātis*) to "exclusion from others" (*apohas*), one can end up with lesser *types* of such things. The Buddhists might, however, say that according to them, negative entities or negative facts need not be admitted for explaining negation—the so-called negative facts are either imaginary constructions or are identical with our awareness of the absence of some entity in something else. This, however, is not admitted unanimously, and thus we have to settle another point of dispute in favor of the Buddhists if we have to admit *Apohavāda*. Again, even if we grant for the sake of argument that it is not easy to explain how universals are related to their instantiators, one may still ask, how does one explain the relation between apoha and the entities that are characterized by it? Jayanta Bhaṭṭa and others have pointed out that the criticisms leveled against the admission of universals on this score by Dharmakīrti and his followers are equally applicable to the admission of apoha (NMJ II.25–26). One may try to avoid this difficulty by saying that while universals are supposed to be real entities, apohas are imaginary constructions, and hence, the question as to how the apohas *actually* enter into relation with other entities does not pose a serious problem. But what I fail to see is how imaginary constructions like apoha can enter into *any* relation (like the causal relation) with other entities any more than eternal or abstract entities like universals (*sāmānyas*), nor can I see why eternal entities should, on account of their eternality, fail to be related to some other entity.

Let us also recall at this point the fact that the universals (*sāmānyas*) admitted by the Mīmāṃsakas or Naiyāyikas are quite unlike platonic universals, which are supposed to be denizens of a third realm that is beyond the ken of sense organs. The Naiyāyikas also do not admit that all common or abstract properties are universals. Thus, while all universals are common properties, the reverse is not true. For the Naiyāyikas, a genuine universal must be an eternal (*nitya*) and unanalyzable property (*akhaṇḍa dharma*), which resides in its locus through the relation known as inherence (*samavāya*). Moreover, only substances, qualities, and movements can be the loci of universals. Thus, while the property of being a cow (*gotva*) is a genuine universal, the property of being a black cow (*kṛṣṇagotva*), which is a complex (and hence analyzable) property, is not a universal (*jāti*)—it is only an imposed property (*upādhi*). Likewise, while substancehood (*dravyatva*) is a genuine universal, the property of being an absence (*abhāvatva*), which does not inhere anywhere, is only an imposed property and not a genuine universal. Again, the property of being a cook (*pācakatva*), which characterizes all cooks, is reducible to activity that is conducive to cooking (*pākānukulakriyā*), which is noneternal, and hence, it cannot also be a genuine universal. In his *Kiraṇāvalī*, Udayana has pointed out many such "impediments to universals" (*jātibādhaka*), that is, conditions under which a common property cannot be treated as a universal. Later Naiyāyikas like Mathurānātha Tarkavāgīśa go a step further and maintain that properties that are common to artifacts (e.g., *pothood* [*ghaṭava*]) are not genuine universals. Prābhākara Mīmāṁsakas maintain that *only* perceptible substances can be the loci of universals. Thus, the admission of universals need not always lead to unbridled reification—which is the nightmare of nominalists like the Buddhists. Moreover, for the Naiyāyikas, universals like *cowness* are not objects of reason alone—nor are they concepts in our heads that are formed by us through the process of abstraction. Most of them are *perceptible*, and unless their perceptibility is admitted, it is difficult to explain how in our perceptual cognition an animal is apprehended *as* a cow rather than as a walrus or as a polar bear. When we thus identify things in this manner, we feel that we are aware of something *positive* that is characterizing the thing before us, and the doctrine of apoha is not compatible with this fact. The same is true of our utterance and understanding of nonnegative sentences. When Udayana remarked that while asking somebody to bring a pot, one does not say "do not desist from bringing what is a non-nonpot,"[34] he was perhaps being facetious, but I think he had a point.

Finally, one word about the illusions under which we are supposed by the *Apohavādins* to live throughout our existence. I am not denying the fact that we are often mistaken about what we claim to know, and I am also willing to admit readily that the inbuilt structure of our cognitive apparatus may vitally influence the manner in which we apprehend things. But does that necessarily mean that the entities around us are in no way responsible for the way in which we apprehend them? Let us admit that we are mistaken in attributing color to the things that we perceive. But can we attribute any color to anything whatsoever at will? The fact that we cannot do so seems to suggest that there must be some properties in those things that make our neurophysical mechanism attribute some specific color to those things. If, under normal conditions, we see five roses as red, then why should it not be admitted that all these roses have some common property, which compels us to ascribe red color to the roses? Unless we admit some such property, it will be difficult for us to explain why the same thing is not seen at the same time as green, yellow, blue, and brown by different persons who are neither color-blind nor sufferers of jaundice (or any other disease that distorts our vision). Once some such common properties in things are admitted, the main thrust of nominalism seems to become blunted. Nominalists cannot get away with simply saying that "the five red roses are only similar," because in the absence of any common property, similarity also becomes inexplicable. Besides, an experience may be declared to be illusory if it is sublated or contradicted by some subsequent experience. But if we admit that our cognitive apparatus is such that each and every one of us will always have a totally distorted and baseless view about the things around us, and that the real nature of things will never be revealed to us in any mundane or veridical experience, then how can we choose between two doctrines that are opposed to each other and also to our commonsense view of the world? Consider, for example, the view of the Advaita Vedāntin that the ultimate reality is one, unchanging, and without any internal distinction; but due to beginningless ignorance (*avidyā*) and the residual impressions (*saṃskāras*) left by the experiences of earlier lives, we see a multitude of things of different types, which are also subject to incessant change; and this illusion will persist unless one has a direct realization of the ultimate truth. This view is equally unacceptable to the Naiyāyikas and the Buddhists. The question is: even if one is willing to give up the worldview as proposed by the Naiyāyikas, how is one to choose between the two other views?

It is absolutely necessary to admit the possibility of error in human beings, because that can act as an effective antidote to dogmatism and fanaticism, which are dangerous. But the admission of universal and incorrigible errors can be equally dangerous, because it may lead to an extreme form of pernicious skepticism. One way out of such a situation may be to suggest that such errors are not, after all, incorrigible, because they may be overcome by an omniscient being, who can see through such illusions by virtue of mystic vision or superhuman powers. But such an answer can hardly be considered to be satisfactory, because once such claims are granted, one may formulate any outlandish theory that is not supported by our experience, and then say that our mundane or ordinary experiences are mistaken and that only a chosen few can have direct access to the ultimate truth. That would obviously be the end of philosophy as a rational discourse. One may profitably recall here Kant's statement in *Critique of Pure Reason*: "[If the concept of cause rested only on subjective necessity] I would not then be able to say that the effect is connected with the cause in the object, that is to say, necessarily; but only that I am so constituted that I cannot think this representation otherwise than as connected. This is exactly what the skeptic most desires" (B168).

The fact that the Buddhist doctrine of apoha suffers from some defects does not necessarily prove that the exclusion theory of meaning is intrinsically wrong. One may still maintain this theory, but it will be incumbent upon the supporter of such a theory to give alternative arguments in favor of this thesis and to show that it can overcome the shortcomings of the rival views without involving any major defect. At the same time, it would also have to give satisfactory explanations of uniform cognition (*anugatabuddhi*) and the application of the same term to different things (*anugatavyavāhara*) without postulating the existence of universals. I hope that supporters of nominalism will come up with some such theory, which can claim a sparse ontology as an advantage.

Notes

The author expresses his sincere gratitude to Tom Tillemans, Mark Siderits, Pascale Hugon, Prasanta S. Bandhyopadhyaya, and Arindam Chakrabarti for their constructive criticisms of an earlier draft and to Shōryū Katsura, Francesco Sferra, and Ole Pind for giving some extremely important information from Tibetan sources. Due to the kind help of Prabhat Ghosh, the text of *Dvādaśāranayacakra* with commentary became available to the author, and this is acknowledged with thanks.

1. "buddhyā vivicyamānānāṃ svabhāvo nāvadhāryate / ato nirabhilapyāste niḥsvabhāvāś ca deśitāḥ" (LS 116).
2. "aparapratyayaṃ śāntaṃ prapañcair aprapañcitam / nirvikalpam anānārtham etat tattvasya lakṣaṇam" (MMK XVIII.9). (b) Some portions of the *Prasannapadā* (PP) of Candrakīrti on this verse are as follows: "tatra nāsmin parapratyayo 'stīty aparapratyayaṃ paropadeśāgamyaṃ svayam evādhigantavyam ity arthaḥ, . . . etac ca śāntasvabhāvam ataimirikakeśādarśanavat svabhāvavirahitam ity arthaḥ. nirvikalpakaṃ ca tat. vikalpaś cittapracāraḥ, tadrahitatvāt tattvaṃ nirvikalpakam, yathoktaṃ Sutre—paramārthasatyaṃ katamat? yatra jñānasyāpy apracāraḥ, kaḥ punar vādo 'kṣarāṇām—iti." Elsewhere Candrakīrti also says: "paramārtho hy āryāṇāṃ tūṣṇīmbhāvaḥ" (PP 57). (c) *Catuḥśataka* (CS) of Āryadeva, k. 194 (ed. Jain Bhaskar; Skt. quoted PP 370, see Suzuki 1994, 143) and *Bodhicaryāvatāra* (BCA) of Śāntideva, verse IX. 33, 35 (ed. Śānti Bhikṣu Śāstrī 1983; see also Vaidya 1960).
3. For this twofold division of reality, see MMK XXIV.8–9 and PP thereon.
4. Dharmakīrti also has stated this view very clearly in PV III.1–3.
5. "na pramāṇāntaraṃ śabdam anumānāt tathā hi tat / kṛtakatvādivat svārtham anyāpohena bhāṣate" (PS V, k. 1, Skt. in Jambūvijayajī 1976, 607).
6. Matilal 1971, 44–45.
7. MB (ed. Shastri and Kuddala) 52–53. The views of Vājapyāyana and Vyāḍi have also been identified by Kātyāyana in two of his Vārttikas on Pāṇini's rule *sarūpāṇām ekaśeṣa ekavibhaktau* (Pāṇini 1.2.64).
8. The arguments of Diṅnāga have been presented in these verses in a very condensed form, and a brief account of these arguments is being presented here with the help of the exposition of these verses as found in Diṅnāga's *Svavṛtti* (PSV) of these verses and the commentary *Viśālāmalavatī* (PSṬ) of Jinendrabuddhi thereon, *Nyāyavārttika* (NV) of Uddyotakara, *Nyāyamañjarīgranthibhaṅga* (NMJG) of Cakradhara, and the commentaries of Bhaṭṭaputra Jayamiśra (ŚVṬ) and Pārthasārathi Miśra (NR) on the *Apohavāda* section of *Ślokavārttika* (ŚV) of Kumārila Bhaṭṭa. For the verses and Diṅnāga's PSV, see Pind (in this volume) and the references contained therein.

9. "tasmāj jātiśabdaḥ kathañcid api bhedasāmānyasambandhajātimadvācako na yujyate, tenānyāpohakṛc chrutiḥ" (PSV, Skt. in Jambūvijaya 1976, 630).
10. "tathāpi sacchabdo jātisvarūpamātropasarjanaṃ dravyam āha, na sākṣāt, iti tadgataghaṭādibhedānākṣepaḥ. atadbhedatve sāmānādhikaraṇyābhāvaḥ. na hy asatyāṃ vyāptau sāmānādhikaraṇyam" (PSV, Skt. Jambūvijaya 1976, 607). "na hy asatyāṃ vyāptāv ityādi. vyāptir ākṣepaḥ. yathā rūpaśabdena madhurādīnām anākṣepo 'tadbhedatvān na taiḥ śabdaiḥ sāmānādhikaraṇyaṃ, na hi bhavati 'rūpam amlam' iti. vyāptau tu bhavati 'rūpaṃ nīlam' iti" (PSṬ, Skt. Jambūvijaya 1976, 607).
11. Incidentally, it may be noted here that the discussion of this alternative is not found in the verses of *Pramāṇasamuccaya* (PS) mentioned earlier, but it has been considered by Uddyotakara in his NV ad NS 2.1.66. This may have been taken by Uddyotakara from some other verse of PS or from some other work of Diṅnāga.
12. Professor Shōryū Katsura and Mr. Ole Pind have kindly informed me that verse 50 in chapter 5 of PS and the PSV thereon (in Tibetan translation) contain the argument that is supposed to show that "arbitrary words" (*yadṛcchāśabdas*) also express only apoha. In brief, the argument is that arbitrary words such as *Ḍittha*, which can be employed without any fixed "ground of application" (*pravṛttinimitta*), are actually words that are supposed to designate collections (*samudāya*); there is no significant difference between words that are supposed to designate universals (*jātiśabda*) and words that are supposed to designate collections (*samudāyaśabda*). Hence, the arguments that have been used in the case of "universal words" (*jātiśabdas*) are equally applicable to "arbitrary words" (*yadṛcchāśabdas*). But this argument is not at all convincing, because words like "forest" (*vana*) or "army" (*senā*), which stand for certain collections, are used only under some specific conditions; and hence, they can never be treated on a par with purely arbitrary words like "*Ḍittha*," "*Caitra*," "John," or "Joseph."
13. "atrāsmābhir vyaktyākṛtijātayaḥ padārtha iti pradhānopasarjanabhāvasyāniyam ena vyavasthāpitam, ekāntavādinaś caite doṣā iti" (NV 308). For the English translation of the relevant portions of Nyāyavārttika, see "Diṅnāga as Interpreted by Uddyotakara" in Matilal 2002, 231–254.
14. "yo hi sacchabdena nirviśeṣaṇāni dravyaguṇakarmāṇy abhidhīyanta ity abhidhatte, taṃ praty eṣa doṣaḥ. asmākaṃ tu dravyaguṇakarmāṇi sattāviśeṣaṇāny abhidhīyante. yatra yatra sattāṃ paśyati tatra tatra sacchabdaṃ prayuṅkte. ekā ca sattā. tatpratyayasyānivṛtteḥ. tasmād bhedānām ānantyaṃ na doṣaḥ" (NV 309).
15. "na punar ayaṃ sacchabdaḥ svaviṣayavyatirekeṇāsati vartate. tasmād asthāna iyaṃ vyabhicāracodaneti" (NV 310).
16. Verses 115–116 of the *Apohavāda* section of *Ślokavārttika* (ŚV, ed. Shastri) are as follows: "apohamātravacyātvaṃ yadi cābhyupagamyate / nīlotpalādiśabdeṣu śabalārthābhidhāyiṣu // viśeṣaṇaviśeṣyatvasāmānādhikaraṇyayoḥ / na siddhir na hy anīlatvavyudāse 'nutpalacyutiḥ / nāpi tatretaras tasmān na viśeṣyaviśesaṇe." According to Pārthasārathi Miśra (NR), these verses contain a rejoinder to PS V, k. 2, which reads as follows: "na jatiśabdo bhedānām ānantyād vyabhicārataḥ / vācako yogajātyor vā bhedārthair apṛthakśruteḥ." This is evident from the following remarks of Pārthasārathi Miśra: "yat tu bhikṣuṇā jātiśabdānāṃ

bhedavācitvam ānantyavyabhicārābhyāṃ nirākṛtya yadā yogasya jātimatsambandhasya jātimātrasya vā vācakatvam, tadāpi viśeṣaṇavacanair nīlādiśabdair utpalādīnāṃ viśeṣyatvasya tathā 'pṛthakśrutaiḥ sāmānādhikaraṇyasya cānupapattir uktā . . . tad apohamātrābhidhāne 'pi tulyam ity āha siddhir ity antena" (NR 422).

17. Verse no. 120 of the *Apohavāda* section in ŚV (ed. Shastri 1978) reads as follows: "athānyāpohavad vastu vācyam ity abhidhīyate / tatrāpi paratantratvād vyāptiḥ śabdena durbalā." According to Pārthasārathi Miśra, this is a rejoinder to PS V, k. 4, which reads as follows: "tadvato nāsvatantratvād upacārād asambhavāt / bhinnatvād buddhirūpasya bhṛtye rājopacāravat." This is evident from the following remarks of Pārthasārathi Miśra: "jātimadabhidhāne'pi 'tadvato nāsvatantratvāt' ityādinā sāmānadhikaraṇyānupapattir bhikṣuṇā darśitā, sā 'pohavādābhidhāne'pi tulyeti darśayati" (NR 423); and "tasmād yatrobhayor doṣaḥ parihāro 'pi vā samaḥ / naikaḥ paryanuyoktavyas tādṛgarthavicāraṇe" (ŚV, Śūnyavāda v. 252, ed. Shastri).
18. See *Dvādaśāranayacakra* (NC), Jambūvijaya 1976, 607–630, where the relevant portions of *Śarkarikā* (ŚVṬ) have been quoted in the footnotes by the learned editor.
19. Corresponding to the verses "na jātiśabdaḥ bhedānām ānantyād vyabhicārataḥ," etc., in the PS of Diṅnāga, we have the following verse in the NC of Mallavādin: "na jātiśabdo bhedānām ānantyād vyabhicārataḥ / vācako niyamārthokter jātimadvad apohavān." Siṃhasūri, while commenting on this verse, remarks as follows: "jātiśabdo viśeṣārthaniyamokter bhedānām avācakaḥ, vyabhicārād ānantyāc ca. jātimato vācakatve ca ye doṣās te 'nyāpohābhidhāne 'pīti pratijñā." Again, after a few pages, we find the following two verses: "nāpohaśabdo bhedānām ānantyād vyabhicārataḥ / vācako yogajātyor vā bhedārthair apṛthakśruteḥ // tadvato nāsvatantratvād bhedāj jāter ajātitaḥ / arthākṣepe 'py anekānekāntaḥ sattādyartho 'py ato na saḥ." While commenting on these verses, Siṃhasūri says: "vidhivādimatavad apohavādimate 'pi catuṣṭaye sthite samāna evātrāpi vicāro granthaś ceti tathaivāha—*nāpohaśabdo bhedānām* ityādiślokadvayam" (Jambūvijaya 1976, 606).
20. Quoted in NR 428. A similar passage has been ascribed by Kamalaśīla to the *Hetumukha*, a lost work of Diṅnāga: "nanu *Hetumukhe* nirdiṣṭam—'ajñeyaṃ kalpitaṃ kṛtvā tadvyavacchedena jñeye 'numānam'" (TSP, ed. Shastri, 385).
21. Professor Francesco Sferra has kindly informed me that it has indeed been claimed in PSV 5.36 that apohas are characterized by oneness (*ekatva*), eternality (*nityatva*), pervasive occurrence (*vyāpyavṛttitva*), etc., and all these features are supposed to characterize universals.
22. "anypohāparijñānād evam ete kudṛṣṭayaḥ / svayaṃ naṣṭā durātmāno nāśayanti parān api" (TS, ed. Shastri, 1002). Śāntarakṣita has tried to obviate many of the difficulties that are alleged to plague the apoha theory of meaning by distinguishing between two types of negation (namely, *prasajyapratiṣedha* and *paryudāsapratiṣedha*), and then drawing a parallel distinction between two types of apoha. Since this issue has been discussed by Hale, Gillon, and Siderits (in this volume), it will not be discussed here.
23. PV I, k. 149–55, PV II, k. 70. These criticisms of Dharmakīrti compelled Naiyāyikas like Jayanta Bhaṭṭa, Bhāsarvajña, Vācaspati Miśra, and Udayana to clarify the pre-

cise sense in which a universal requires a substratum and how it can be multiply located at the same time. But their answers to Dharmakīrti are not of immediate concern to us.

24. The main thrust of Dharmakīrti's arguments for the apoha doctrine may be found in the following verses in PV I, Svārthānumāna (ed. Gnoli 1960):

yad yathā vācakatvena vaktṛbhir viniyamyate /
anapekṣitabāhyārthaṃ tat tathā vācakaṃ vacaḥ // 66
. . . .
pararūpaṃ svarūpeṇa yayā saṃvriyate dhiyā /
ekārthapratibhāsinyā bhāvān āśritya bhedinaḥ // 68
tayā saṃvṛtanānārthāḥ saṃvṛtyā bhedinaḥ svayam /
abhedina ivābhānti bhāvā rūpeṇa kenacit // 69
tasyā abhiprāyavaśāt sāmānyaṃ sat prakīrtitam /
tad asat paramārthena yathā saṃkalpitaṃ tayā // 70
vyaktayo nānuyānty anyad anuyāyi na bhāsate /
jñānād avyatiriktaṃ ca katham arthāntaraṃ vrajet // 71
tasmān mithyāvikalpo 'yam artheṣv ekātmatāgrahaḥ /
itaretarabhedo 'sya bījaṃ saṃjñā yadarthikā // 72
. . . .
jvarādiśamane kāścit saha pratyekam eva vā /
dṛṣṭā yathā vauṣadhayo nānātve 'pi na cāparāḥ // 74
aviśeṣān na sāmānyam aviśeṣaprasaṅgataḥ /
tāsāṃ kṣetrādibhede 'pi dhrauvyāc cānupakārataḥ // 75
tatsvabhāvagrahād yā dhīs tadarthevāpy anarthikā /
vikalpikā 'tatkāryārthabhedaniṣṭhā prajāyate // 76
tasyāṃ yadrūpam ābhāti bahyam ekam ivānyataḥ /
vyāṛttam iva nistattvaṃ parīkṣānaṅgabhāvataḥ // 77
arthā jñānaniviṣṭās te yato vyāvṛttirūpiṇaḥ /
tenābhinnā ivābhānti vyāvṛttāḥ punar anyataḥ // 78
ta eva teṣāṃ sāmānyasamānādhāragocaraiḥ /
jñānābhidhānair mithyārtho vyavahāraḥ pratanyate// 79
sa ca sarvaḥ padarthānām anyonyābhāvasaṃśrayaḥ /
tenānyāpohaviṣayo vastulābhasya cāśrayaḥ // 80
. . . .
saṃsṛjyante na bhidyante svato 'rthāḥ pāramārthikāḥ /
rūpam ekam anekaṃ ca teṣu buddher upaplavaḥ // 87
bhedas tato 'yaṃ bauddhe 'rthe sāmānyaṃ bheda ity api /
tasyaiva cānyavyāvṛttyā dharmabhedaḥ prakalpyate // 88
. . . .
abhinnapratibhāsā dhīr na bhinneṣv iti cen matam /
pratibhāso dhiyāṃ bhinnaḥ samānā iti tadgrahāt // 107
kathaṃ tā bhinnadhīgrāhyāḥ samāś ced ekakāryatā /
sādṛśyaṃ nanu dhīḥ kāryaṃ tāsāṃ sā ca vibhidyate // 108
ekapratyavamarśasya hetutvād dhīr abhedinī /
ekadhīhetubhāvena vyaktīnām apy abhinnatā // 109

sā cātatkāryaviśleṣas tadanyasyānuvartinaḥ /
adṛṣṭeḥ pratiṣedhāc ca saṃketas tadvidarthikaḥ // 110

. . . .

yad rūpaṃ śābaleyasya bāhuleyasya nāsti tat /
atatkāryaparāvṛttir dvayor api ca vidyate // 139

Cf. other readings of these verses in Sāṃkṛtyāyana 1943 and Miyasaka 1971–72.

25. This view has been expressed beautifully in the following verse, which has been quoted in many works: "vikalpayonayaḥ śabdā vikalpāḥ śabdayonayaḥ / kāryakāraṇatā teṣāṃ nārthaṃ śabdāḥ spṛśanty amī."
26. "svākāram abāhyaṃ bāhyam adhyavasyan vikalpaḥ svākārabāhyatvaviṣaya iti; yathāha—'svapratibhāse 'narthe 'dhyavasāyena pravṛttir iti.'" (NVTṬ 441). For some critical comments on this passage, see *Nyāyabhāṣyavārttikaṭīkāvivaraṇapañ jikā* (NBhVP 41).
27. This can be understood from a verse from his tract dealing with apoha. The verse has been quoted in part by Vācaspatimiśra and in its entirety by Aniruddha in his NBhVP, as well as by Cakradhara in his NMJG (who has specifically ascribed it to Dharmottara). This verse reads as follows: buddhyā kalpikayā viviktam aparair yad rūpam ullikhyate / buddhir no na bahir yad eva ca *vadan nistattvam āropitam* / yas tattvaṃ jagato jagāda vijayī *niḥśeṣadoṣadviṣāṃ* / vaktāraṃ tam iha praṇamya śirasā 'pohaḥ *sa* vistāryate." NBhVP 41 gives *padam* for *vadan*, *niḥsattvam* for *nistattvam*, *ābhāsate* for *āropitam* and *niḥśeṣadoṣadviṣo* for *niḥśeṣadoṣadviṣām*; NMJG 14 gives *sma* for *sa*. Vācaspatimiśra has quoted the portion "buddhyā. . . . bahiḥ" in NVTṬ and Nyāyakaṇikā (see VV).
28. NK 750–60: "athocyate—yādṛśam eko govikalpo bāhyātmatayā svapratibhāsam āropayati, govikalpāntaram api tādṛśam evāropayati, vikalpāś ca pratyekaṃ svākāramātragrāhiṇo na parasparāropitānām ākārāṇāṃ bhedagrahaṇāya paryāpnuvanti, tasyobhayagrahaṇādhīnatvāt. tadagrahaṇāc ca vikalpāropitānām ākārāṇām ekatvam āropya 'vikalpānām eko viṣaya'—ity ucyate. tad eva ca sāmānyaṃ bahirāropitebhyo vikalpākārebhyo 'tyantabhedābhāvena abhāvarūpaṃ svalaksaṇa-tadākārāropitaiś caturbhiḥ sahobhiḥ samasya 'ardhapañcamākāra' ity ucyamānam āropitabāhyatvaṃ śabdābhidheyaṃ śabdasaṃsargaviṣayaḥ. tadadhyavasāya eva svalaksaṇāvasāyaḥ, tadātmatayā tasya samāropāt."
29. NK 360–61:

> anyavyāvṛttisvabhāvaṃ bhāvābhavasādhāraṇaṃ cedam, 'gaur asti,' 'nāśti' iti prayogāt. bhāvātmakatve hy asya 'gaur asti'—iti prayogāsambhavaḥ, punaruktatvāt. 'nāsti'—iti ca na prayujyate, virodhāt. evaṃ tasyābhāvātmakatve 'nāsti'—iti punaruktam, 'asti'—iti virudhyate. yathoktam—

ghaṭo 'stīti na vaktavyaṃ sann eva hi yato ghaṭaḥ /
nāstīty api na vaktavyaṃ virodhāt sadasattvayoḥ //

> etasmād eva ca bhinnānāṃ api vyaktīnam ekatāvabhāsaḥ. idaṃ hi sarveṣām eva vikalpānāṃ viṣayaḥ, asya aikyād vikalpānām apy ekatvam. teṣām ekatvāc

ca tatkāraṇānāṃ pratipiṇḍabhāvināṃ nirvikalpakānām apy ekatvam. teṣām ekatvāc ca tatkāraṇānāṃ vyaktīnām ekatvāvagamaḥ.

30. NMJ II.17. For more on Jayanta Bhaṭṭa's criticism of the Apoha doctrine, see the contribution of Masaaki Hattori to this volume.
31. Prajñākaragupta, the author of the commentary *Pramāṇavārttikālaṁkāra* (PVBh) on *Pramāṇavārttika*, sticks however to the extreme view that words cannot perform any positive function like "referring" to its object. See PVBh 262–266.
32. See PVSVṬ 248–254. and TSP ad TS 1094–1095.
33. This position has been clearly stated in the following passages: (i) "śabdais tāvan mukhyam akhyāyate 'rthas— / tatrāpohas tadguṇatvena gamyaḥ / arthaś caiko 'dhyāsato bhāsate 'nyaḥ / sthāpyo vācyas tattvato naiva kaścit" (JNĀ 203). (ii) "yathā hi nīlotpale niveśitād indīvaraśabdāt tatpratītau nīlimasphuraṇam anivāryam, tathā gośabdād apy agavāpoḍhe niveśitād gopratītav ago'pohasphuraṇam anivāryam." (JNĀ 203). (iii) "yat tu goḥ pratītau na tadātmā parātme ti sāmarthyād apohaḥ paścān niścīyata iti vidhivādināṃ matam, anyāpohapratītau vā sāmarthyād anyāpoḍho 'vadhāryate iti pratiṣedhavādinām matam. tad asundaram. prāthamikasyāpi pratipattikramādarśanāt" (RNA 59).
34. NVTP 300: "na hi ghaṭam ānaya iti vaktavye anaghaṭaṃ na mā naiṣīr iti vaktāro bhavanti."

9

Apoha as a Naturalized Account of Concept Formation

Georges Dreyfus

Since its promulgation by Dignāga (480–540),[1] the apoha theory has evinced passionate responses from Indian thinkers. Hindu realists have forcefully rejected it as an attempt to mask the problems created by the Buddhist nominalist denial of abstract entities. Buddhist antirealists have responded to these criticisms by insisting on the value of this theory, which in their eyes provides the resources for furthering the nominalist project of explaining thought and language in a world of particulars. The fact that this debate went on for a long time and is still largely unresolved is a testament to the intricacies and ingenuity of its arguments and the importance of its philosophical stakes. It also, however, makes the comprehension of this theory that much more difficult. In this essay,[2] I do not try to settle the issue and evaluate the arguments and counterarguments provided by both sides, a task that may well be impossible. Rather, I locate the nature of the contribution made by the apoha theory, which I see as a naturalized account of our cognitive capacities to form concepts rather than as an attempt to use negative formulations to articulate an original semantics. My account focuses on the views of Dharmakīrti (600–660), who provides the formulation around which almost all later discussions have revolved.

The gist of the apoha or exclusion theory is well known and is agreed upon by most scholars. It is the idea that thought and language do not relate to real things by capturing real properties but by excluding particulars from contradictory classes. For example, the word "tree" acquires its signification not by capturing the putative property of *treeness* but by excluding some particulars from being nontree. The assumption behind this theory is that only particulars are real and that the commonality presupposed by thought and language is only fictional. In fact, the apoha theory can be seen as an attempt to explain how thought and language can function in a world bereft of real abstract entities. Hence, it is a form of nominalism, "the widely shared intuition that the real is always particular" (Siderits 1999, 342). But this minimal consensus[3] about the theory quickly breaks down when one tries to define more precisely the nature of this nominalism.

In the following pages, I briefly consider the nature of Dharmakīrti's nominalism and its possible connection to conceptualism. I explore the problems commonly associated with these positions and show how Dharmakīrti avoids some of their pitfalls through a cognitive account of concept formation. My analysis focuses on the role of similarity, which I argue is to be understood as being part of a cognitive account rather than as a doomed ontological attempt to replace real properties by real resemblances.

Resemblance can then be seen to play a pivotal role in Dharmakīrti's apoha theory, linking perception to fully formed concepts and explaining how we can move from the perception of bare particulars to the realm of thought and language. In exploring Dharmakīrti's account, I also examine the role that mental content plays. Perceptions lead to the creation of mental representations based on assessments of similarities and dissimilarities. These mental representations are then taken to stand for fictionally constructed commonalities. In this way, Dharmakīrti can claim to show that thought and language are grounded in reality and hence nonarbitrary, despite the rejection of real universals. Finally, I deal with the role of error in the apoha theory according to which thought and language are necessarily mistaken because they proceed dichotomously. This radical conclusion is, however, quite at odds with our intuition that some thoughts are more correct than others. Toward the end of this essay, I explore the resources that Dharmakīrti and his followers can muster to respond to this objection.

In dealing with Dharmakīrti's views, I try to remain as close as possible to his own formulations. I believe that it is problematic to deal with the apoha theory in abstraction from any historical location. Nevertheless, historical precision has its limits, especially when dealing with the philosophical reconstruction of a theory, as is the case here. Thus, it should be clear that in dealing with Dharmakīrti's views, I do not try to capture the ways in which Dharmakīrti understood his work as a commentary on and a defense of Dignāga's apoha theory. Rather, I attempt a philosophical reconstruction of his theory, presenting it as a viable attempt to defend the nominalist project of showing how thought and language can be accounted for in a world of particulars. Hence, at times, I feel free to succumb to the temptation of supplementing his views with those of Tibetan commentators, particularly those belonging to the Sakya (*sa skya*) tradition. Since these thinkers have attempted to remain close to Dharmakīrti's spirit, I can claim that my interpretations capture the gist of his thought, though at times my formulations may be more Dharmakīrtian than those of Dharmakīrti. In this way I hope to present a reconstruction of Dharmakīrti's apoha theory that is philosophically compelling while remaining historically responsible.

NOMINALISM AND CONCEPTUALISM REVISITED

Some contemporary thinkers inspired by Nyāya realism have depicted the apoha theory as an extreme nominalism that denies that thought and language can have any objective basis. Dravid (1972, 345), for example, says about Dharmakīrti: "However, it must be granted that the Buddhist is the most thorough-going nominalist in the history of thought." For him, Dharmakīrti holds that the real is completely unknowable and that, therefore, the only kind of knowledge we can have is the kind we construct. Concepts are the life of knowledge but have no objective counterparts. They are, according to Dravid, the products of creative thought without any objective support in the real world onto which they are arbitrarily and a priori projected (1972, 345).

Although this presentation of the Buddhist view is not without some basis, it misses some of the distinctive elements of Dharmakīrti's apoha theory, which is not just a clear formulation of Dignāga's view but a systematic attempt to show that the rejection of real universals does not entail the arbitrariness of thought and language. Although thought and language presuppose commonalities that can only be fictional, this does not mean that they are deprived of any objective basis. Thought and language are causally related to our experiences of things and hence are grounded in reality. Thus, they have an objective and natural basis and are not just the reflections of our human interests. Hence, Dharmakīrti is not an extreme nominalist in Dravid's sense of the term.

Other scholars have recoiled from such an extreme description and have offered more nuanced accounts of Dharmakīrti's views. Siderits, for example, has described Dharmakīrti's position as a radical nominalism, that is, being committed to the view that "in the case of a kind term like 'cow,' all that is common to the individuals called cows is the name 'cow'"

(1999, 341). At the outset, let me say that there is a great deal of truth in this characterization. Indeed, Dharmakīrti is quite radical in his rejection of abstract entities and his critique of Hindu realists. He lambastes them for their ultrareified view of abstract entities and forcefully rejects real non-individual entities, as is made very clear throughout his works, particularly in the *Pramāṇavārttika*, where he examines and dismisses several versions of realism.[4] But this is not the only issue raised by Siderits's description of Dharmakīrti as a radical nominalist.

An easy objection to Siderits's formulation would be that the commonality that is required to account for thought and language cannot reside simply in our use of the same term to designate discrete individuals. For, upon analyzing the notion *sameness of terms*, we are forced to conclude that identifying individual terms as being the same presupposes the concept of sameness of meaning, in relation to which the individual terms can be identified as being the same. For example, we identify different tokens of the word "cow" as being the same because they identically signify *cowness*, not just because they sound identical (in fact, they often don't). Thus, commonality cannot be due simply to the use of a common term. This is not, however, how I understand Siderits, for I take him to make a different and more cogent point, namely, that we can explain our kind terms by pointing to the agreed-upon fictions that we form through education and language acquisition. Hence, for Siderits and many other nominalists, thought is to be explained in terms of the agreed-upon fictions that we construct linguistically as reflections of our human interests.

It is precisely this one-sided emphasis on the preponderant role of language and human interests that I find problematic. For it leaves out central elements of Dharmakīrti's account such as his reliance on mental content and the natural constraints under which this content is formulated. These elements indicate a somewhat different position. In the following pages, I explore the central elements of this account and show how Dharmakīrti can be thought to offer a naturalized account of thought and language based on a theory of concept formation. This is what I perceive to be Dharmakīrti's distinct contribution: the attempt to respond to the realist critique by providing an account of causally constrained concept formation. Hence, far from being arbitrary, thought and language rest for him on natural processes and have a basis in reality, despite the absence of any real commonality.

The centrality of the process of concept formation in Dharmakīrti's account suggests that his position could be described not inaccurately as a form of conceptualism. In an earlier contribution, Siderits (1999, 341) describes conceptualism as the view that "what all cows have in common is just that they fall under the same concept, where concepts are thought of as mental contents of some sort or other." I think that this description, which Siderits unfortunately rejects, is not an inaccurate characterization of Dharmkīrti's position. This does not mean that Dharmakīrti is not a nominalist, for he clearly rejects the existence of real abstract entities and is quite comfortable with its sparse ontology. But since he relies on mental content and the constraints under which this content is formulated to explain the works of conceptuality, I think one could describe him as a conceptualist as well (a similar point could be made about Ockham). I do not think, however, that we need to emphasize this term as long as the specificity of Dharmakīrti's project is recognized, namely, the attempt to further the nominalist project of explaining how nonarbitrary thought and language can be possible in a world of particulars by appealing to a theory of concept formation and the constraints that such an account entails.

The description of Dharmakīrti's position as conceptualist is not without its own set of problems and misunderstandings. First, conceptualism is often assimilated to the position articulated by British empiricists and as such is not given much consideration. Second, conceptualism is also thought to be a confused attempt to find a third way between realism and nominalism by appealing to the reality of concepts. But as has often and rightly been argued, such an attempt is doomed from the outset. There is no middle ground between realism and nominalism. Either abstract entities exist or they don't. In either case, various possibilities may be open, but a middle ground is not one of them. Moreover, the appeal to the notion of *concept* is problematic inasmuch as it involves a notion that is as vague as it is widespread. Hence, conceptualism is often dismissed as an attempt to mask the problems of nominalism by confusing the issues through vagueness and equivocation. Finally, conceptualism is also accused of reintroducing properties in the guise of resemblances. But as many thinkers have argued, such an attempt cannot succeed, for if we accept real similarities, we might as well accept real properties. Thus, it looks like conceptualism cannot be taken seriously, for it ends up reintroducing by the back door what it sought to exclude. Can Dharmakīrti's apoha theory do better and offer a viable account compatible with some of the conceptualist insights, particularly those concerning the importance of resemblances? This is the question I answer first as a way to indicate the nature of Dharmakīrti's apoha theory.

IS DHARMAKĪRTI A RESEMBLANCE THEORIST?

As already mentioned, conceptualists are often described as defending a resemblance theory of universals according to which concepts are based on similarities from which properties are construed. This view is contrasted with recurrence theories of universals, which hold that the nonarbitrary application of general terms to individual things can only be explained if these things share some common property. Throughout the history of philosophy, many thinkers have asserted resemblance views of universals. The most famous of these in the West are Hobbes, Berkeley, and Hume. Although their views are not identical, they share the idea that only particulars are real and that general terms are to be explained in relation to concepts derived from similar uses of real particulars resembling each other. For example, Hume (1960, 20) says: "when we have found a resemblance among several objects that often occur to us, we apply the same name to all of them, whatever differences we may observe in the degrees of their quantity and quality, and whatever differences may appear among them." For Hume, the basis for our uses of kind terms such as "tree," "medicine," etc., is to be found in the similarities among the objects we encounter. But as many critics have argued, the appeal to similarities faces the question of the status of these similarities. Are they real, or are they merely the reflections of our judgments? If the former, it seems hard to see how one could avoid the conclusion that there are real properties. This is so because two things cannot be similar absolutely but only in relation to a third term. For example, two different samples of red are similar only in relation to their being red. But what is the status of this being red? If it is not real, it is hard to see how there could be real similarities. If on the other hand it is real, then there is no reason to introduce real similarities in the first place. We have real properties and do not need anything else to account for the commonalities of thought and language.

Dharmakīrti's theory may at first appear to follow the same line of argumentation as the one defended by British empiricists. He describes, for example, the similarities (*sādṛṣya*, *'dra ba*) among things in terms of their having "a [common] effect" (PV I, 108c). Similarity between cows, for example, comes from their capacity to perform similar functions, not from the inherence of a universal cowness.[5] When we see large beings that provide milk, carry loads, produce manure, etc., we are struck by the similarities and construct the unitary concept of *cowness* to account for these similarities. It is these similarities in the ways in which we understand our experiences that provide the basis for the construction of properties. In a well-known passage, Dharmakīrti says: "Certain [things] are naturally [*svabhāvena, rang bzhin gyis*] determined as producing common results such as being evaluated as the same [object], being identified as having the [same] function, etc., despite the fact that they are distinct, just as sense-bases, etc. For example, certain medicines, taken separately or together, despite their being distinct, are seen to cure fever, etc., [whereas] other things do not."[6]

Dharmakīrti argues that despite being distinct, things can produce similar results on the basis of which abstract properties are constructed. To illustrate his point, he provides well-discussed examples that do not need to detain us here. When we see that various medicinal plants such as the *harīkata* and the *abhaya* can cure fever, we may be tempted to assume that they share some common essence. But this is a mistake, for the only commonality they share is that they are similarly therapeutic. There is nothing over and above this resemblance regarding their medicinal effects that is required to explain how these plants are all remedies.

All this seems to suggest that Dharmakīrti's view is not substantially different from the one classically associated with British empiricism and hence that it suffers from the same problems. But it is here that our thinker offers a much more cogent account, for the similarities that he posits are not objective but exist only in our experiences or, rather, in the ways in which we spontaneously conceptualize our experiences. We experience things and are struck by similarities and dissimilarities between our experiences. We then form intuitive senses or evaluations of similarity (*ekapratyavamarśa*)[7] and dissimilarity on the basis of which we develop fully formed concepts. It is these intuitive assessments of similarity that provide the nonarbitrary basis for the elaboration of fully formed concepts, for they are the results of the causal capacities of objects as we experience them through nonconceptual perceptions. We perceive things and assess them to be similar or dissimilar and on this basis elaborate our concepts.

This recognition of the conceptual nature of similarities may appear to make his account of resemblances circular and to collapse two levels of discussion: the level at which objects have functional similarities and the level at which we conceive them as having these similarities. I would argue, however, that this is not the case at all.

Dharmakīrti's theory is actually quite coherent and may well provide the most defensible nominalist account of similarities. His point is that similarities do not exist "out there," but are the products of our interactions with the world. There is no such thing as real similarities between shapes or colors,[8] as we tend to think, but only evaluations of similarity between shapes or colors that provide the basis for our conceptual elaborations.

These evaluations already involve conceptualization. It is not the case that objects are perceptually given to us as similar or dissimilar. Rather, it is we who conceive of them as such. But these conceptual assessments are of a lower order than full-blown predicative judgments. They are the spontaneous evaluations that we come up with naturally when we encounter objects. They are not the products of a higher level of conceptualization but are directly caused by our experiences. This does not mean that they reflect reality, for they are conceptual. But they are as close to reality as conceptuality can ever come, the first level out of which higher-level conceptualization is elaborated.

It is this cognitive role of similarity that is missed when Dharmakīrti's account is accused of being circular. This assumes that Dharmakīrti's intention is to point to similarities among judgments to posit objective similarities. That would obviously be circular, but if my account is correct, this is not what Dharmakīrti is up to. Rather, he points to our naturally induced sense of similarities and dissimilarities to respond to the realist argument that without real properties shared by individuals there is no way to justify our conceptual judgments, which rely on these properties. Dharmakīrti's answer is not to find some objective similarities that would substitute for the missing properties but to argue that there is no need for such substitutes. All we need are our evaluations of similarities and differences as they arise causally from our encounters with reality. These assessments provide the nonarbitrary basis on which we further construct fictional commonalities that we project onto reality. Hence, Dharmakīrti's account does contain a resemblance theory of a kind: not one, however, that substitutes objective similarities for properties, but one that emphasizes the cognitive role that our conceptions of similarities and differences play as the basis for further conceptual elaborations.

This discussion of the conceptual nature and cognitive role of similarities eliminates some of the qualms concerning Dharmakīrti's apoha theory. It shows that for Dharmakīrti this theory is not an attempt to use a negative formulation to create a semantic theory escaping the problems created by the absence of real properties, as if speaking negativese would magically solve the problem of universals, but a cognitive account of the process through which we construct commonalities on the basis of similarities and dissimilarities. This is what Dharmakīrti means in his summary of the apoha theory when he says: "because all things essentially abide in their own nature, they support the exclusion from similar and dissimilar things."[9] The commonality necessary to explain thought and language does not exist in reality. Hence, it must be constructed by excluding things from similar and dissimilar classes. For example, the statement "this is a golden pot" involves the predication of a property, being a golden pot, of a particular object made of gold. But since this property is fictional, it is not given in experience but must be constructed through the exclusion of the particular object made of gold from the class of similar things (brass pots, clay pots, etc.) and the class of dissimilar things (cows, airplanes, etc.).

This much is well known. What is often less noticed in this statement is that Dharmakīrti is not just speaking about the exclusion of things from the classes they do not belong to, but also connecting this exclusion to the way things are. It is because things abide in their own nature that we can exclude them from the classes they do not belong to. It is this connection that is ignored when his theory is accused of providing an account of thought and language as arbitrary projections onto the real world. For Dharmakīrti, the exclusion of things from the classes they do not belong to is not random but is based on how we experience real things through the mediation of our spontaneous evaluations of similarities and dissimilarities. It is to capture the cognitive nature of this process of exclusion that Dharmakīrti describes the classes of things that are to be excluded as "the similar and dissimilar things." In Dhamakīrti's formulation, the apoha or exclusion theory is not a semantic formulation in which the use of negative terms solves the problems created by the rejection of universals, but a three-tiered cognitive theory of the process of differentiation in which fully formed concepts emerge from quasi-immediate assessments of similarities and dissimilarities brought about by perception.

A NATURALIZED ACCOUNT OF CONCEPT FORMATION

A picture of the apoha is emerging as a cognitive account of the process of concept formation. This cognitive account is in evidence in the *Pratyakṣa* chapter of the *Pramāṇavārttika,* an important chapter that has not yet received the full attention that it deserves in the scholarly literature existing in Western languages. As a result, most scholars have tended to focus on the semantic side of the apoha theory, neglecting the cognitive account of concept formation. This cognitive account is found in a crucial passage, which delineates the main lines of articulation of the apoha theory and as such may have served as a basis for later developments of the apoha theory such as Śāntarakṣita's distinction among three types of negation. In this passage, Dharmakīrti seems to be responding to an objection to his assertion that the conceptual process is grounded in perceptual experience. The objection is that the assertion of a causal link between things and concepts contradicts the basic Buddhist epistemological tenet that words do not relate to real things but only to fictional entities. Indirectly, the adversary is also questioning how the Buddhist epistemologist can explain the referential function of language. How can language apply (*pravṛtti, 'jug pa*) to the real world if all language has access to is agreed-upon fictions? In response to this objection, Dharmakīrti says:

> Although a word does not rest on an elimination [found in] external [things] that has the capacity [to perform a function], it is related to the [object's] reflection [provoked] in conceptual thought [by words], which rests on this [elimination existing in external objects]. Therefore, since it rests on an elimination of others, a word signifies an elimination of others. That which pertains to the reflection of the object appears to the cognition [arising] from words similar to the elimination [of others], but it is not the real object, [for] it arises from mistaken latencies.[10]

In his response, Dharmakīrti grants that language and conceptuality do not bear directly on reality and insists on the mediating function of the conceptual representation, the concept as a mental event, here described as reflection (*pratibimba, gzugs brnyan*). This is the central element linking conceptuality to reality. Thought proceeds by constructing conceptual representations, which are taken to stand for agreed-upon fictional commonalities and projected onto discrete individuals.

This passage raises a multitude of questions, the first of which I cannot address here. It concerns the textual interpretation of Dharmakīrti's reference to "an elimination [found in] external [things] that has the capacity [to perform a function]." Does Dharmakīrti hold that eliminations can be real? Some later commentators thought so and proposed the concept of objective elimination (*arthātmaka-svalakṣaṇānyāpoha, don rang mtshan gyi gzhan sel*), to refer to a real negative entity, the real object as it is differentiated from the other real objects.[11] Other scholars have rejected this description, which in their eyes compromised the integrity of the nominalist rejection of real universals.

But this passage also raises other, and at this point more directly relevant, questions. What is the role of the notion of reflection referred to in this passage in the apoha theory? What is the nature of this reflection? Is it a genuine exclusion or is it a real mental event? Here, as often, Dharmakīrti is not as explicit as one may wish, and as a result his commentators have differed on the relation between reflections, real things, and exclusion. Tibetan commentators in particular have struggled with these questions. Here, I briefly introduce some of their insights, for I believe that they introduce concepts that clarify these delicate points.

Tibetan discussions of these questions often center on the concept of object universal, which is the conceptual representation or reflection mentioned earlier by Dharmakīrti. The term *don spyi*, that is, *arthasāmānya*, which I translate as "object universal," is not a Tibetan invention, for it is found in Dignāga's writings, where it plays an important semantic role (see Pind's essay in this volume). It seems, however, to have played little role among later Indian commentators, perhaps as a consequence of its neglect by Dharmakīrti, whose semantic theory differed quite radically from Dignāga's. For reasons not entirely clear, Tibetan commentators revived its use to explain the ways concepts operate, but their interpretations of this term are far from being unified. Here, I will follow the insights of the Sakya commentators, particularly Goramba (*go ram pa bsod nams seng ge*, 1429–1489 A.D.), who offers a clear account of the question that follows closely the views of Sakya Paṇḍita (*sa skya paṇḍita*, 1182–1251 A.D.). In discussing the nature of the object universal or conceptual representation, Goramba distinguishes two aspects of conceptual representations: the representation or appearance of an object to the conception as a real mental event and the content of that appearance. He says:

> Now, is the objective aspect of the conceptual cognition a [real] appearance[12] or is it an [unreal] elimination? Here, [I would like to distinguish] two factors: a cognitive factor and a factor superimposed onto the external jar. Among those two, the former is a [real] appearance because it is the object that is taken as an object of self-cognition (*rang rig, svasaṃvitti*) of a conceptual thought. The latter is an elimination because it is an imputation.[13]

The conceptual representation (Dharmakīrti's reflection) of the object is a real mental event and, hence, not an actual exclusion. As such, it is to be sharply distinguished from its unreal content. Or, to put it slightly differently, concepts as real mental events are to be distinguished from concepts as content, which can be variously characterized as exclusions, universals, and properties. All these descriptions are equivalent and come to the same view of conceptuality. Thought proceeds by constructing universals or unreal properties, which are assumed to be instantiated by individuals. This construction is based on a "standing for" relation in which mental representations are made to stand for the agreed-upon commonality that individuals supposedly share.

Thus, we can see that there are two equivalent ways to formulate the apoha theory. Universals can be understood as the content of concepts or as properties instantiated by things. For example, *cowness* can be taken as the content of the concept of *cow* or it can be described as the property shared by cows. In the first case, universals are taken as the mistaken identifications of the representation and the property that the representation is assumed to stand for. This is the object universal of *cow*. In the second case, the universal is understood as a property assumed to be instantiated by things. Since this property can only be validated in a purely negative way, it can be reduced to the mere elimination of the contrary of the object considered. This is the universal as pure elimination.

We now realize the degree to which Dharmakīrti's account relies on the process of concept formation. His entire apoha theory can be constructed from the point of view of a process of representation in which concepts qua mental events are mistaken to stand in for agreed-upon commonalities. For example, the content of the concept of being a cow is created by identifying the conceptual representation of a cow with the fictional property that the representation is assumed to stand in for. This representation is constructed out of a primitive space of similarities and dissimilarities in which objects are not yet categorically organized but consist of features distinct from other features. We then conceive of colors and shapes and distinguish them from other colors and shapes. These distinctions in term of similarities and differences provide the basis for the full-blown categorization involved in the predicative judgment "this is a cow." We then conceive of real objects not just as fleeting features but as fully organized stable objects, such as cows, falling within well-determined categorical schemes.

We may wonder, however, what is the nature of this primitive space of similarity and dissimilarity? We have described it as a quasi-immediate and spontaneous sense of similarity and dissimilarity, a pre-predicative form of evaluation or assessment. For Dharmakīrti, this form of evaluation cannot but be conceptual. It is not given in perception, which is nondetermined (*nirvikalpaka*) and hence provided only with an unarticulated content. Any determination of content comes from the process of construction in which we engage immediately upon encountering the world through our senses. It is this process of construction that Dharmakīrti describes as conceptual, that is, as involving an internal process of articulation through differentiation. In his later *Pramāṇaviniścaya*, Dharmakīrti gives this definition of conceptualization (*kalpanā*): "conceptualization is that consciousness in which a representation [lit., appearance] is fit to be associated with words."[14] When we conceive of an object, for instance a jar, we do not apprehend this jar directly but through the mediation of its conceptual representation, which we identify by associating it with a linguistic sign. This is how fully formed conceptualization works. But what about our pre-predicative evaluations of similarity? Are they also associating representations with linguistic signs?

It is here that some commentators have made a creative use of Dharmakīrti's notion of fitness (*yogyatā*, *rung ba nyid*) contained in the definition of conceptualization. Mokṣākaragupta, for example, argues that babies have simple nonlinguistic concepts that allow them to perform basic functions, much like animals do (Kajiyama 1966, 40–43). Similarly, Śāntarakṣita holds that the word "fit" is added by Dharmakīrti to include prelinguistic conceptions. Their concepts, which are perhaps best described as protoconcepts, are fit to be verbalized in that they provide the basis for later verbalization, though they are not actually associated with words.[15] It should be clear, however, that this is not the main point that Dharmakīrti has in mind when formulating his definition. His main target is quite different, namely, the Mīmāṃsā view that there is an inherent fitness (*yogyatā*) between words and meaning. Against this view, Dharmakīrti argues that the fitness between words and meaning is not inherent but conventional. Nevertheless, I believe that these commentators are not wrong in making this definition account for the pre-predicative level of conceptualization necessitated by Dharmakīrti's three-tiered account. Spontaneous assessments of similarity are prelinguistic, though they are conceptual in the Dharmakīrtian sense of the word.

This prelinguistic level of conceptualization leads to a fully articulated conceptualization based on the formation of conceptual representations that are made to stand for the commonalities we construct out of the more primitive space of similarities and dissimilarities. As real events, these representations are causally produced, being based on our encounters with realities, though they do not directly reflect such encounters. This causal link is essential to understand Dharmakīrti's response to the Hindu realists. For him, our thoughts are based on our experiences and the sense of similarities and dissimilarities that we derive from such experiences. As such, they are importantly constrained by our natural constitution. We think, for example, that certain colors are similar not just because we share certain language games and have been educated in certain ways, but also and very importantly because of the kind of being that we *naturally* are. That is, we are, to a large extent, naturally constrained when we see certain things as similar or dissimilar, and education, culture, or language can do little to this. These natural constraints can be spelled out in several ways. From a purely Buddhist perspective, we can point to karma and the tendencies left by beginningless lives as explaining the ways we see and conceive the world.[16] From a more modern standpoint, we can point to our evolutionarily produced perceptual apparatus as importantly determining the way we conceive of things. This apparatus largely determines the primitive sense of similarities and dissimilarities out of which full-blown categorization emerges. Although humans disagree on the ways in which they conceptualize full-blown objects, they largely agree on the relevant features that first emerge when they encounter objects. This is so though there is a very large number of features that could be potentially considered. But here we are largely constrained by our constitution, which leads us to consider only a limited range of features such as color and shape. Such agreement is based on our natural constitution and provides the primitive space of similarities and dissimilarities discussed here. Although such a space does not fully determine the way we categorize the objects of our experiences, it significantly constrains the posterior process of concept formation. This is why we can understand each other relatively easily, particularly when we refer to concrete, middle-sized objects. Language games do not operate in a vacuum but exist in an embodied situation that significantly constrains them.

This is what is distinctive in Dharmakīrti's thought: a naturalized account of how we are led to certain judgments. For example, when we perceive certain shades of blue or green as being similar, we are not just making judgments on the basis of our education and the language we speak but we are largely following the dictates of our perceptual apparatus. We are the kind of beings who naturally see these shades as similar, and for Dharmakīrti this is all there is to say. Or, almost all, for not all humans see all colors similarly. There are individual and cultural variations. It is well known, for instance, that several Asian cultures do not seem to distinguish blue and green in the same ways as we do in modern English. Tibetans, for example, will say of grass that it is *sngon po*, the word that we would usually translate as "blue." Thus, it is clear that culture and language do play a significant role in the formulation of the similarities and differences that underpin our conceptual practices. But the role of culture and language is more limited than the depiction of Dharmakīrti as a radical nominalist would suggest. For although some cultures do seem to parse green and blue differently, I do not think that there is any culture in which yellow and blue or white and black are conflated. This is so because our sense of similarities and differences is importantly constrained by our senses. This is what Dharmakīrti means when he states that "certain [things] are naturally [*svabhāvena, rang bzhin gyis*] determined as producing common results." The common result he envisages is the primitive space of similarities and dissimilarities described earlier. Out of this space, which is to a large extent naturally constrained, emerge full-blown conceptual representations, which stand for the commonalities that we assume objects instantiate.

This three-tiered account provides, to use Tillemans's useful formulation (in this volume), a bottom-up account of the ways in which thought and language relate to reality. Once this account is in place and conceptualization is shown to emerge from experiences and our spontaneous assessments of similarity, it becomes possible to think about the apoha theory as a top-down account of the meaningfulness of language and the various operations that this allows. This semantic account is also part of the apoha theory, as is evident in other parts of Dharmakīrti's works, as, for example, when he shows how the apoha theory can account for the informativeness of certain forms of inference.[17] For Dharmakīrti, this semantic account plays an important role in his works, but it presupposes an explanation of how the scaffolding of conceptualization is constructed from the bottom up. Only after this account is in place can the nominalist feel entitled to use the apoha theory to explain how general concepts can be used without having to pay the exorbitant ontological price that the realist is trying to extract. This is how the apoha theory works as the central piece of the nominalist project of explaining how thought and language can be possible in a world of particulars.

THE MISTAKEN NATURE OF CONCEPTS

One of the unavoidable consequences of this cognitive account of the apoha theory is that conceptualization is inherently distorting and hence necessarily mistaken. In a world of individuals, commonality among different objects does not exist in reality but is constructed and superimposed (*adhyāropa, sgro 'dogs*) onto individuals. This commonality emerges when a representation is mistaken for a property constructed on the basis of a radical dichotomization of the universe of discourse and assumed to be instantiated by individuals. Although this property is taken to exist in reality, it does not, for there is no commonality outside of our imagination. Nevertheless, we subjectively assume that commonality between objects exists and on this basis we relate to objects in our practices. It is only on the basis of this confusion that we are able to relate to the world.

This assertion of the radically dichotomous and necessarily mistaken character of thought raises obvious questions. For if conception is mistaken, how can it then determine its object? And is it not the case that any determination will be equally off the mark? Dharmakīrti answers that the validity of conceptual cognitions derives not from their ability to reflect reality accurately, but from their causal connection with reality via perception. An erroneous conception, such as the apprehension of smoke as permanent and an inference of fire on the basis of smoke, do not differ in their ability to provide an adequate representation of reality, for in this respect they are both mistaken. Where they differ is in whether they tap into the causal regularities of reality or not. The conception of smoke as permanent is the result of inner mistaken tendencies rather than that of the experience of causal regularities, and hence, it cannot lead to successful actions in the world, except by chance. By contrast, the inference of the existence of fire upon seeing smoke is formed on the basis of the experience of causal regularities and leads to further success. Hence, it is nondeceptive and valid, though equally mistaken. Dharmakīrti illustrates his point by a double example: the apprehension of the light of a lamp being taken as a jewel and the apprehension of the light of a jewel being taken as a jewel. This double example is well known and does not need to detain us here.[18] But what needs to be emphasized is that the point of these examples is to stress that all conceptions are equally mistaken. The apprehension of a jewel with respect to a jewel's glitter is no less erroneous than a similar apprehension with respect to a lamp's light. The only factor that differentiates the conception we hold to be factual from the other is its practical success.

This stress on the radically mistaken character of thought is well known and does not need to be belabored any further. What must be emphasized is the centrality of this view of conceptuality as mistaken. The fact that thought is radically dichotomous and necessarily mistaken is at the center of the apoha theory. And yet, this is one of the aspects of the apoha theory that is often glossed over by commentators, traditional or modern. The temptation to downplay this radical aspect of the apoha theory is quite understandable. It is discomforting to assert that the inference of fire from smoke and the view that smoke is permanent are equally mistaken. Our intuition is that one is more correct than the other and a complete theory should be able to account for this difference.

Tibetan commentators have been quite conscious of these difficulties. Although they have not ignored Dharmakīrti's repeated assertion of the inherently mistaken nature of conceptuality, they have been tempted at times to downplay its radical nature. Geluk (*dge lugs*) thinkers are well known for their moderate realist attempt to differentiate the degree of mistakenness in conception. Accordingly, they describe the inference of fire from smoke as mistaken with respect to its appearing object (*snang yul*, which is the object universal of smoke identified as real *smokeness*) while being unmistaken with respect to its apprehended object of application (*'dzin stangs kyi yul*, the fact there is fire on the smoky hill). This distinction, which was first introduced by Chaba (*phya pa chos kyi seng ge*, 1109–1169), is not necessarily committed to a moderate realist position,[19] but it is easy to see how it could easily lead to such a view. Correct thinking is not mistaken with respect to its apprehended object because it is inferring the property of *fireness* over the fiery hill, and such a property exists, albeit not in the way thought conceives of it. Hence, inference can be distinguished from erroneous thinking, which is mistaken with respect to both appearing and apprehended objects.

The Sakya tradition has struggled with the same issues but has come up with different strategies, which remain closer to Dharmakīrti's radical rejection of realism. For Sapaṇ and his commentators, the problem is to maintain Dharmakīrti's assertion of the radically mistaken nature of thought while satisfying our phenomenological demand for distinguishing between correct and incorrect conceptions. This is particularly important in the domain of reasoning where we want to distinguish the kind of reasonings that lead to inferential knowledge from spurious ones. To solve this quandary, Sapaṇ and his followers make a distinction between *practical application* (*'jug pa*) to the object and *critical explanation* (*'chad pa*) of such application.[20] To clarify this technical distinction, which is important to explain the status of reasoning in an antirealist framework, let us take the case of the famous inference: the sound is impermanent because it is produced, just as a jar. When we state this reasoning, we speak from a conceptual and hence mistaken point of view. The mistake is that we take the subject (*chos can, dharmin*) of the argument, a sound, as being the actual object and proceed on this basis to infer the predicate (*chos, dharma*) or property of being impermanent. The realist takes this predication to show that properties and universals cannot be dispensed with. The proponent of the apoha theory responds that predication does not connect two entities but excludes an individual object from being what it is not. When I say "the sound is impermanent," I am only seemingly relating an individual to a property instantiated by this individual, but in reality I am excluding an individual sound from being nonimpermanent.

This well-known explanation seems to give a convincing antirealist account of predication and reasoning. There is, however, a problem with this account, namely, the assumption that the real individual is the subject of predication. This is highly problematic in Dharmakīrti's system, where thinking or language can never relate directly to reality. How then can predication be described as the exclusion of an individual from being what it is not if such an individual is not available to the conceptual process of exclusion? The Sakya commentator Shākya Chokden (*śākya mchog ldan*, 1428–1509) responds by agreeing with the objection. Conceptuality is limited to nonexistent constructs that we imagine to be real. In the case of the subject of our predication of impermanence, what we are dealing with is a conceptually understood substance (lit., "the substance considered from the point of view of exclusion," *sel ngo'i rdzas*). This pseudosubstance is the actual basis of predication and, hence, is called the "actual subject" (*song tshod kyi chos can*), but it is nothing but an elimination and hence a conceptual fiction. When this subject is mistaken as real, it is designated as the "assumed subject" (*rlom tshod kyi chos can*), that is, the subject (falsely) assumed to be real.

This distinction between the actual (*song tshod*) and imagined (*rlom tshod*) terms of the reasoning explains the way in which reasoning can work in a world where we have no direct access to real individuals. The actual subject (*chos can*, *dharmin*), the predicate of the probandum (*bsgrub bya'i chos*, *sādhyadharma*), and the reason (*rtags*, *liṅga*) are constructs and hence fictions. But reasoning proceeds by confusing these concepts and applying them to reality. We do not abstractly deduce constructs from constructs, but, rather, we do so by applying them to reality. For instance, in our example we infer impermanence from the property of being produced. The actual reason (*song tshod kyi rtags*) is the fictional property of being produced and from it we infer an equally fictional property of being impermanent by mistakenly applying this construct to what we imagine to be the subject. That is, we think that sound is impermanent because it is produced, ignoring that we are really dealing with useful fictions.

For Shākya Chokden, this distinction between actual and imagined terms of the reasoning is not meant just to emphasize the mistake at the heart of the conceptual process. It also intends to find a place, albeit a limited one, for our intuition that there is a difference between right and wrong reasonings. Erroneous conceptions miss the mark whereas correct ones capture some aspects of reality. Sakya thinkers such as Shākya Chokden can be seen to accommodate this intuition by distinguishing actual from imagined terms of the reasoning. Although in reality the terms of a reasoning are fictional, in our imagination they are not, for we take our concepts to be real and proceed to reason on this basis. Hence, our reasonings do capture reality in some sense of the word, but this capturing occurs only in our imagination and hence is largely illusory.

This illusion turns out, however, to be useful. Since our concepts are born from experience, they tap into causal regularities and hence their use can lead to success. Moreover, since our interlocutors share the same mistaken tendencies to confuse the conceptual and the real, the reasonings we direct at them work as well. It is only when we step back and critically examine the terms of our reasonings that we come to differentiate our fictional constructs from the real entities of the world. When this distinction is made, the mistaken nature of thinking is exposed, and thinking comes to a halt. This distinction between precritical application and critical examination provides a basis for an antirealist explanation of inference. For Buddhist logicians, inference does not deal directly with reality. It does not establish or refute anything in direct reference to a real substance (*rdzas la dgag sgrub mi byed*).[21] Accordingly, reasoning only deals with conceptual creations, which are mistaken for real entities. Thus, it is only in our imagination that we can conceive of reality. But this imagination is neither free-floating nor arbitrary, for it reflects our education and the social process of language acquisition we have gone through. It is also largely determined by the natural constraints under which we operate. We come to conceive of things as sharing certain properties on the basis of our assessment of similarities and dissimilarities not just because we have been socialized in certain ways and have particular interests, but also, and importantly, because we operate under the natural constraints that our embodied situation entails. For Dharmakīrti, this is the bottom line. As a consistent antirealist, our thinker has to recognize that in a world of individuals, thought, which proceeds by generalities, cannot capture the real and is limited to the fictional domain. Hence, all that we have are our imagined constructs, but since those are naturally produced they work, and this is all we need to explain the successes of our epistemic practices.

Notes

1. As far as dates are concerned, I am not making any original claim but merely following the largely accepted consensus given by Hattori 1968, 4.
2. This essay is a revised version of the paper presented at the Apoha conference at Crêt Bérard, Switzerland. As such, it has greatly benefited from the ideas of all the participants, but I want to acknowledge more particularly Amita Chatterjee, Jonardon Ganeri, Pascale Hugon, Mark Siderits, and Tom Tillemans for their feedback and suggestions.

3. An important exception to this consensus is the kind of moderate realist interpretation developed by many Tibetan thinkers. Geluk (*dge lugs*) commentators in particular maintain that the apoha theory excludes only the universals advocated by the extreme realists of the Nyāya type and hence is compatible with the existence of real properties existing in things. See Dreyfus 1997. In this essay, however, I will ignore their views and focus on the mainstream of Buddhist epistemological tradition.
4. See, for example, PV I, 88 (ed. Miyasaka), where Dharmakīrti seems to target the Sāṃkhya view, and I, 89–90, and III, 25–26 (ed. Miyasaka), where he refutes more particularly the Nyāya view.
5. PVSV (ed. Gnoli) 46.11.
6. "gcig rtogs don shes la sogs pa / don gcig sgrub la 'ga' zhig ni / tha dad yin yang rang bzhin gyis / nges te dbang po la sogs bzhin // dper sman kha cig tha dad kyang / lhan cig pa'am so so yis / rims la sogs pa zhi byed par / mthong gi gzhan gyis ma yin bzhin // ekapratyavamarśārthajñānādyekārthasādhane / bhede 'pi niyatāḥ kecit svabhāvenendriyādivat // jvarādiśamane kāścit saha pratyekam eva vā / dṛṣṭā yathā vauṣadhayo nānātve 'pi na cāparāḥ" (PV I, 73–4 [Miyasaka 1971]).
7. This term is not easy to translate. It refers to a quasi-immediate though conceptual sense of similarity. I first used the term "judgment," but this wrongly suggests that we are dealing here with a fully predicative structure. I have chosen "evaluation" or "identification" as a way to capture this pre-predicative sense of similarity and difference. Tibetans have translated this term as "conception of oneness" (*gcig tu rtogs*).
8. Some Hindu thinkers hold that there is such a thing as a similarity (*sādṛṣya*). The Prābhākara school of Mīmāṃsā holds similarity to be a basic category of reality. The Bhāṭṭa holds a middle ground between the Buddhist denial of objective similarity and the Prābhākara emphasis of similarity as a fundamental building block of reality. See Bandyopadhyay 1982.
9. "gang phyir dngos kun rang bzhin gyis / rang rang ngo bo la gnas phyir / mthun dgnos gzhan gyi dngos dag las/ ldog pa la ni brten pa can // sarve bhāvāḥ svabhāvena svasvabhāvavyavasthiteḥ / svabhāvaparabhāvābhyāṃ yasmād vyāvṛttibhāginaḥ" (PV I.40. Mookerjee and Nagasaki 1964, 91).
10. "phyi rol nus pa rnam gcod la / reg pa med kyang sgra de ni // rnam par rtog pa'i gzugs brnyan ni / de yi mthar thug rnams dang 'brel / des na gzhan sel mthar thug phyir / mnyan pa gzhan sel byed par brjod // ldog pa bzhin du sgra dag las / shes la gzhan gyi gzugs brnyan snang / gang de'ang don gyi bdag nyid min / de 'khrul bag chags las byung yin // bāhyaśaktivyavacchedaniṣṭhābhāve 'pi tacchrutiḥ // vikalpapratibimbeṣu tanniṣṭheṣu nibadhyate / tato 'nyāpohaniṣṭhatvād uktānyāpohakṛc chrutiḥ // vyatirekīva yaj jñāne bhāty arthapratibimbakam / śabdāt tad api nārthātmā bhrāntiḥ sā vāsanodbhavā" (PV III, 163.cd–165).
11. For a discussion of the different types of apoha, see the paper by Katsura in this volume; see also Dunne 2004.
12. The contrast drawn here is between appearance (*snang ba*), which is real, and elimination (*sel ba*), which is not.
13. RTKN 65a 2–4: "'o na bum 'dzin rtog pa'i bzung rnam de snang ba yin nam sel ba yin zhe na / de la shes pa yin pa'i cha dang / phyi rol bum par sgro btags pa'i cha

gnyis las / snga ma ni snang ba yin te / bum 'dzin rtog pa rang gi ngo bo la rang rig mngon sum du song ba'i yul yin pa'i phyir / phyi ma sel ba yin te / sgro btags yin pa'i phyir."

14. "rtog pa ni brjod pa dang 'dres rung ba snang ba'i shes pa ste / abhilāpasaṃsargayogyapratibhāsā pratītiḥ kalpanā." See PVin I (ed. Vetter) 40.6–7, NBṬ ad NB I.5 (ed. Malvania), Tib. of PVin, P. 5710, *Ce*, 252b 4.
15. See TS 1214–1226, transl. Jhā 1986.
16. See, for example, PV III, 29–30.
17. See PV I, 41–51.
18. It is found in PV I, 80–81. See also Dreyfus 1997, 316–319.
19. For a discussion of Chaba's views, see Hugon 2008.
20. The locus classicus of this distinction is Sapaṇ's statement in RT 8b6: "[Be] wise by differentiating when explaining, and successful by mistaking [real things and concepts] when engaging in practical activities" ("'chad tshe rnam par phye bas mkhas / 'jug tshe gcig tu 'khrul bas thob").
21. RTGG, vol. X, 113.6.

10

Apoha, Feature-Placing, and Sensory Content

Jonardon Ganeri

BRIDGING THE GAP BETWEEN SENSATION AND THOUGHT

The thesis that sensory experience presents something other than ordinary physical objects has much to recommend it, but any philosopher who endorses it must eventually attempt to bridge the gap it opens between sensory and cognitive (or nonconceptual and conceptual) content.[1] The Buddhist notion of "exclusion" (*apoha*) serves here as a functional term for whatever additional explanatory resource is needed to bridge the gap, and (as Tillemans observes in his contribution to this volume) the explanation might proceed either by "working up" from sensory experience or else by "working down" from conceptual content. In this chapter, I will concentrate on the "bottom-up" strategy. I do so first by examining some recent work in the philosophy of perception to see whether and where a notion of "exclusion" might be operative. I will then try to correlate these observations with Buddhist versions of the "bottom-up" approach.

"Bridging the gap" means, more accurately, showing how perceptual experience can supply normative constraints on belief and judgment—how it is that belief and judgment are accountable to and constrained by experience (see Millar 1991). It need not imply a commitment to a stronger thesis—call it the Construction Thesis—that concepts can be "constructed" or "built" out of nonconceptual contents.[2] As we will see, it seems that the most that can be constructed out of nonconceptual contents is what might be labeled "protoconcepts," mental constructions that share some but all not of the attributes possessed by concepts.

The thesis I began with has long been associated with sense-data theory, but that theory's commitment to the claim that the immediate objects of perceptual experience are disembodied and immaterial sensory elements (elements that, moreover, actually have the properties experience ascribes to ordinary things) has proved more mystifying than illuminating. The recent work I will refer to claims instead that what sensory experiences present are spatially located instances of phenomenal qualities. The twitches and itches, chirps and cheeps, flickers and flashes that comprise the manifest image are represented in sensory experience as a twitch here, an itch there, a flash in the distance, a cheep just off to the left, and so on. This thesis has been most fully defended by Austen Clark in his *A Theory of Sentience* (2000), but his formulation, as he himself notes, is substantially in agreement with Christopher Peacocke's influential work on nonconceptual "scenario content" (Peacocke 1992a, 1992b, 2001). Clark's ideas are also the subject of a special issue of *Philosophical Psychology* (2004).

Let me note at the outset that Clark sees bridging the gap between sensation and thought as an important ambition of the theory:

> Working upwards from sensory capacities, and downwards from subject and predicate in logic and grammar, we meet at feature-placing. The representation of features in space is arguably the most sophisticated of sensory capacities. From my point of view, looking up from the muck and goo of the simplest sentience, such spatial representation is a complex and sophisticated achievement. Yet here it is sitting just below the least sophisticated of linguistic capacities, those sufficient for a feature-placing language. And if we can make that one tiny step, sensation and thought can at last commune. They can share contents. (Clark 2000, 151)

I will argue that there are three places where a notion of "exclusion" enters the theory of sentience. One is in the construction of what Clark calls a "quality space" and the associated relational account of qualities; a second has to do with the partition of phenomenal appearance into qualitative features and their apparent locations; the third is at the point of transition from feature-placing to full object perception and reference.

Quality Space: The Structure of Phenomenal Appearance

Stimuli—particular occasions of transducer irritation—cause sensations, and to those sensations are ascribed two sorts of properties: the properties they represent things as having and the properties in virtue of which they so represent things. A red sense impression is not itself red, but there is some property of it because of which it is a sensation presenting red. The term "qualia" blurs this distinction: qualia are thought to be properties represented by one's sensory experiences and yet also to be properties of the sensory experiences themselves. Following Galen Strawson (1989), we might label the two sorts of property "qualitative" and "phenomenal." Phenomenal properties are those that characterize how things appear: if the apple looks red, then red is a phenomenal property of the sensation. Qualitative properties are those properties of sensations in virtue of which things appear the way they do. Confusion arises because often our only way to refer to a sensation presenting red is as I have just done: the phenomenal properties of sensations are used to name their qualitative ones (Sellars therefore called the latter "red*").

In this chapter, I am interested primarily in the content of sensory experience, and I will follow Clark in using the terms "quality" and "qualia"—as in "the quality red"—to refer to the phenomenal properties of a sensation. The term "qualitative property" is reserved for content-explaining properties of the sensation itself.

Any catalogue of the contents of sensory experience in a given sense modality must include a description of the qualities that are represented, and it has long been known to psychologists of perception that there is no simple correlation between the stimulus that produces a sensory experience and the quality (that is, phenomenal property) that the sensory experience presents. In the case of vision, for example, there are several different combinations of wavelengths that will all produce a sensation of any particular hue.[3] Perceiving subjects have, moreover, discrimination thresholds below which distinct stimuli are perceived as presenting the same quality. And again, it is possible for the same stimulus to present different qualities to different perceivers.

But if there is no correspondence between stimuli and presented qualities, how should the qualities that those stimuli present be defined? Clark argues that the relations of matching, discriminability, and relative similarity are used to construct what he calls a "quality space," an "ordering of the qualities presented by a sensory modality" (2000, 1; see also his 1993). Parameterized, in the case of vision, by hue, brightness, and saturation, the quality space places two qualities near to each other if perceivers tend to find instances of them similar, further away to the degree their instances are perceived as different. Thus:

> To the question "What are the occupants of quality space?" the natural answer is "qualities," and what these considerations show is that while

> we might be able to *label* a point in quality space with some stimulus specification—some class of stimuli which happened to present that quality—we cannot *identify* the quality with that class. A finite class of occasions might help to pick out the quality, but it cannot be used to define the identity of that quality. So from the very beginning, the relations of matching, discriminability and relative similarity among classes of stimuli are used to order something other than those stimuli themselves. Discriminations among stimuli serve to order the qualities that the stimuli present. (Clark 2000, 6)[4]

This ordering is what generalizes across subjects: even if two people see the same light source as having a different color, they will concur in placing that color between others presented by different stimuli (2000, 14). Even if the sea looks "blue" to me and "red" to you, we might agree that it has the same hue as the sky and a different one from the sun.

A quality space is like a structural description of a skeleton, which maps the invariant relationships between the various bones, even though in actuality no two skeletons have bones of the same shape or size. The occupants of quality space are defined, not in terms of stimuli or classes of stimuli, but through a *structure description*:

> If we are to define a term [for a quality], we cannot mention any stimuli. We can mention only the structural properties that give the quale its place in the quality space. "Orange" cannot be defined as "the color of ripe oranges" or in any similar way, no matter how sophisticated. It can only be defined as something like "the color midway between red and yellow, and more similar to either than to turquoise". The terms "red," "yellow," and "turquoise" would all receive similar analyses. (Clark 2000: 16)[5]

Here, then is a first place where the notion of "exclusion" is employed in the description of phenomenal appearance. Phenomenal properties are defined by their relational position within an order rather than in terms of the stimuli that cause them to be presented. Paradigmatic stimulus instances cannot play any role in the definition of such terms, because of the problems of paradigm existence and contingency: if "orange" is defined to be the color of *this* ripe fruit, then this ripe fruit must exist for anything to look orange; and it is no longer a contingent matter that this fruit is indeed orange in color. But neither conclusion is true.

Finally, there has to be a corresponding ordering among the properties of the internal states themselves (the "qualitative properties") in virtue of which this ordering among phenomenal properties obtains: for vision, perhaps this lies in the functioning of so-called opponent processes (12–13). These provide the neurophysiological basis of the quality space.[6] In that sense, the ordering is an inherited part of our natural endowment. Finding one color more similar to a second than a third is not a matter of convention or social practice: it is built into the processes of vision.

The Feature-Placing Hypothesis

Phenomenal quality is, however, not the only dimension of variation in the content of sensory experience: "there is more to sentience than sensory qualities" (1). Even by the standards of a creature that had no higher cognitive abilities than sentience alone, one would be radically impoverished if there were nothing more to sentience than this:

> Consider a humble animal whose consciousness stops at sentience. One *imagines* its mental life to consist of nothing but a flux of sensory qualities. In a widely repeated and ancient image, its stream of mental processes is filled by variegated qualia, which over time pop up, bob along, combine, recombine, and ultimately sink back down into the muck. A mental life of pure sensation would be nothing but a stream, flux, a flow of such stuff. . . . But this picture, ancient and widely repeated as it is, radically underestimates the sophistication needed by even the simplest animal. An animal whose mental life is a pure flux of qualities . . . could not distinguish matte red next to glossy green from matte green next to glossy red . . . [The ability to do this] marks a significant threshold in the complexity of one's psychological organization. To pass it one needs somehow to focus the attribution of qualities, so that one can distinguish a scene containing a red square from one containing something that is red and something else that is square. (Clark 2000, 79)

The problem here referred to is a version of what Frank Jackson (1977) called the "Many Properties" problem, a problem about the way features get to be bound or integrated. A creature with only the capacities to discriminate red from green and matte from gloss could not distinguish between the two scenarios: both represent the same conjunction of qualities. Unless a sentient creature could solve this problem, it would also not be able to distinguish a single chirp from two simultaneous and qualitatively identical chirps coming from different locations. In fact, of course, the capacity to do this with falls well below the capacity to employ concepts and language.

It might seem that the problem could be solved by the introduction of spatial qualia ("local signs"), but Clark shows why that cannot work in a series of metaphors that will be familiar to those who work in Buddhist Studies:

> Merely adding more qualities will not help: they will be lost in the flux with all the others. In a similar way, the ancient image of a thing as a bundle of such qualities—concretions settling out of the flux—*smuggles in* more organization than one might suppose. If the qualities are sticks, we need some distinct principle by which to bundle the sticks together. A piece of string serves admirably, but notice that it serves a rather different function than that served by additional sticks. Tossing in more sticks leaves one just as disorganized as before; they will soon be bobbing down the stream, undifferentiated from all the rest. Even special sticks, labeled "spatial qualia," are soon lost. . . . We require some distinct principle by which to create bundles. Lacking string, the simplest way to count our things—our piles or bundles—is by location. Here is one bundle, and there is another . . . I suggest that the threshold of the Many Properties problem is the point at which we add to the flux of sensory qualities a distinct capacity for sensory reference. (Clark 2000, 79)

The argument here is that all but the simplest sentient creatures (perhaps a creature with only olfactory sensory abilities would be simpler) do more than merely *enjoy* qualia. They must also have a distinct capacity to *place* those qualia: hear a sound as coming from a certain direction and at a certain distance. Places become objects of what Treisman and Gelade call "focal attention": "Any features which are present in the same central fixation of attention are combined to form a single object. Thus focal attention provides the "glue" which integrates the initially separable features into unitary objects" (1980, 98). The conclusion drawn is that any schema describing the contents of sensory experience must be partitioned into two dimensions of variation: variation in location and variation in the qualities at those locations (Clark 2000, 60). It is important to this proposal that places are themselves able to function as loci of focal attention, and not only as the objects that may or may not be located at those places. Referring to the experimental results of Michael Posner on the spatial cueing of attention, Clark (2004a, 457) argues that "feature-placing can indicate or pick out a place by spatial coordinates derived from the operation of the sensory system itself; it does not need some object to glom onto—some object to which to attach its referential force."

The "feature-placing" hypothesis asserts that this pair of capacities is strongly akin to the capacities ascribed by Peter Strawson to the user of a feature-placing language. A feature-placing language does not have resources for reference or predication; it exists below the level of the subject-predicate distinction. What it does have are resources to identify places and to locate features: thus the sentences in such a language are of the form "red here" or "raining now." Speakers of such a language can judge "cow here" and "cow here again," but they cannot distinguish where this is the case because of two cows and where the case involves the same cow a second time. That is, they do not yet have the capacity to reidentify enduring particulars, the capacity fundamental to reference and so to predication. According to the feature-placing hypothesis, sensory experience analogously requires the capacity to identify phenomenal properties in quality space and also the capacity to place those qualities at locations on or near the body of the sentient creature (locations with respect to which the sentient creature can stand in an ongoing information link).[7] This latter capacity is what Clark slightly misleadingly calls the capacity for sensory reference. Singling out places is rather a protoreferential analogue to reference, and in placing features there is a protopredicative analogue to predication, in a feature-placing language.

I want to argue that in the idea of "feature-placing" we have a second point at which the notion of "exclusion" is involved in the description of sensory experience. Qualities belong at the level of generality, places at the level of particulars. One reason for this is that to a single place can be attributed many qualities, but a given quality has at most just one place (this is, indeed, a further argument against "spatial qualities": they are not qualities). But Strawson points to another, yet more telling asymmetry. According to Strawson, the asymmetry between particulars and general characteristics has at its source the fact that, while both supply principles for the "collection" of other particulars and general characteristics, the nature of the respective principles they supply is different. As Strawson puts it, there is

> a certain asymmetry which particulars and general characteristics of particulars have relative to each other, in respect, as I put it, of the possession of incompatibility ranges and involvement ranges. General characters typically have such ranges in relation to particulars, particulars cannot have them in relation to general characters. For every general character there is another general character such that no particular can exemplify them both at once; but for no particular is there another par-

ticular such that there is no general character they can both exemplify. Again, for many a general character there is another general character such that any particular which exemplifies the first must exemplify the second or vice versa; but there is no pair of particulars so related that every general character the first exemplifies must be exemplified by the second or vice versa. (1974, 126)

The "incompatibility range" of a general characteristic is the group of other general characteristics which cannot also be exemplified by a particular which exemplifies it. Thus, if green is in the incompatibility range of red, then no place can be both red and also green. On the other hand, particulars do not have such incompatibility ranges: if a particular exemplifies a given general characteristic, that does not tell against any other particular doing so as well. Similarly with "involvement" ranges: scarlet is in the involvement range of red, because any particular that exemplifies scarlet must also exemplify red. Again, there is no analogue for particulars. These asymmetries are the reason we must distinguish the linguistic devices of identifying reference and predication, "such linguistic and other devices as will enable us both to classify or describe in general terms and to indicate to what particular cases our classifications or descriptions are being applied" (1966, 47).[8]

It is in the asymmetry between qualities and their apparent spatial locations that another application of the notion of "exclusion" is available. The idea behind the thought that the former are general and the latter particular, and that the former are "located in" the later, is that the former alone have incompatibility ranges: that is, for each quality, there are other qualities whose exemplification at a place is excluded by its own.[9] If the dimensions of variation in phenomenal appearance were not so partitioned into two factors, one protoreferential, the other bearing the hallmarks of generality, then no amount of logical manipulation of the qualia could get us above the level of mere flux of unrestituted sensory elements.

Quasi Objects and Protoconcepts

Feature-placing by itself isn't enough for the introduction of objects. For that, we need sortal concepts, concepts that provide identity conditions for the things that fall under them. In a recent paper (Clark 2004a), Clark argues that feature-placing *is* enough to locate, count, and keep track of things, and those abilities permit the introduction of quasi-objectual entities that lack the robust identity conditions of objective particulars but can easily be mistaken as such:

> Feature-placing can give us the wherewithal to locate, track and count some putative entities, such as the waves on the ocean as they come crashing onto the beach. But waves are still quite a ways away from being individuals or "objective particulars." Consider: Two waves can fuse; one wave can split into two; "a" wave has entirely different parts at different times; and if all those parts are arranged in just the same way at a different time, it is a purely verbal question whether we have the same wave again, or a different one that is qualitatively identical. Waves are often mentioned by Buddhist thinkers who want to point to something that appears to be an object but is not. In this I think they are right. (Clark 2004a, 465)

These entities appear to be objects but are not. The reason is that it is a constitutive part of the idea of a physical object that it can change its location. With feature-placing capabilities alone, including an ability to bind features together in "focal attention," sensory content can be made to reach up as far as the cognition of quasi-objectual entities that can be counted, located, and tracked, but not reidentified as the same again as they undergo movement in the visual field. And it is a mistake, albeit one that is easy enough to make, to regard those entities as objective particulars. (According to Clark, the missing link is shape perception.) Let us call the content of such cognitions "protoconcepts." Protoconcepts are constructible for nonconceptual sensory content, but there remains a gap between them and fully fledged concepts. How is that gap to be bridged?

Object Perception and Reference

A creature can solve the "Many Properties" problem if it has the capacity to discriminate between matte red next to glossy green and matte green next to glossy red. It might still not have the ability to perceive enduring physical objects. If what we have said so far is correct, the further step is the same as the move up from a feature-placing language to a language of individual reference and predication. This step involves a new ability, the capacity to discriminate between red thing here and red thing there, on the one hand, and red thing here and same red thing there, on the other. For that, one needs sortal concepts and the associated ability to discriminate between distinct items of the same sort. Here again, the notion of "exclusion" has a role, for the relevant ability might be redescribed as an ability to distinguish between what an object is and what is excluded from or other than the object. The idea of what is "other than *x*" and the idea of what is "identical to *x*" are clearly complementary to one another. Notice that this concept of "exclusion" is different from either of the preceding two, for it obtains between individual objects and not occupants of quality space. If what we mean by "conceptual thought" is the employment of the full intentional machinery of reference and predication, then possession of this further ability is the criterion that distinguishes thinkers from the merely sentient.

I would like to suggest, in fact, that at this level the notion of "exclusion" has an even more important role: it can function as a semantic device for converting a sortal into a feature term. "That which is nonwater-pot" is a term that spreads its reference over many things of various sorts; it possesses no criterion of identity of its own. So its negation cannot have one either. In that sense, it is like "gold" or "red" in marking only the presence of a feature. A double negation transforms the subject-predicate sentence "this is a pot" into the feature-placing "no absence of pot here." The capacities required of someone who understands this second sentence fall short of an ability to reidentify particulars. This semantic device serves to excise the sortality from a sortal concept: what remains is a feature-placing construct, a protoconcept.

DHARMAKĪRTIAN BUDDHIST THEORY

Thanks to a considerable body of recent work, we now have a quite detailed understanding of Dharmakīrti's "bottom-up" approach. If I might attempt an extremely compacted summary, Dharmakīrti's view is that particulars, which arguably lack extension either in space or in time, individually or jointly cause sensory impressions, each of which carries an "aspect" or "image" (*ākāra*) that is particular to itself, and which in turn and as a result of impressed mental tendencies (*vāsanā*) cause another cognition (*ekapratyavamarśajñāna*), this time carrying an "aspect" or "image" that is general, in the sense that specific differences have been excluded and "overlayed with nondifference," an "image" that is then imputed to the original particular itself as a result of an "unconscious error" that overlooks the lack of identity between the general image and the particular (see Dreyfus 1997; Dunne 2004, 84–144, 159–161; Katsura 1993; Tillemans 1995, 1999).

Dharmakīrti endorses the thesis with which I began this chapter: perception has a content; yet this content is something other than the perception itself, but is not an ordinary physical object or part of one. The content consists of the "particular" (*svalakṣaṇa*) and the "aspect" (*ākāra*). The content is "nonconceptual" (*nirvikalpaka*) in the sense that it is not describable in terms of concepts and language available to the perceiver themselves (PV III.1–2). Of all the recent work in the philosophy of perception, Dharmakīrti's theory has most affinity with that of Clark and Peacocke. Although Dharmakīrti does not himself speak of the spatiality of perceptual content, his twelfth-century expositor Mokṣākaragupta does so, saying that the "particular" is manifested as "determined in space, time, and aspect" (*deśa-kāla-ākāra-niyataḥ puraḥ prakāśamānaḥ*; TBh_{ST} 21.10–11). The same definition is repeated by the Kashmiri Śaiva philosopher of language Abhinavagupta in the course of a discussion of the Buddhist conception (ĪPV I 86, 4–8 on ĪPK 1.2.1–2). Dharmottara too refers to time (*kāla*), space (*deśa*), and aspect (*ākāra*), according to recent work from the Tibetan by H. Krasser. On that conception, sensory experience presents particulars with uniquely individuating spatial, temporal, and aspectual characteristics. The manifest image presents aspects at spatiotemporal places (those places being, it seems, extensionless).

THE TWO "ASPECTS"

An initial puzzle here concerns the particularity of the first "aspect," the aspect that is associated with particular sensory impressions. I take it that this refers to something more than merely the fact that no two sense impressions are exactly alike (something that is, in any case, only contingently true). On the other hand, these aspects cannot be qualities, for qualities are defined by relations of discriminability. If I have sense impressions from two objects and cannot discriminate them in hue, then ipso facto the two sense impressions present the same color. So qualities are not particular to sense impressions; indeed, as we have seen, sensory qualities belong at the level of the general. We might, however, recall our earlier distinction between phenomenal properties and qualitative properties and identify the *ākāra* with those specific properties of the particular sensation itself (qua mental event) in virtue of which it represents what it does.[10] (NB: Dignāga's *svākāra* and perhaps Dharmakīrti's *grāhakākāra* pertain to the higher-order question: what is it like for the experiencer to be experiencing something red? See Ganeri 1999a; cf. Clark 2004b, 564–565.)

The role of impressed mental tendencies in Dharmakīrti's account might be seen to reflect the fact that quality spaces are the result of relations of matching, discriminability, and relative similarity, and that those relations are grounded in the neurophysiology of perceivers. That one sees the color of this and the color of that as the same or as more similar to each other than to a third is a fact about the hardware of our visual sensory systems, about the wavelengths and intensities required to get the retinal transducers to fire, and so on.[11]

The second *ākāra* in Dharmakīrti's theory—an "image" that is qualified by an exclusion and "overlayed with nondifference"—is most closely akin to the "protoconcept" referred to earlier, something that represents a quasi-objectual entity but lacks full identity conditions (Buddhists use the example of the apparently circular whirling firebrand).[12] John Dunne (2004, 158–161) describes this cognition as a "judgment," but perhaps "protojudgment" would be a better term:

> To solve these problems, Dharmakīrti appeals to the notion of a "judgment of sameness" (*ekapratyavamarśajñāna*). On this argument, the claim that all the entities in question have the same effect rests ultimately on the fact that they eventually produce a second-order cognition—a judgment—in which the individuals in question are identified as the same type of entity. All the entities we call "blue," for example, produce perceptual images that, when the proper conditions are in place, will lead to the judgment, "This is blue". . . . [I]t is "by their nature" (*svabhāvena, prakṛtyā*) all entities that we indirectly call a "water-jug" have the same effects. For the apoha-theory, perhaps the most important "same effect" that each of the entities in question produces is the aforementioned "judgment of sameness" (*ekapratyavamarśajñāna*).

This marks the transition from sensation to a primitive sort of *seeing as* or typed perception, a sort that does not yet involve or presuppose any linguistic abilities or the related ability to combine, apply, and work with concepts (the ability to combine concepts into larger conceptual complexes being partly definitive of what a concept is and what it is to possess one). My use of the label "protoconcept" for the more rudimentary perceptual capacity under discussion is closely related to Michael Dummett's notion of a "protothought":

> Perhaps the least difficult case for the characterization of protothoughts, at least as we engage in them, is the purely spatial one. A car driver or a

> canoeist may have rapidly to estimate the speed and direction of oncoming cars or boats and their probable trajectory, consider what avoiding action to take, and so on: it is natural to say that he is engaged in highly concentrated thought. But the vehicle of such thoughts is certainly not language: it should be said, I think, to consist in visual imagination superimposed on the visually perceived scene. It is not just that these thoughts are not in fact framed in words: it is that they do not have the structure of verbally expressed thoughts. But they deserve the name "protothought" because, while it would be ponderous to speak of truth or falsity in application to them, they are intrinsically connected with the possibility of their being mistaken: *judgment*, in a nontechnical sense, is just what the driver and the canoeist need to exercise. (Dummett 1993a, 122)

Dummett's point is that the pioneers of analytical philosophy such as Frege went wrong in claiming that all perceptual recognition and *seeing as* has to be understood as concept involving; this left them without the resources to explain the role of prelinguistic perceptual capabilities, not to mention the nonlinguistic perceptual abilities of animals. Here, it seems to me, Dummett is joining sides with the Buddhists against their Grammarian and Naiyāyika opponents.

Error: Sensory and "Unconscious"

Dharmakīrti's acknowledgement of the phenomenon of sensory error indicates that he admits that sensation is intentional. Possibilities for misplacing a quality at a location open up if sensory content has a feature-placing structure. These are, as Dharmakīrti acknowledges, genuinely *perceptual* errors. They are "nonconceptual errors" (Dunne 2004: 88, 130; cf. Clark 2000: 191–197). There is also, as the passage from Dummett suggests, the possibility of error in those feature-placing complexes we are calling protoconcepts: I might see something as a circle of fire when in fact it is not.

What of the "unconscious error" that overlooks the fact that these general "images" are perceptual products and not real properties of external particulars (cf. Dunne 2004, 139–144; Tillemans 1995)? Well, that is exactly what Clark's relational account says too:

> A consequence of this account is that qualitative character is a relational affair. Qualitative properties seem to be intrinsic properties, but they are not. . . . Perhaps it is part of our ordinary conceptual framework that qualia are intrinsic. And perhaps that part is a portion of our folk inheri-

> tance that we must renounce. Empirical inquiry suggests that the facts of qualitative character have at root a relational form. (Clark 2000, 19)

Again,

> I take the experiments of Wallach and Gilchrist (and also now of Whittle) to be quite revealing. They strongly suggest that our intuitions that chromatic qualities are intrinsic qualities are just wrong. (Clark 2004b, 566)

In Wallach's experiment, a projector is used to project a circle of light onto a screen, while a second projects an encircling ring. Wallach found that the apparent brightness of the central circle can be altered without adjusting the first projector, but simply by changing the brightness of the annulus: the brighter it is, the darker the center appears. What seems to be an intrinsic property, the brightness of the central disk, is in fact a relation between two levels of luminance. It is indeed an "unconscious error," a false bit of our folk inheritance, that leads us to think of the visual qualities (hue, saturation, brightness) as intrinsic monadic properties of the things they purport to qualify. That we are inclined to view the deliverances of our senses as ascribing intrinsic properties to external things rather than as involving properties whose relational character is something for which the structure our visual apparatus is responsible is, quite possibly, an error we are naturally selected to make.

The Causal Efficacy of Particulars and Only Particulars

A last bit of Dharmakīrti's theory that seems amenable to this sort of interpretation is his thesis that only the particulars are causally efficacious, while the constructed generalities lack causal efficacy. Here again, the relational account seems to concur. If that account is correct, then it is a mistake to try to define qualities in terms of their causal functional roles, whether "long-arm" roles that "reach out into the world of things" or "short-arm" roles that are "purely internal" (Block 1997). The reason is that the basic relations that define the structure of quality space are not causal ones:

> Rather than accept a theory that requires us to specify some particular "causal niche" that is always and only occupied by sensations of orange, we should abandon the theory. There is no "causal niche" filled characteristically and uniquely by sensations of orange. But a relational structure can be built up using relations other than "causes" . . . if our goal

> is to describe qualitative character, the root relations will be those of qualitative similarity. (Clark 2000, 18)

What defines orange is its falling between yellow and red and so on. Likewise, the "constructed universals" of Dharmakīrti's system are defined by formal relations and not causal ones: unlike the particulars, they do not have causal roles. Once again, paradigmatic particulars have no role in the definition of quality terms. (It is, however, also possible to argue that Dharmakīrti's reference to "sameness of effect" points rather in the direction of "short-arm" causal roles.)

Positivist and Negativist Interpretations of Apoha

Dharmakīrti's introduction of the mental "image" is a response to the objection that a purely exclusionary ("negative") account of sensory content does not get the phenomenology right. Thus Dunne (2004, 137–138):

> Dharmakīrti and his interlocutors all maintain that affirmative conceptual cognitions (i.e., those that are not negations) present some positive content as an object in cognition. For Dharmakīrti, this positive content would be an appearance or image in the mind. If, however, the universal is a mere negation, then a conceptual cognition—i.e., one that has a universal as its object—would involve no positive content at all, for how can a negation be presented in a positive form? . . . On Śākyabuddhi's interpretation, Dharmakīrti accounts for the positive content of such cognitions by pointing out that the cognitive image (*ākāra*) "that which excludes other," is indeed part of what constitutes a universal formed through exclusion.

Let us compare this objection to the apoha theory with objections that have recently been pressed against Clark's attempt to give a relational account of qualia:

> Sensory qualities have an "absolute value" as well as a relative one; a determinate identity over and above their set of similarity relations . . . it isn't pure difference alone that is presented by border regions. We don't just say *this is different from that*, but we describe the difference in terms of a quality—our visual system says "redder here than there." But what's *being redder*? (Levine 2004, 546–548).

> Since the same difference can occur anywhere in the color map, this leaves it mysterious why a certain pattern should look brown-and-orange rather than turquoise-and-olive (assuming that the differences within these pairs are the same). (Matthen 2004, 517).

Clark's response is that a relational account will associate each quality term with a unique definite description as long as the relational structure is asymmetric and so noninvertible (Clark 2004b, 569). But that does not seem to speak to the objection, the force of which is that sensory experience presents qualities as having a "positive value," so that a purely relational account is not adequate to the phenomenology of appearance. (The problem of distinguishing between the appearance of a square and a diamond is perhaps related; Peacocke 1992a.) Contemporary work in the philosophy of perception is divided over whether the best way to proceed is broadly Fregean—that is, introducing *ways* of perceiving or *manners of presentation* of sensory contents—or broadly Russellian—that is, introducing further, more finely grained, sensory properties. Post-Dharmakīrti Buddhist discussion over the status of the "image" (*ākāra*) might be seen as exploring similar and other possibilities.[13]

Bottom-Up vs. Top-Down Approaches to Apoha

One striking aspect of Dharmakīrti's discussion is the complete absence of the distinction between quality terms like "red" and sortal or count terms like "water jug." But if what I have been arguing is right, then this ought not surprise us. For the concept of exclusion is seen to be operative in describing the content of phenomenal appearance, that is, the first and second identifications I mentioned earlier. The transition from a feature-placing language to a language of reference and predication is thus not a primary concern for Dharmakīrti's apoha theory. One might think that one could proceed to a definition of a "sortal space," somewhat akin to an Aristotelian ordering of universals, and in doing so one would appeal to much the same resources that were involved in the definition of a quality space, but responsive to facts of quantitive rather than qualitative character. That, indeed, might have been Dignāga's use of the apoha doctrine.[14] However, it seems that natural kind terms are much less like color terms and much more like proper names; indeed, that they rely on paradigms and causal definition in precisely the way the space of qualities allegedly does not. I have suggested instead that apoha functions from the top-down as a semantic device for converting sortals into feature terms. In fact, if the distinctive aspect of natural kind terms is that the "conventions" they are introduced by are paradigm-invoking baptismal events, then what this device does is to excise the semantic dependence of the term on some one particular.

Is Dharmakīrti an Idealist?

A striking feature of Dharmakīrti's account is his decision, at certain key places, to surrender what John Dunne calls his "External Realism" (and Dreyfus, his *Sautrāntika*) in favor of an "Epistemic Idealism" (or *Yogācāra*). That is, the presented contents of experience are no longer thought to have external correlates. It is a retreat from the view that we have ongoing information links with our external surroundings and so from the idea of sensory (proto)reference. I am not sure, but it would seem to me that this involves a retreat to the doctrine of spatial qualia in place of thinking of experience as genuinely intentional; or perhaps it is a move away from identifying apparent locations with real ones. That is, however, extremely conjectural. On a related note, also conjectural, is the identification of the particulars in Dharmakīrti's system. Even if they lack spatial extension (there is not yet scholarly agreement about this: Keyt 1980; Dreyfus 1997, 85–86; Dunne 2004, 98–113), it would not follow that they are not spatial at all. For perhaps a *svalakṣaṇa* is a (minimally discriminable) spatiotemporal point characterized by phenomenal properties; at least, this might be so when Dharmakīrti is wearing his External Realist hat. Perceiving spatial regions, boundaries, and shapes then falls in with the perception of physical objects as requiring something more than sentience alone.

CONCLUSION

To bridge the gap between sensation and thought, we must work up from the "muck and goo" of basic sentience and also down from the sophisticated world of identifying reference and predication. I have tried to show that there are various ways in which a notion of exclusion might have a role to play in these two movements. Going up, by providing a relational account of the sensory features and an explanation of that in which the generality of features consists. Moving down, by converting sortals into feature terms. Feature-placing is then the middle ground where sensation and thought meet in the formation of protoconcepts. And if, as it seems,

Dharmakīrti's theory of perception can attribute to sensory content greater richness than a sense-data analysis does, then the apoha theory helps us to see how it is possible for sense experience normatively to constrain the content of our beliefs.

Notes

1. "The central idea behind the theory of nonconceptual mental content is that some mental states can represent the world even though the bearer of those mental states does not possess the concepts required to specify their content" (Bermúdez 2003). It seems that the Buddhists—but not the (Navya-)Naiyāyikas—endorse an Autonomy Thesis about nonconceptual (*nirvikalpaka*) content. The Autonomy Thesis asserts that a creature "can be in states of nonconceptual content despite not possessing any concepts at all" (ibid.; cf. Peacocke 1994).
2. Again, it seems that Buddhists were more willing to endorse, and Navya-Naiyāyikas firm to reject, the Construction Thesis. See Matilal 1986, 309–354.
3. Such combinations are known as "metamers." Perhaps one can see them as a consequence of that fact that sensory transducers are "broadly tuned": a transducer designed to respond optimally to photons whose wavelength is 430 nanometers will also respond less strongly to light of 480 nanometers. They conflate wavelength and intensity.
4. Concerning the qualities that are here described as the occupants of quality space Clark later says, "these are, precisely, phenomenal properties, or properties of appearance" (2000, 7).
5. In the Appendix, Clark shows how this is done by defining a Ramsey sentence for the whole structure and assigning quality terms their Ramsey correlates
6. John Dunne (in this volume) refers to J. Edelman's theory of neuronal group selection as fulfilling a similar role.
7. Compare Peacocke's idea of "scenario content," well summarized by Bermúdez: "Peacocke suggests that a given perceptual content should be specified in terms of the ways of filling out the space around the perceiver that are consistent with the content's being correct. For each minimally discriminable point within the perceiver's perceptual field (where these are identified relative to an origin and axes centered in the perceiver's body) we need to start by specifying whether it is occupied by a surface and, if so, what the orientation, solidity, hue, brightness and saturation of that surface are. This specification gives us the way in which the perceiver represents the environment. The content of that representation is given by all the ways of filling out the space around the perceiver in which the minimally discriminable points have the appropriate values. The representation is correct just if the space around the perceiver is occupied in one of those ways." (Bermúdez 2003)
8. Strawson applies the Aristotelian criterion to explain the asymmetry between subject and predicate, but not the asymmetry between places and features; indeed, he regards it as distinctive of the former. I do not see, however, why it should not apply at the lower level too. Bob Hale has argued that Strawson's

criterion is unable to ground the subject-predicate distinction or give a criterion for singular terms; see Hale 1979, 1996. That is perhaps a reason to think that it should in fact be invoked *only* at the feature-placing level, for only at this level can we be sure that there are no "expressions which serve for the expression of generality but are . . . grammatically congruent with singular terms" (Hale 1996, 456). There is nothing analogous to "something" in a feature-placing language.

9. It seems possible to define involvement ranges in terms of incompatibility ranges—see Ganeri (2001, 104–114).
10. On this aspect, Dunne comments, "when a perception occurs, a sensory object acts as a contributing cause for the production of the cognition in which an 'image' (*ākāra*) of that sensory object appears." (2004, 84). And, "the singularity of the perceptual image is not congruent to (i.e., has not isomorphic correspondence with) the singularity of its physical causes. Instead, the singularity of the image *correlates with* a singularity of causal function: multiple external causes are producing a single effect, the image" (2004, 108).
11. This would be consistent with Siderits's suggestion (Siderits 2006, 98) that Buddhist nominalism relies on the assumption of a "rudimentary similarity space" that might be a result of processes of natural selection, but not with the significance he assigns to conventions and social practices; at most, there may be some role for nurture or acculturation. Pascale Hugon (in this volume) points out that the processes of selection are done "*before* the setting of the convention." Amita Chatterjee's proposal (in this volume) that the relevant mechanism can be understood in terms of Gibsonian "affordances" is another way to make the point.
12. George Dreyfus (in this volume) explores the role of protoconceptualization in post-Dharmakīrtian Tibetan theory.
13. See Parimal Patil's work on Jñānaśrīmitra and Ratnakīrti: Patil 2003 and this volume.
14. See Katsura 1979; Ganeri 2001, 104–114. Dignāga's "top-down" apoha theory is examined in detail by Prabal K. Sen and Ole Pind (in this volume).

11

Funes and Categorization in an Abstraction-Free World

Amita Chatterjee

A theory of meaning can contribute to an account of human cognition if and only if it is also a theory of understanding. Apoha semantics, even in its earliest formulation, was something more than a mere theory of reference. Dignāga in his *Pramāṇasamuccaya* definitely attempted to teach us how to pick out and talk about individuals that inhabit our world, but at the same time he showed us how to understand the meaning of predicative expressions without resorting to realism about universals. Dignāga's theory therefore qualified as a theory of understanding too. I have a different reason for being enthusiastic about apoha semantics. My recent explorations in cognitive science have convinced me of the potential of the apoha theory as a theory of human cognition. Cognitive scientists have always considered cognition a bridge between perception and action. However, while the first-wave cognitive scientists emphasized the representational and computational nature of cognition, the third-wave cognitive scientists favor a noncomputational coupling between perception and action of a dynamic embodied system. A debate is raging between the moderates and the radicals regarding the dispensability of mental representations in explanation of higher-order cognition. A clinching argument in favor of any of the contending parties has yet to be found. It is my hunch that Dharmakīrti's account of apoha construed as a theory of mental representation will help us to imagine the contours of a noncomputational theory of mental representation, which will add to the plausibility of the moderate position. My goal in this paper is therefore to develop

Dharmakīrti's account of apoha as a theory of mental representation. My goal being what it is, I shall not enter into certain well-known debates associated with the apoha theory, namely, whether the double-negation involving apoha semantics is compositional in nature or not, whether admitting a quasi universal or apoha makes a Yogācāra Buddhist a nominalist or a conceptualist, etc. I shall only try to show how contemporary cognitive science and the Buddhist theory of apoha can be mutually supportive.

Let me begin my exercise by telling you the story of Funes the Memorious by Borges (1964). It's a fascinating story with significant implications.

> Funes fell off his horse, and from then onward he could no longer forget anything. He had an infinite rote memory. Every successive instant of his life experience was stored forever. His memory was so good that he gave proper names or descriptions to all the numbers. Each was a unique individual for him. But as a consequence, he could not do arithmetic: he could not even grasp the concepts of counting and number. . . . The same puzzlement accompanied his everyday perception. He could not understand why people with ordinary, frail memories insisted on calling a particular dog, at a particular moment, in a particular place, in a particular position by the same name that they call it at another moment, in a different time, place and position. For Funes, every instant is infinitely unique and different instants were incomparable and incommensurable. (Harnad 2005, 30–31)

So Funes wanted a language in which he would be able to name not only every individual object but also every distinct experience.

The story of Funes must sound familiar to you. Just like Funes's world after his accident, the Buddhist world is inhabited by causally potent (*arthakriyākārī*) unique individuals (*asadṛśasvalakṣaṇas*), which are ineffable (*śabdasya aviṣayaḥ*) but can be directly apprehended by a sensory-motor system. This worldview always appeared fantastic to their realist opponents, maybe because a person like Funes whose world is too fine-grained cannot possibly survive. For in his abstraction-free existence, Funes would be unable to count, to speak, to communicate, or to understand the physical world. He would not be able to correlate different features of the external world, and therefore each thing he would have to learn anew. Consequently, he would fail to distinguish between his friends and foes, food and poison, shelter and trap. He would at best be a tacit sensory-motor system, blown here and there by environmental vagaries, like a dried leaf buffeted by rain and wind.

As Siderits (2005) rightly points out, the Buddhist nominalists, like ostriches, were knowledgeable in survival techniques. They could distinguish between things-as-they-are and things-as-they-appear-to-us, thus creating, as Tillemans (in this volume) reminds us, a scheme-content dichotomy. But otherwise, how would they survive in the splitter's world?[1] So they grant that living beings are endowed with imagination (*kalpanā*), which enables them to have general thought contents, though the real world comprises only unique individuals. Thus, they avoided all the misfortunes that visited Funes. They had language, the number system, means of reasoning, and concepts of objects to navigate around the world. But they never lost sight of the fact that the natural kinds and artifacts that surround them in their everyday world are what they are because organisms as sensory-motor systems have learned to lump the unique particulars in a certain way. Universals, according to them, are nothing but *façons de parler* (*vārttāmātra*).[2]

If our given world is one of unique particulars, then how do we cognize it? By virtue of possessing our sensory-motor system, we are able to sense the unique individuals. Cognition, however, necessarily presupposes categorization. This is a problem that baffled the ancient Buddhists and the contemporary cognitive scientists alike. The problem assumes a grave proportion because at one level we are unconscious of the principle by which we organize our world of unique individuals, yet we must explain the exact mechanism of categorization. While the Buddhists relied mainly on speculation supported by philosophical arguments in this task, modern cognitive scientists are trying to discover the same principle(s) of organization with support from brain scientists. The latter group also resorts to philosophical speculation where evidence from brain sciences is lacking. The Buddhists developed their apoha theory to solve many problems including the problem of categorization at a very basic level. So let me quickly run through a resumé of an account of categorization from the perspective of cognitive science,[3] interspersing it with some Buddhist insights wherever relevant.

Organisms with a sensory-motor system are receiving innumerable signals from the external world every moment of their waking existence. These signals or information atoms come from some distal "things" and our sensory surfaces receive their features. Though these received features change in each moment, our sensory systems can detect and extract some invariant representations of color, shape, size, feel, sound, taste, etc. These representations are called invariants because though the size, shape, etc., of the shadows of individuals on the retina change as we move in relation to them or as these move in relation to us, our visual system can detect some constancy in them. For example, a unique feature *blue* interacts with my visual sense organ momentarily and produces an ever-changing blue awareness (*pratibhāsa*). Immediately after my visual sense organ receives this blue awareness, an invariant representation of blue (*nīla-ākāra*) is produced in the second moment, which Dharmakīrti calls *manovijñāna*.[4] In the second moment too, I am purely in the receiving mode and have not contributed anything to this representation. Thus, I go on accumulating sensory invariants or representations. These representations also leave their traces behind. The unique features of the world by virtue of their causal power affect me differently. So I start reacting differently to different stimulations. However, if I am not to live from moment to moment, my adaptive sensory-motor system must have "systematic differential interactions" with the world. That is, the same kind of input must result in the same output and a different output must be the outcome of the different kind of input. The moment I look for "kinds," I enter the domain of categorization. We must realize that this level of categorization is an implicit one of which we cannot be introspectively aware.

How does our sensory-motor system learn the tacit process of categorization? It is easier to answer what we learn to do than how we learn to categorize. Philosophers put forward a number of hypotheses and cognitive scientists are still looking for the exact mechanism, algorithmic or heuristic, underlying our process of categorization. But all of them agree on the point that while categorizing we selectively abstract certain features and overlook differences among others. Which features does our sensory-motor system choose? We can't say a system begins by detecting the similarity of features in representations because, all representations being of unique features, it is unlikely that the system will find some point of intersection between representations of any of the distal features. Fodor therefore suggested that all our categories are innate. We cannot learn to acquire categories. But declaring the process of categorization innate makes the process an unexplainable mystery. Instead, cognitive scientists try to find the basis of the human skill of categorization.

Our sensory-motor system usually does not give equal weight to all features. In the course of evolution, it learns to consider some features more salient than others. In many cases the clue comes from the context or the environment, which the Buddhists can easily identify with *saṃskāras*, "dispositions of the system." To explain in the contemporary parlance, these dispositions may be due to natural selection. When, within a large number of unique representations, the system pays selective attention to a single representation, say *x*, it proceeds by contrasting *x* with the rest and achieves a dichotomy of *x* and non-*x*. This is a simple difference, which has objective grounding in different representations of real individuals. Here *x* stands out prominently against the background of non-*x*. So this figure-ground differentiation is the first stage of abstraction, or the first interpretive step taken by the system. Within the putative class of non-*x*, it might make further divisions, say, of *y* and non-*y*, and within non-*y*, of *z* and non-*z*, and so on. But how does the system generate a class *y* within the cluster non-*x*? This is probably the most difficult hurdle to cross, since ex hypothesi each member of non-*x* is different from others. However, the system because of its disposition fails to distinguish among different stimuli, finds them compatible with one another, and starts reacting to them in the same manner. It is like a frog's reaction of swallowing moving black dots, be they flies or stone pellets. The process of natural selection endows the frog with this power of overgeneralization for increasing its chance of survival. Thus the system of the frog learns to overlook the distinction between flies and moving stone pellets and to grasp only "the black moving dot forms" as their food items. A human sensory-motor system also learns to conflate some differences and superimpose a single form on distinct representations. That is, while the comparison process among distinct representations is on, the system finds that some representations can be conjoined with certain other representations. For instance, the frog system finds that the black form, the dot form, and the moving form are not incompatible with one another and can be combined to the benefit of the system, and therefore the system retains this combination in its memory. Thus, the system may go on combining different invariants and configure them by adding some internal coding. In this way, out of different representations of distal features, a sensory-motor system constructs some "objects." These generate some affordances (Gibsonian) for the system, which enable it to react to its environment appropriately. By this time, as is evident from the earlier example, depending upon the context, the system has learned to suppress or overlook some minor differences and consider some clusters of representation similar to some other cluster *C*. Similarity judgments (*ekapratyayavamarśa*), in most cases, depend on the similarity of experience/reactions evoked by two or more clusters of invariants and not on the real similarity between them. At this point, we can smoothly merge the narrative of cognitive science with the Buddhist narrative. It is possible to hold without any inconsistency that the clusters of representations that affect us like a previously chosen cluster *C* share the negative character of being not non-*C*, since they are functionally different from non-*C*. This, however, does not imply that $C_1, C_2 \ldots C_n$ share some positive class character of *C*. Yet we lump all *C*s together for our practical convenience, subject to our cognitive constraints, and this is what Dignāga and Dharmakīrti meant by the process of ascribing a putative universal (*jāti-yojanā*). Thus we can get the following equation: *jāti* = *apoha* = *atadvyāvṛtti*. (Universal = exclusion = distinctness from what is not that.) Understanding this process, of course, requires grasping negations at two levels: the first is the simple *difference* among representations and the second is negation at the functional level realized through *the act of canceling* the differences by overlooking them. Here ends the process of implicit categorization or protoconcept formation. We have learned to lump different invariants, to grasp some forms, and to recognize them. All these require the capability to abstract. Ontologically speaking, all these abstracted elements, be they representations of properties, of class characters, or of objects, are on a par, since they are constructions of our sensory-motor system and cannot be found in the external world. This tacit categorization, however, has not been done once and for all. Depending on the background provided by the context, the system might opt for some different schemes and cut up the world differently. Moreover, there may be many correct categorizations for the very same set of representations, correctness being a function of success in everyday life.

Once the process of categorization gets stabilized and some objects-in-general (*artha-sāmānya*) have been fixed by bringing representations under some protoconcepts and by superimposing externality, the system associates a name with each "object." Naming or the use of words (*nāma-yojanā*) paves the way for explicit learning and hastens the process of manipulating thought contents. Organisms become more and more skilful in abstraction. We can, for example, handle concepts of primroses and prime numbers with equal felicity because we can communicate with one another and navigate in our environment more efficiently by using verbal descriptions, which is much more advantageous than proceeding through trial-and-error experience. But even at this stage, we cannot claim that we have cut up the world at its natural joints or that our putative kinds are real natural kinds, since the external constraints come to us only in the form of affordances at the sensory-motor level. Our linguistic descriptions are somehow grounded in our sensory-motor invariants. Without implicit categorization, explicit categorization or learning cannot take place, yet that is no guarantee for grasping the world as it is. Dharmakīrti and some of his followers therefore rightly recognized the role of significant errors behind successful human practices. Cognitive scientists are not at all bothered by this kind of antirealism, because they do not presume to do ontology at all. If they do ontology, then they restrict their ontic claims to their terms of art. As Harnad (2005, 40) says, "at bottom all our categories consist in the ways we behave differently toward different kinds of 'things,' whether it be the 'things' we do or do not eat, mate with or flee from, or the things that we describe, through our language, as prime numbers, affordances, absolute discriminables, or truths. And isn't that all that cognition is for—and about?"

Like the cognitive scientists, the Buddhists in the Dignāga-Dharmakīrti tradition did not claim that their processes of classification, naming, or description have any correspondence with the external world. They did not believe that there is any natural relation between a word and its referent or its meaning. Rather they attempted to refute the relation between the word and its referent as admitted by the Naiyāyikas.[5] The tortuous explanation that the apoha theory offers for our reference success or reference failure might somewhat undermine its excellence as a theory of word/sentence meaning, yet as a theory of mental representations and concept acquisition and concept possession, apoha theory appears highly plausible. Such a construal of apoha may not appear too improbable, if we remember that even in fourteenth-century Europe, Ockham and his followers maintained that universals are "names" rather than external entities. Names, according to them, were units of language of thought and were called *conceptus.* So universals were not associated with a mode of being but with mental representations. Besides, in the previous section we have shown why, to introduce abstractions in a world of unique particulars, we need to admit an implicit process of categorization in which is rooted our verbal categorizations.

The Buddhist worldview centering on ineffable particulars was entirely alien to their opponents, the Indian realists, who simply could not understand how any sane person could ever ascribe categories, names, and descriptions to representations or awareness, which are ascribable to external objects only.[6] So the Buddhists did not spare any effort to explain why they thought their scheme of categorization fell in the realm of appearances and not in the realm of reality. Of course, once explicit verbal categorization was introduced, they could not avoid talking about word meaning because they were forcibly drawn to the linguistically designated determinate world of cows and horses, of pots and pans, of fire and water, which was the universe of any realist discourse. Faced with the debate on the nature of what a word stands for—a universal (*jāti*), a particular (*vyakti*), a particular shape formed out of the configuration of limbs (*ākṛti*), or a particular characterized by a universal (*jāti-viśiṣṭa-vyakti*), or all of them together—Dignāga replied that a word expresses its object through the exclusion of other things just as an inferential mark establishes the object to be inferred through the exclusion of that which does not possess that inferential mark. When someone infers fire on a hill, seeing smoke there, he does not cognize any particular fire on the hill. He is only sure about one thing, that the hill does not lack fire. Similarly, when somebody hears the word "fire," the word acts as a sign for the object that is not nonfire, that is, "knowledge of the word 'fire' leads to our knowing the object of reference as excluded from nonfire." That means the sign leads to the knowledge of the signified, if we know that the former is excluded from whatever excludes the latter. It is obvious that the relation between the designatum and the designated is as indirect as that of a sign and a signified in an inferential situation. Hence we need some computational moves to link a designatum with its designation. If Dignāga's model were extrapolated to the realm of mental representations, then we would most probably have ended with a computational theory of mental representations. Dharmakīrti's account, however, is not overtly computational and cognitive scientists of a noncomputational persuasion may profitably exploit it. The reason for this is that while Dignāga tried to explain the relation between the name and the named logically, Dharmakīrti's causal story came very close to the naturalistic explanation of the cognitive scientists. A neo-Gibsonian stance at this juncture will help us explain the last point.

J. J. Gibson was the first person in contemporary cognitive science to raise and answer the very important question: how does one obtain constant perceptions in everyday life on the basis of continually changed sensations? But he never talked about concept acquisition or possession. Rather, he avoided all concept talk by introducing a new idea of "affordance," which a sensory-motor system can directly pick up from the environment. Gibson (1979, 127) wrote, "the affordances of the environment are what it offers the animal, what it provides or furnishes either for good or ill. The verb 'to afford' is found in the dictionary, but the noun 'affordance' is not. I have made it up. I mean by it some thing that refers to the complementarity of the animal and the environment . . . .Affordances are properties taken with reference to the observer. They are neither physical nor phenomenal."

Gibson's theory of perceptual information pickup suggests that perception depends entirely upon information in the stimulus array. When a perceiver moves around in an environment, some aspects of the perceptual array change (transformations), while others do not (invariants). The invariant information is what he calls the "ambient optic arrays"[7]—texture, gradient, flow patterns, the horizon ratio, shadows, color, convergence, symmetry, layout—and these determine what is perceived. Gibson holds that these are the features of the visual environment or the retinal image. These invariants are available to be picked up by the perceivers. Higher-order invariants determine what may or may not be picked up by a particular perceiver, that is, the perceptual flow of the perceptual environment. Perception thus requires an active agent.

According to Gibson, perception is designed for action and not for having some inner experience or representation. Affordances are perceivable possibilities for action. They are powers in relation to which an organism can act. Perceiving a chair may afford a human perceiver a place to sit on, perceiving a door affords opening by pushing. To pick up this information is to pick up the affordances, that is, to perceive something in a particular way. This is a direct operation and not an inferential process. Affordances may be individuated in terms of what they afford a perceiver, namely, a particular behavior.

The ecological philosophy of perception suggests that perception itself is noninferential and that it does not involve any mediation. Perception gives knowledge of things in a way that is not the product of reasoning by the perceiver, nor is it the result of computations performed within the perceiver. Since affordances are directly picked up, the possibility of cognitive error is completely ruled out. An organism can only fail to pick up the available affordances and thus fail to respond to its environment successfully. I think nobody can fail to observe some resemblance between Gibson's theory and Dharmakīrti's theory of nonconceptual perception. It is also obvious that the Buddhist theory is richer because it can also offer an explanation of our determinate concept-laden perceptions, for which Gibson's explanation is not satisfactory. Because of these inherent weaknesses, Gibson's successors, Maturana and Varela gave us a slightly different narrative. To quote from Maturana (1980, 5), "living systems are units of interactions; they exist in an environment. . . . When an observer claims that an organism exhibits perception, what he or she beholds is an organism that brings forth a world of actions through sensory-motor correlations *congruent* with perturbation in the environment in which he or she sees it to conserve its adaptation" (58). Thus perception, to the neo-Gibsonians, is an act of the whole animal, the brain-body complex, in continuous interaction with its changing environment. To use neo-Gibsonian jargon, meaningful perception is a result of structural couplings of the organism and its environment. This is known as the enactive approach to perception. There is no need to enter into the details of this enactive theory. I shall just highlight the important points relevant to our discussion. The enactive approach makes it clear that the world does not contain stable objects with specific properties as our commonsense worldview tells us. Objects in the world are created in stages by dynamic interaction between organisms and the world. This is also true about simple sensual properties. There is no single type of environmental property, say, the property *blue* shared by all blue-producing features of the environment. So it does not make any sense to say that the function of our color vision is to detect that single type of property. Color properties emerge due to the enactment by the perceptuo-motor coupling of animals with their environments. So different animals have different phenomenal color spaces, though organisms belonging to the same species because of the similarity in their embodiment have similar phenomenal color spaces.

Now I would like to draw your attention to some important implications of this enactive theory: (1) Though we are, in a sense, the architects of the objects of our determinate common sense experience, these objects have some objective counterpart in the ever-changing particular features of our environment. (2) As the construction of our world is dependent on the dispositions of our embodiment, organisms belonging to the same species are constrained to perceive the world in similar ways. That is why we tend to endow the world with some "prespecified" features relying on the sensory-motor capacities of human beings or higher primates, which may not apply to the world of organisms different from us.

These observations bring me to the last point that I want to make, concerning the scheme-content dichotomy. The Gibsonian and neo-Gibsonian worldview is compatible with Putnam's theory of internal realism because all three of them deny the existence of a prestructured world as upheld by the metaphysical and commonsense realists, yet also avoid the cultural relativist position that the external world appears the way it does due to some cultural convention. The followers of Dharmakīrti would also admit Putnam's (1987, 12) slogan that "the mind and the world jointly make up the mind and the world." Of course, by "the world" here the Buddhists would mean the conventional realm. If the Buddhists were ready to be confined to the conventional world, they could have avoided the scheme-content dichotomy, for, though the perceptuo-motor dispositions and cultural-linguistic legacy of the human system constrain their perception of the world in a certain way, objects of the commonsense world are not mere projections of their beliefs and interests. Their beliefs and interests enter in a big way in the "construction" of the commonsense world, yet they do discover the so-called conventional truths. It is even possible to infer through back calculations what caused the affordances. Probably the Tibetan interpreters of Dharmakīrti who upheld a theory of causal isomorphism between the particulars and their respective representations hinted at this point (see Dreyfus in this volume). But as long as Dharmakīrti and his followers continued to hold on to the distinction between the *svalakṣaṇa*s and the *sāmānya-lakṣaṇa*s[8] and preferred to grant the latter an inferior ontic status, the scheme-content dichotomy would persist. For soteriological considerations Dharmakīrti would leave the scheme-content gap wide open—beginning with and ending in an abstraction-free world.

Notes

1. This terminology of "splitter" and "lumper" I have borrowed from Bloom 2004.
2. Vide Jayantabhaṭṭa's NMJ.
3. In constructing this account I have depended on Cohen and Lefebvre 2005, relying mainly on the papers of Harnad, Rey, Sowa, Papafragou, Panaccio, and Lalumera.
4. "svaviṣaya-anantaraviṣayasahakāriṇendriyajñānena samanantarapratyayena janitaṃ tan manovijñānam," NB I.9.
5. "anyeṣāṃ ca svalakṣaṇādināṃ bāhyānāṃ vācyatvenāyogasya pratipāditvāt," TS 1218, TSP.
6. "nāmajātyādināñ ca yā yojanā . . . sā arthagato dharmaḥ na jñānasya," TS 1222, TSP.
7. Gibson concentrated his research on visual perception and hence all examples are from the visual domain. The scope of his theory has been expanded by his successors.
8. These are not only diametrically opposed in nature; their modes of access are also different, the former is accessible by nonconceptual perception and the latter by inference.

12

Apoha Semantics

SOME SIMPLEMINDED QUESTIONS AND DOUBTS

Bob Hale

Some Buddhist Nominalists thought, or can at least plausibly be interpreted as claiming,[1] that a commitment to universals can be avoided by explaining the meaning of general predicates or kind terms by exclusion, appealing to their theory of apoha. In broad terms, their idea seems to have been that whereas there is a strong temptation to think that different particulars to which a general predicate such as "animal" can be correctly applied must have something in common, there is no such temptation to suppose that different particulars to which such a general predicate does *not* apply must do so, so that if we can understand the positive general predicate as conveying no more than that the particulars to which it applies are not nonanimals—as "excluding the other" (*anyāpoha*)—we can see that there is no commitment to shared universals after all.

I shall pursue two main questions: (1) How is the central idea—of avoiding ontological commitment to universals by recourse to doubly negative predications—to be understood, and how is it supposed to work? (2) What difficulties does it face, and is it possible to overcome them?

WHAT IS THE DOCTRINE?

I want to begin by identifying what I see as the main elements of the Buddhist Nominalist solution to the problem of universals. As I understand it, the doctrine of apoha involves the following leading ideas or claims:

1. *The meaning of a kind term* K *is to be explained in terms of negation as: not non-*K.

As Mark Siderits (2006, 95) puts it:

> The meaning of a kind-term is the "exclusion of the other" (*anyāpoha*). This builds on the idea that since a given predicate determines a bipartition of the world, mastery of that predicate may be expressed either as the ability to tell when the expression does apply or the ability to tell when it does not. So . . . the meaning of "crow" may be given as: not non-crow. What all the things that are called crows have in common, then, is that they are not in the class of non-crows.

2. *By giving the meaning of kind terms in this way, one can avoid commitment to universals.*

The thought here (cf. Siderits 2006) is that while we might be tempted to think that there has to be a single shared character that all crows have in common, there is little or no temptation to think that there must be a single shared character that all noncrows—as Siderits observes, a very disparate assortment, including "ostriches, teapots, the number 7, and the wizard Gandalf"—have in common.

3. *The two negations in* "x is *not non-*K" *are importantly different*—"*non-*K" *is the nominally bound negation of the kind term* "K," *whereas* "not" *is verbally bound negation.*

The terms "nominally bound" and "verbally bound" are taken from B. K. Matilal (cf. Siderits 2006). I think we can state the intended point by saying that the prefix "non-" applies to nouns to form a kind of term negation, whereas "not" is used as a sentential operator, since if we think of negation as attaching to the verb in a simple sentence, this is equivalent to taking it to be the sentence as a whole (or the proposition expressed by it) that is negated. Crucially, "*x* is non-*K*" and "*x* is not *K*" are to be distinguished. The latter is understood like negation in classical logic, in which the Principle of Bivalence is assumed, so that "*x* is not *K*" is true if and only if "*x* is *K*" is not true, and the classical logical Law of Excluded Middle holds (i.e., "$p \vee \neg p$" is true for all choices of *p*). But while "*x* is non-*K*" is incompatible with "*x* is *K*," it is *not* its contradictory. So "*x* is *K* or *x* is non-*K*" is *not* a logical law—it may be that neither disjunct is true, just as Jones may be neither kind nor unkind.

The difference is important because while "*x* is not not *K*" (i.e., "$\neg\neg$ *x* is *K*") collapses, classically, to "*x* is *K*," "*x* is not non-*K*" does not. This may seem to enable defenders of the doctrine to avoid the objection that our ability to determine what belongs to the class of noncrows, for example, is parasitic upon our ability to determine which are the crows. The objection would be good if both negations were classical, because then—applying the principle of Double Negation Elimination—"*x* is not non-*K*" would simply reduce to "*x* is *K*." But "the application of verbally bound negation to a predicate formed through the nominally bound negation of some predicate *P* does not yield a straightforward assertion of *P*. . . . To say of something that it is not non-*P* is just to refuse the characterization of the thing in question as non-*P* without commitment to any positive characterization" (Siderits 2006: 95).

These, then, are the main ideas or claims I shall discuss.

SOME DOUBTS AND DIFFICULTIES

Most of the difficulties I want to discuss focus upon one general doubt about the central idea of apoha semantics—that the meaning of a kind term *K* is to be explained as not non-*K*—or arise in connection with responses to it. This general doubt turns on the widely accepted idea that some sort of principle of compositionality has to be respected by any satisfactory semantical theory for a particular language. But I'll begin by briefly airing another doubt, about the thought that Siderits presents as underlying this central idea.

Positive and Negative Application Conditions

The thought, as Siderits puts it in the quoted passage, is that predicate mastery can be taken equally to consist *either* in the ability to determine when the predicate *applies or* in the ability to determine when it *does not apply*. In general terms, this idea has a good deal of plausibility. But this initially plausible idea—or at least its application in the present context—seems to me to be undermined by the introduction of nominally bound negation.

The point is this: when a predicate *P* effects an *exhaustive* partition of the universe of objects into two exclusive classes, one containing the objects to which *P* applies and the other those to which it does not apply, then if we have learned to determine when *P* does *not* apply, and we know that *P* applies to all and only those objects *x* such that it is *not* the case that *P* does not apply to *x*, we are in position to know just which objects are the ones to which *P does* apply. So we can arrive at an understanding of *P* indirectly, in this way, just as well as directly, by learning *P*'s positive application conditions. To put the essential point another way: if I am told which objects do *not* belong to the extension of *P*, I can tell, for any object whatever, whether it belongs to the extension of *P* or not. For given any object *x*, I can ask myself: is *x* one of the objects that I have been told does *not* belong to the extension of *P*? By hypothesis I know the answer. If, and only if, the answer is that *x* is not one of those objects, I can infer that it belongs to the extension of *P*.

In the case with which we are concerned, however, things are more complicated. We are, as in the case just described, to arrive at an understanding of the application conditions for *P* indirectly, by way of first grasping the application conditions for non-*P* and then taking *P* to apply just in case non-*P* does not. But while non-*P* effects an exhaustive partition of the universe of objects into two exclusive classes, just as *P* does, it does not apply to all and only those objects to which *P* does not apply, only to some proper subclass of them. Table 12.1 illustrates.

TABLE 12.1

P	*Not P*	*Not P*
Not non-P	*Not non-P*	*non-P*

It is a good question what ability can constitute my grasp of the meaning of non-*P* in this case—whether it must consist in my being able to determine which objects lie in its extension, or whether it might instead consist in an ability to determine which objects are the ones to which non-*P* does *not* apply. But the question that most concerns us is not this one, but what my grasp of the meaning of *P* could consist in. In particular, could it consist in my ability to determine which objects are not non-*P*? And it seems to me the answer must be that it can't. For if all I know is that an object is not non-*P*, I am in no position to infer that it is *P*—as my picture indicates, it may just as well *not* be *P*, even though it is not non-*P*.

In short, the very feature of nominally bound negation that distinguishes it from verbally bound or sentential negation appears to render it unfit to serve as a starting point for an indirect route to a mastery of a positive predicate *P* by exclusion (i.e., *x* is *P* if and only if *x* is not non-*P*). If this is right, there is an awkward—and perhaps fatal—tension in the doctrine we are examining. For the interpretation of "excluding the other" in terms of a combination of sentential and term negation—rather than just double sentential negation or double term negation—is required to give plausibility to the idea that by understanding general predicates in terms of exclusion we can avoid commitment to universals. If both negations are of the same type, the doubly negative predication just collapses back into the positive predication, and it is hard to see how the latter's unwanted apparent commitments are supposed to be avoided. But if "*x* is not non-*P*" isn't equivalent to "*x* is *P*," and so to "$\neg\neg$ *x* is *P*," how can one explain the meaning of the positive predications by means of the doubly negative one?

I shall not pursue this difficulty further just now, but will return to it briefly later. First I want to press what appears to be, if anything, an even more serious difficulty.

Compositionality

An initial—and at least initially serious—cause for concern arises from the theory's apparent disregard for considerations of compositionality. It is very widely agreed that a competent speaker's knowledge of the meanings of most of the expressions of her language isn't, and couldn't be, acquired through a separate process of training in the use of each expression piecemeal. Rather, her understanding of complex expressions in general results somehow from her knowledge of the meanings of their parts together with a grasp of the semantic significance of the way they are put together. In broad terms, the worry is that any attempt to understand or analyze a positive predication "*x* is *P*" as some sort of doubly negative predication "*x* is not non-*P*" offends against the extremely plausible idea that the meaning of a complex expression has to be understood in some such way—as a function of, or as determined by, or somehow composed out of the meanings of its parts. If, for example, the claim is that "*x* is *P*" means, or can be analyzed as, "*x* is not non-*P*," the objection will be that the contained sentence "*x* is non-*P*" and the whole containing "it is not the case that *x* is non-*P*" cannot be understood save on the basis of a prior and independent grasp of the meaning of the constituent predicate *P*—so there can be no question of explaining what "*x* is *P*" means by recourse to the more complex paraphrase. This would just get the meaning-dependency relations back-to-front.[2]

Two quite radical responses to this worry would be: (i) To *accept* the principle of compositionality for *senses* or *meanings*, but *reject* compositionality for *reference*, and argue that the proposal does not require claiming that "*x* is *P*" has its sense or meaning given by "*x* is not non-*P*"—and thus only requires rejecting compositionality of reference. (ii) To accept *both* forms of compositionality, but *reject* the idea that the proposal requires any claim of *equivalence*, either in terms of sense or in terms of reference/existential commitment. Instead, the idea is that the nominalist could agree that positive predications carry commitment to universals, but can argue that no important loss of information is incurred by using "*x* is not non-*P*" *instead* of "*x* is *P*," thereby avoiding the acknowledged commitments of "*x* is *P*."

I call these *radical* responses because they both involve jettisoning what seems to me the distinctive idea of apoha semantics, which *is* a claim about what positive predications *mean*. It is mainly for this reason that I am going to set them aside, at least for now. Instead I want first to explore the question whether the central idea of apoha semantics can be presented and developed in such a way that—so far from conflicting with plausible principles of compositionality for sense or reference—it actually constitutes a kind of compositional theory of meaning that explains how general predicates have meaning in a nominalistically acceptable way, that is, without implying or presupposing the existence of shared characteristics belonging to the objects to which they apply.[3]

NOMINALISTIC COMPOSITIONAL SEMANTICS

In any compositional semantics, the expressions of the language are divided into two classes: there will be a *base class* of simple expressions that are, in a certain sense, primitive, together with a class (the *derived class*) of complex expressions, whose meanings can be seen as built up out of those of simpler expressions and ultimately from those of the simplest expressions. Normally, the base class will be finite, and normally the class comprising all other expressions belonging to the language will be, at least potentially, infinite. Expressions in the base class will typically be syntactically simple, but anyway are always *semantically* simple in the sense that their meanings are not further decomposed, and so cannot be worked out on the basis of the meanings of other expressions and therefore have to be learned directly in some way, such as ostensive training. Other expressions will typically be syntactically complex but need not be, since syntactically simple expressions may be semantically complex, being introduced by definition on the basis of other expressions. Such definitions may be explicit—as illustrated by the definition of "vixen" as meaning "female fox"—or implicit—as illustrated by the usual recursive definition of "+" by means of the recursion equations: $a + 0 = a, a + b = (a + b)$. Some—but not all—of the expressions in the base class may be associated with constructions by means of which complex expressions are formed from simpler ones—their meanings will be given by rules that in some way fix the meanings of the resulting complex expressions in terms of those of their constituents. A particularly important case of this kind is the explanation of the (classical) logical constants by stipulations or rules that give the truth conditions of complex sentences formed by their means.

From our present point of view, the interesting question is what (kinds of) expressions may be taken as lying in the *base class* for a *nominalistic* semantic theory. I shall not try to answer this question fully, but will give what I hope is a plausible partial answer—this will be enough for my purposes. Two points—one negative, the other positive—are, I think, pretty obvious, but important enough to justify emphasizing them.

The negative point is that the nominalist's base class *cannot*—with crucial exceptions to which I shall soon attend—include *general predicates*. His goal is to explain the meanings of these expressions in a way that avoids any assumption of a common characteristic belonging to all and only the entities to which such expressions apply. This is to be accomplished by showing how their meanings result, in a nominalistically acceptable way, from those of expressions in the base class.

The positive point is that there is no reason why the base class should not include *names* of particulars of one sort or another, and it surely will include semantically simple such names. Thus if one took Plato to be a particular and took the use of the proper name "Plato" to be learned by direct association with its bearer, "Plato" might be taken to belong to the base class. But a complex name for the same particular—such as "the author of *Theaetetus* and *The Sophist*"—would not do so, because it involves a general relational predicate.

Clearly, however, the nominalist's base class must comprise some expressions of some other kinds, since from simple names alone, no sentences can be formed, only lists of particulars. Without some further expressions, it will not be possible to *say* anything using only expressions belonging to the base class. But if sentential operators and other expressions are to be explained in a way that is at all plausible, there must be semantically basic sentences—atomic sentences—to anchor those explanations.

I claim that there must be at least one general predicate. But there is a strong constraint governing the admission of predicates into the base class: any admissible basic predicate must be such that it is evident, or can easily be seen, that its application does not require the existence of any shared characteristic belonging to the objects to which it applies.

I think that perhaps only one predicate—the general predicate of identity—can plausibly be taken to satisfy this constraint, and I shall assume that the nominalist can include it in his base class. In this case, one might argue as follows: Obviously each object is identical with itself and with no other object. Let a and b be distinct objects. Then each of them is identical to itself—$a = a$ and $b = b$. But clearly the truth of these two statements does not require that a and b should have anything in common—they can be as different as you like. And if we consider the whole universe of objects, then there is a true statement of self-identity for each of these objects, but this plainly does not require that there be some shared characteristic that all objects whatever have in common.

It might be thought that a similar argument could be given to justify including the opposite of the identity predicate—the distinctness predicate—in the base class. One would need to argue that where $\{a,b\}$ and $\{c,d\}$ are disjoint pairs of distinct objects, there need be nothing in common between the two pairs (i.e., for it to be true that $a \neq b$ and that $c \neq d$). Maybe one could argue for that—although I'm not sure how—but in any case the nominalist will be able to introduce the distinctness predicate anyway, provided he can include (sentential) negation. For he can then just define $x \neq y$ to mean $\neg x = y$.

I can see no reason why the nominalist can't have *sentential* negation, along with the other standard sentential operators (conjunction, disjunction, etc.) in his base class. But as we know, a proponent of apoha semantics requires another kind of negation operator, which can be applied to *nouns* rather than complete sentences (or, equivalently, in effect, to verbs). There is nothing to prevent the nominalist from introducing a form of term negation as opposed to sentence negation. So far, the only nouns in our nominalistic language are proper names of particulars. So it will be these to which term negation is applied, at least in the first instance, at the basic level, rather than common nouns like "crow," "man," etc. With this point in the open, it is also clear that more needs to be said about how term negation is supposed to work. If term negation ("non-") is applied to a *general* term, such as "man," we naturally assume that the resulting compound "nonman" is another general term—so in this kind of case, "non-" forms general terms from simpler general terms. But this can't be how it works at the basic level, where it is *proper names* rather than common nouns that are negated. Here we take a proper name, "Socrates," say, and form a new expression, "non-Socrates." Pretty clearly the output expression has to be of a different logico-syntactic category from the input—"non-Socrates" isn't a proper name. Whatever negating a proper name gives us, it won't be another proper name. Obviously the idea has to be that, at this basic level, applying "non-" to a proper name results in a general term or predicate.

There is nothing objectionable in that, provided that the general predicates thereby produced are nominalistically acceptable—that is, don't require that the objects to which they apply have some shared character. But it seems clear that this condition is met—"non-Socrates," for example, is true of each and every one of those objects that is distinct from Socrates, and it is perfectly clear that there need be nothing that all of these objects have in common.[4]

But there is a snag. Suppose *n* is any name. Then "*x* is non-*n*" will be true if and only if "*x* is not *n*" (i.e., $\neg\, x = n$) is true. So "*x* is not non-*n*" will be equivalent to "*x* is not not *n*" (i.e., $\neg\neg\, x = n$), and so to "*x* is *n*" (i.e., $x = n$) after all. That is, the complex predicate ". . . is not non-*n*" can, it seems, be true of one and only one object, namely *n* itself—so our attempt to exploit the combination of sentence and term negation to get a general predicate that clearly assumes no shared character but is potentially applicable to many objects seems to have fallen flat.

Siderits confronts this snag but is undaunted, because he thinks there is a way around it. What is needed, evidently, is to explain non-*n* in such a way that it is *not* equivalent to the general predicate "distinct from *n*" (i.e., $\neq n$). Indeed, we need non-*n* to be not merely different in meaning from $\neq n$, but to diverge from it extensionally—more precisely, we want non-*n* to be true of no more than a proper subset of the universe of objects apart from *n*, i.e., $\bigvee - \{n\}$. How is that to be achieved? Siderits proposes that we associate the original name *n* with a "paradigm image p_n which is formed so as to be manifestly incompatible with some but not all of the remaining particulars in the universe" (2006, 96). So—to continue with his example—if *n* is a particular crow, we might associate with *n* an image that is incompatible with the perceptual images we get from ostriches, tumbleweeds . . . and so on for many other kinds of thing, but is not incompatible with the perceptual images we get from other crows. Then—or so, I take it, the thought runs—non-*n* will apply more selectively, in the way desired, not to every object distinct from *n*, as before, but to just those objects that are incompatible with the image p_n. And the class of objects that are not non-*n* will comprise, not just *n* itself, but with it any other objects that are not incompatible with p_n—all the crows, in the example.

FURTHER DOUBTS

Does the paradigm-image maneuver work? In the remainder of this paper, I shall try to do two things. First, I'll try to argue that if it *does* work, it can turn the trick for the nominalist *without reliance* on any special doctrine of *doubly negated predicates.* Second, I'll raise some independent doubts about whether the paradigm-image maneuver can work.

In fact, I can't see that much argument is needed for the first of these claims. The point is quite simple. If, by associating with a particular object *n* a certain paradigm image, we can ensure that the negative term non-*n* applies, not to everything in the universe other than *n* itself, but only to some of the objects distinct from *n*—all the noncrows, say—then what is to prevent us from directly introducing a nonnegative general term n^+ with the stipulation that it is to be true of exactly those objects that are compatible with the paradigm image p_n associated with *n*? The play with the not non-*n* construction is just an idle wheel—all the real work is done by the paradigm-image maneuver. Another way to put this point is that non-*n* will differ in extension from $\neq n$ only if the term to which "non-" is prefixed applies to more objects than just *n* itself, and so is *already* (functioning as) a general term. If so, we don't need the doctrine that meaning of a kind term is the exclusion of the other to get general terms without universals.

But does the paradigm-image maneuver work? An obvious objection to it is that it simply assumes that *images* can possess, unproblematically, the kind of generality of application that the nominalist agrees he needs to *argue* words can have. But why suppose there is any *less* of a problem in seeing how an *image* can fit many things without their having some shared quality or qualities than there is in seeing how *words* can apply to many things without shared qualities? Doesn't the maneuver simply pass the buck from words to images?

As I read him, Siderits confronts essentially this problem in the central pages of his paper and proposes an ingenious solution to it. Here is his formulation of the problem:

> How does one learn to form [the paradigm image]? . . . What we want to know is how one can learn to see ostriches but not other crows as distinct from a particular crow if the distinctness in question is not qualitative distinctness. . . . How can one learn to conform one's linguistic behavior to that of others if there are no objective features to form the basis of this discriminative capacity? (Siderits 2005, 96)

And here is his solution:

> The appeal to social convention does bring out the point that the ability in question depends on practices that are responsive to human wants

> and interests. . . . For it turns out that there is something all the non-s_n may be said to share, when "non-s_n" is formed in accordance with the relevant convention, namely that they all fail to satisfy a certain desire, say the desire to eat crow. Obtaining an ostrich or tumbleweed will fail to satisfy that desire; the desire persists in their presence. Obtaining s_n satisfies the desire. But the same hold for s_{n+1}, s_{n+2}, and s_{n+3}, so these also belong to the class of things correctly asserted to be not-non-s_n. This is why all four things are said to be crows. (96)

So, the ingenious move, as I understand, is this: to the realist's insistence that there has to be a shared characteristic, the nominalist can, and should, respond: "There is, but what is shared in *not* an *objective* feature, but a common relation *to us*. There are no objective common features—so no universals—but that doesn't mean there can't be nonobjective common features, grounded in the relations between classes of particulars and their different relations to our wants and interests. These nonobjective common features are acceptable to the nominalist, and they suffice to account for our use of general terms."

Siderits goes on to offer an equally ingenious defense of this move against the objection that different particulars could not all equally serve to satisfy a certain desire if they themselves did not have a shared nature—in essence, the reply is that the capacity of different particulars to satisfy some single desire no more requires that they have some common nature than does the capacity of different drugs to relieve fever. In general, different things may play the same causal role without there being something in common between them all in virtue of which they do so.

There is room, I suspect, for further rounds in this debate. It might be claimed, for instance, that while a functional property may well supervene on different base properties, it is in virtue of particulars' possession of *general properties* that the functional property supervenes. It is at best unclear how different particulars could all serve in the same role if their capacity to do so were not underpinned by *some* general properties, albeit not necessarily the same ones in all cases. But I shan't pursue this further now. Instead I want to comment briefly on some other aspects of the proposed solution and to air a different cause for concern about appealing to the capacity of different particulars to satisfy a single desire, such as the desire for crow pie (or, generalizing, to stand in some single relation to some want or interest).

First, then, it seems to me that although the play with the capacity of different particulars—for example, different noncrows—to fail to satisfy a certain desire is introduced to answer a question about how we form suitable paradigm images, in fact the images just drop out of sight. They play no discernible role—and certainly no essential one—in explaining how we achieve conformity in linguistic behavior, or how we come to apply a single word to many different particulars. Once again, if the explanation works, it seems that it can work perfectly well *without* any introduction of intermediary *images*—so they may as well be ditched.

Second, although Siderits puts his proposal in terms of a nonobjective shared feature common to all the objects which are non-s_n (noncrows)—namely, they all fail to satisfy the desire for crow—thereby leaving work to be done by equating the positive with the doubly negative, it again appears that, if the proposal can work at all, it can be implemented directly, appealing to a different nonobjective similarity among all the particulars that are s_n (i.e., the ones to which we apply "crow")—namely, that they all satisfy the desire for crow. So once again, the distinctive doctrine of apoha semantics—that a kind term *K* means not non-*K*—is doing no essential work.

But *does* the proposed solution work? My final worry about it concerns the legitimacy of appealing to desires such as the desire to eat crow at this fundamental level of semantic theorizing. Roughly, my fear is that appealing to any such general desire—that is, desire with a general content that has to be specified in general terms, if it is to be satisfiable by many different particulars—simply reintroduces our old worry about circularity. We can all agree that the nominalist must aspire, in his semantics, to explain how we can meaningfully ascribe to ourselves desires and other attitudes that are general in the sense that their objects are not particulars—they are desires, not for this or that particular, but for something of a certain general kind. For example, I want a pint of beer, but there is no particular pint of beer of which it can correctly be said that it is the pint of beer I want—I just want it to be the case that I have a pint of beer, and I don't mind which (so long as it tastes good, etc.). My desire for a pint of beer is like Quine's desire for a sloop—no particular sloop, just relief from slooplessness (Quine 1966). And the nominalist must hope to be able to explain this in a way that avoids commitment to universals. I myself see no reason why he shouldn't do that in just the way Quine advocated—by construing my desire for a sloop, when it is the desire for relief from slooplessness rather than some covetous desire for my neighbor's sloop, as a desire to the effect that a certain existential proposition be true: namely, that $\exists x(x \text{ is a sloop} \wedge \text{I own } x)$. But this solution to that problem highlights precisely my cause for concern over appealing to general desires to explain how different particulars can all fall under some general term. The explanation of how desires can, in the relevant way, be general presupposes that the terms involved in specifying their objects are already general, in the sense of being potentially applicable to many distinct particulars. To put the point another way: a nominalistic compositional semantic theory, for English, say, is going to generate a partial ordering of the sentences of English—an ordering that at least roughly reflects the order in which understanding of the expressions has to be acquired. In that ordering, one would expect sentences like "this is a sloop and I own it" and "this is crow pie and I am eating it" to come well before sentences like "I want a sloop" (with the relief from slooplessness meaning) and "I want some crow pie."

It may be said that the proposed solution can agree that the understanding of sentences ascribing desires cannot precede that of sentences describing what is, in effect, the satisfaction of those desires, but claim that this does not make it illegitimate to appeal to facts about what will or won't satisfy a given desire in explaining how words get to apply to many things. This might be right, if by what satisfies my desire for crow pie, say, one simply means what will bring it about that I no longer want crow pie. But what satisfies my desire for crow pie—in this sense of satisfy—may very well not have parts of something that once had wings and black feathers in it. It may be a glimpse of a far more appetizing roast pheasant, or the news that I've won the lottery, or a bang on the head.[5] So appealing to what satisfies or fails to satisfy desires for *X* is liable to give *X* the wrong extension. But if by what will satisfy a desire, one means what it is a desire for, we are back with the preceding worry.

Perhaps it can be argued that the circularity I have argued afflicts the appeal to desires (and other attitudes) is not vicious, but I have to confess that at this stage, I don't see how.

CONCLUDING REMARKS

I have raised two main kinds of doubt about the doctrine of apoha and the associated defense of nominalism. First, I have suggested that the doctrine of apoha by itself does not provide a means of avoiding commitment to universals, and that when it is supplemented with other ideas in the way Siderits describes, it is the other ideas that do the work, with nothing essential contributed by the apoha doctrine of exclusion of the other.[6] But I have also expressed some skepticism about whether the other ideas can anyway really give the nominalist a way of explaining how we can talk in general terms without commitment to universals.

I should emphasize that the main difficulties I have raised for the apoha doctrine have been developed on the assumption that it will involve explaining the meaning of general terms by the way of "exclusion of the other." In particular, it is this which gives rise to the difficulty over compositionality that I have been mainly concerned to press. So far, it remains possible that a defense of nominalism that preserves something of the spirit of the apoha doctrine might be mounted, adopting one or the other of the two more radical responses to the problem of compositionality that I mentioned earlier in the paper. Roughly put, the idea would be that instead of claiming that one can give a semantics for general terms under which "*x* is *K*" is explained as meaning "*x* is not non-*K*," one might claim that negation can be viewed as a device that cancels existential commitments carried by expressions occurring in what is negated. Perhaps one could then argue that "*x* is not non-*K*" is precisely *not equivalent* to "*x* is *K*." Rather, it is a *weaker*—ontologically less committed—statement, which the nominalist may offer as a *replacement* for the more problematic "*x* is *K*." This would involve granting that general positive predications of the form "*x* is *K*" are fully intelligible without benefit of nominalistic reinterpretation, but claiming that this leaves the nominalist free to introduce "*x* is not non-*K*" as a replacement that avoids the unwanted commitment to universals associated with what it replaces. Whether a stable version of nominalism can be developed along these lines is a good question, but it is not one I can discuss here.[7]

Notes

This paper is based upon a talk given at the conference "Apoha Semantics and Human Cognition" held in Crêt Bérard, Switzerland, May 24–28, 2006. In sharp contrast with all the other speakers at the conference, I have no expertise in this area and am certainly not competent in the exegesis of the Buddhist nominalists whose ideas and doctrines were the subject of the conference. I have therefore relied heavily upon my fellow speakers' contributions and am greatly indebted to them for helping me to acquire some understanding of the intriguing but perplexing ideas I am trying to talk about in my paper. I need hardly say—but I will—that none of them can reasonably be held responsible for any misunderstandings and errors of interpretation that may remain.

1. I have drawn especially upon Siderits 2006.
2. I think what Pascale Hugon (in this volume) calls the charge of circularity/interdependence can be seen as an alternative formulation of the objection I am considering here. She considers, and I think means to endorse, the response she takes to have been Dharmakīrti's—that the realist is equally open to the same charge. But I do not myself think that this ad hominem argument against the

realist works. The realist can agree that when he fixes the convention for the application of the term "tree," for example, he also settles what is *not* to count as a tree. But this does not mean that his theory makes the understanding of "tree" *depend* upon understanding "not nontree." I think formulating the problem for the nominalist in my way helps to make it clear that it is a problem that distinctively afflicts the apoha doctrine.

3. In fact, as we shall see, a good deal of what Mark Siderits says in "Buddhist Nominalism" (2005) fits in very well with such an approach, even though he does not explicitly address questions about compositionality.
4. Obviously, they may be said to have it in common that they are all distinct from Socrates, but there is nothing to unsettle the nominalist in that. The essential point is that there is, or at least need be, no characteristic they share, *in virtue of which* they are each distinct from Socrates.
5. This is the main problem with the theory of desire advanced by Bertrand Russell in Lecture III of *The Analysis of Mind*. This problem with Russell's theory—which identifies whatever it is that causes the feeling of discomfort that he takes to be involved in any desire to go away as what the desire is for—is discussed at some length in Kenny 1963, 101–110.
6. In this respect, what I have said is very much in line with a number of remarks in Tom Tillemans's paper (this volume).
7. I should like to record my gratitude to the other participants in the Crêt-Bérard conference both for their helpful discussion and for their patience with my amateurish efforts in the area of their expertise. I am especially grateful to the organizers, and to Professor Shōryū Katsura for helping me toward a better grasp of the Buddhist nominalist ideas I have tried to discuss in this paper.

13

Classical Semantics and Apoha Semantics

Brendan S. Gillon

As is well known, Indian Buddhist thinkers found universals (*sāmānya*) to be metaphysically repugnant. Yet, many Indian thinkers considered universals necessary to account for how general expressions, such as common nouns, manage to apply to an unbounded set of individuals. Indian Buddhist thinkers, however, believed that a satisfactory account of this linguistic phenomenon could be provided without appeal to universals. In particular, such thinkers, called *apoha-vādins* (proponents of exclusion), held that, through exclusion (*apoha*) and difference (*anya*) of individuals, an *ersatz* universal could be found that would provide an empirically adequate account of the generality of a general expression.

Though no explicit semantic account of these two forms of negation, exclusion (*apoha*) and difference (*anya*), is given by the *apoha-vādins*, it is still tempting to think that they might have had such an account in mind, at least implicitly. This conjecture[1] is plausible for several reasons. To begin with, Indian grammarians had devised a generative grammar and semantics of classical Sanskrit. In addition, Indian grammarians themselves did distinguish two forms of negation, *prasajya-pratiṣedha* and *paryudāsa*, dubbed "verbally bound negation" and "nominally bound negation" by Matilal, though these forms were never given an explicit analysis (Matilal 1971, 162–165). Indeed, what precisely these negations consist in is still not clear (Gillon 1987). Finally, Dharmakīrti, one of the early developers of the apoha theory, himself had given an explicit semantic analysis of the Sanskrit word *eva* (*only*), which plays a crucial role in the statement of the truth conditions of the classical Indian syllogism (Kajiyama 1973; Gillon and Hayes 1982; Gillon 1999).

The aim of this paper is to show that the two most obvious candidates from contemporary logic that one might use to explicate the *apoha-vādin*'s notions of exclusion (*apoha*) and difference (*anya*), namely, internal and external negation, do not provide the *apoha-vādins* with the *ersatz* universals they were looking for. Below, I shall first state what a semantic theory is and rehearse its principal features, availing myself of the semantics of monadic predicate logic (without quantifiers). I shall then set out various semantics for monadic predicate logic using internal and external negation in various combinations. I shall conclude by showing that their combination does not permit the definition of an *ersatz* universal that appeals only to individuals and thereby satisfies the scruples of those, such as Buddhists, who found the positing of universals to be metaphysically repugnant.

WHAT IS A SEMANTIC THEORY?

The idea that provides the basis of contemporary semantic theory dates back to Pāṇini, the great Sanskrit grammarian of ancient India (c. fourth century B.C.E.). His idea can be summarized as follows: each Sanskrit sentence can be analyzed into minimal constituents and the sense of each minimal constituent contributes to the sense of the entire sentence. Thus, a grammar of a natural language should not only generate all and only its acceptable constituents, it should also make clear how meanings, once associated with the simplest constituents, determine the meanings of the complex constituents they constitute. The meanings associated with minimal constituents comprise, among other things, the constituents of situations, namely, actions (*kriyā*) and their participants (*kāraka*).[2] Here, however, is where the grammar fell short, for neither Pāṇini nor his successors had a clear way to give a mathematical treatment of situations and their components. The remedy to this obstacle did not appear until the second quarter of the twentieth century, when began to emerge the one discipline whose business it is to make precise how values associated with constituents determine the value of the constituents they make up, namely, the subdiscipline of logic known as model theory. Its founder, Alfred Tarski (1901–1983), recognized the pertinence of such investigations to the study of how complex constituents in a natural language acquire their meaning from the constituents that constitute them, though he himself doubted that a satisfactory formal account of this property of natural language constituents could be worked out (Tarski 1935, 1944).

Natural language semantics, then, addresses two central questions: What values are to be associated with the basic constituents of a language? And, how do the values of simpler constituents contribute to the value of the complex constituents the simpler ones make up? The utility of model theory is to enlighten us on how to proceed with answering the latter question. If a satisfactory answer is obtained, then it will explain not only how changes in our understanding of complex constituents changes with changes in their constituents but also how it is that humans are able to understand completely novel complex constituents. After all, one's understanding of complex constituents cannot be accounted for by appeal to memorization of a language's constituents, as explicitly noted by Patañjali twenty-three hundred years ago,[3] any more than an appeal to memorization can explain how it is that humans knowing elementary arithmetic can understand previously unencountered numerals. Rather, one is able to understand novel complex constituents since they are combinations of simple ones that are not novel and that are antecedently understood.

MODEL THEORY

For the sake of illustration, let us consider a portion of classical predicate logic: monadic predicate logic without quantifiers. The formulae are obtained from two disjoint sets, the set of individual constants (*CN*) and the set of one-place predicates (PD_1). Atomic formulae (*AF*) are obtained by prefixing a one-place predicate to an individual constant. Thus, if *P* is a one-place predicate and *c* is an individual constant, then *Pc* is an atomic formula.

Definition 1: *Atomic Formulae of Monadic Predicate Logic*

$\alpha \in AF$, the atomic formulae of Monadic Predicate Logic, iff $\alpha = \Pi c$, where $\Pi \in PD_1$ and $c \in CN$.

The set of formulae is the smallest set that includes the atomic formulae and those formulae obtained from prefixing a formula with a unary connective or enclosing a pair of formulae between parentheses and placing between the formulae a binary connective. There is but one unary connective $\neg$ and there are four binary connectives: $\wedge$, $\vee$, $\rightarrow$, and $\leftrightarrow$.

Definition 2: *The Formulae of Monadic Predicate Logic (MPL)*

FM, the set of formulae of monadic predicate logic, is defined as follows:

(1)	If AF $\subseteq$ FM;
(2.1)	if $\alpha \in$ F M, then $\neg\, \alpha \in$ FM;
(2.2.1)	if $\alpha, \beta \in$ F M, then $(\alpha \wedge \beta) \in$ FM;
(2.2.2)	if $\alpha, \beta \in$ F M, then $(\alpha \vee \beta) \in$ FM;
(2.2.3)	if $\alpha, \beta \in$ F M, then $(\alpha \rightarrow \beta) \in$ FM;
(2.2.4)	if $\alpha, \beta \in$ F M, then $(\alpha \leftrightarrow \beta) \in$ FM;
(3)	nothing else is in FM.

A model for this notation comprises a nonempty universe *U* and an interpretation function that assigns to each member of *CN* a member of *U* and that assigns to each member of PD_1 a subset of *U*. Values assigned to symbols will be called their denotations.

Definition 3: *Model for MPL*

Let CN and PD_1 *be the constants and the monadic predicates, respectively. Then,* $\langle U, i\rangle$ *is a model for monadic predicate logic iff U is a nonempty set and i is an interpretation function of CN into U and of* PD_1 *into the subsets of U.*

An atomic formula is said to be true in a model if and only if the individual assigned to the constant is a member of the set assigned to the predicate.

Thus, let *i* be the interpretation function of the model and let *Pc* be the formula, then *Pc* is true in the model iff $i(c) \in i(P)$. Composite formulae are true by dint of the usual definition.

Definition 4: *A Classical Valuation for the Formulae of MPL*

Let $\langle U, i\rangle$ *be a model for MPL.*

(1) ATOMIC FORMULAE:
Let Π *be a member of PD1 and let c be a member CN. Then,*

(1.1)	$v_i(\Pi c) = T$	iff	$i(c) \in i(\Pi)$.

(2) COMPOSITE FORMULAE:
Then, for each α *and for each* β *in FM,*

(2.1)	vi $(\neg\alpha) = T$	iff	$v_i(\alpha) = F$;
(2.2.1)	vi $(\alpha \wedge \beta) = T$	iff	$v_i(\alpha) = T$ and $v_i(\beta) = T$;
(2.2.2)	$v_i(\alpha \vee \beta) = T$	iff	either $v_i(\alpha) = T$ or $v_i(\beta) = T$;

(2.2.3) $v_i(\alpha \rightarrow \beta) = T$ iff either $v_i(\alpha) = F$ or $v_i(\beta) = T$;
(2.2.4) $v_i(\alpha \leftrightarrow \beta) = T$ iff $v_i(\alpha) = v_i(\beta)$.

SIMPLE COPULAR SANSKRIT CLAUSES

Of course, we are not interested in the semantics of monadic predicate logic, rather we are interested in the semantics of natural language, in particular, the semantics of classical Sanskrit. Now the syntax and semantics of a natural language is vastly more complicated than that of monadic predicate logic, even if we confine our attention to simple copular clauses. For the sake of simplicity, let us confine our attention to simple copular clauses consisting of a common noun (e.g., *kāka*, *gau*, *puruṣa*, etc.) and a proper noun (e.g., *Devadatta*, *Yajñadatta*, *Kṛṣṇa*, etc.). In such a case, the model theory developed earlier for monadic predicate logic without quantifiers can be adopted virtually without change to such elementary Sanskrit clauses. An atomic (simple) copular clause, as just stated, comprises a proper noun and a common noun (both in the nominative case). The formation of composite simple copular clauses is straightforward but involves complications regarding the placement of the connectors, which need not detain us here.

A model for this language comprises a nonempty universe and an interpretation function that assigns to each proper noun an individual in the universe and to each common noun a subset of *U*. An atomic simple copular Sanskrit clause is true if and only if the individual in the universe denoted by the proper noun (i.e., the individual assigned to the proper noun by the interpretation function) is a member of the set denoted by the common noun (i.e., the set assigned to the common noun by the interpretation function).

Consider the sentence:

(1) Devadattaḥ puruṣaḥ.
Devadatta is a man.

This sentence is true if and only if the thing denoted by the proper noun *Devadattaḥ* (*Devadatta*) is a member of the denotation of the common count word *puruṣaḥ* (man).

THE PROBLEM OF LEXICAL SEMANTICS

While model theory is enlightening with respect to the fundamental question of how the meanings of simpler constituents contribute to the meaning of the complex constituents the simpler ones make up, model theory tells us nothing about how values are assigned to the simplest constituents. Indeed, model theory is not concerned with this question, for it is interested in the properties of formulae as the interpretations of their minimal constituents are allowed to vary freely, holding the interpretation of the logical symbols constant.

Although, in natural language, it is clear that the assignment of meanings to words is arbitrary, nonetheless, once fixed, the assignment does not vary freely from occasion of use to occasion of use, for without these invariable meanings, communication would be impossible. But what imparts this invariability and how is what imparts it learned?

Let us confine our attention to proper nouns and common nouns. If we set aside the problem of the indeterminacy of demonstration, noted by Wittgenstein, which afflicts both the learning of the meaning of proper nouns and the learning of the meaning of common nouns, we notice another contrast between the learning of proper nouns and the learning of common nouns. Once one associates the individual denoted by the proper noun with the proper noun, nothing more need be said. In contrast, even after one associates an individual of which the common noun is true with the common noun, one still has to figure out how to associate another individual of which the common noun is true with the common noun. In other words, once one knows to whom or to what a proper noun applies, one need know nothing further to apply it again, since there is nothing else to which it applies; however, even after one knows that a common noun applies to a particular individual, one is no further ahead ipso facto with respect to the problem of knowing to what else it applies. The contrast resides not so much in the fact that a single individual is associated with a proper noun, while more than one individual is associated with a common noun; rather, it lies in the fact that what is associated with a proper noun is bounded, whereas what is associated with a common noun is not. Thus, when one learns what is associated with a proper noun, one learns that it applies to one individual and no other. Indeed, a proper noun cannot apply to anything else without losing its meaning. But when one learns what is associated with a common noun, even if one knows what all the things it applies to are, nothing in its meaning rules out its applying to something that comes into existence subsequently, provided the individual that comes into existence is of the right sort. (Of course, we are putting aside the problem of vagueness.)

How, then, does one manage to know to what a common noun applies and to what it does not? Model theory provides no help here. Indeed, the only answers to this problem are answers put forth long before the advent of model theory. One answer is to say that they are, in a sense, just like proper nouns: associated with each common noun is but one thing, a universal, by dint of which association, one knows, for any new individual, whether it applies or not. But, what is a universal and how does one grasp it? Many of those who have felt that these questions have no satisfactory answer have been driven to seek answers elsewhere.

APOHA MODEL THEORY

Buddhists maintained that *anya-apoha* (exclusion of what is different) provides an answer to the question of how one knows the meaning of a general expression, without having recourse to universals. The idea is that one can use two kinds of negation: *anya* (what is different) and *apoha* (exclusion). Unfortunately, these thinkers never specified what precisely these two negations are. An obvious and natural suggestion is the negations found in contemporary three-valued logic, internal negation and external negation. Let T, F, and N be the three values of three-valued logic. The two unary operators are defined as follows:

$$
\begin{array}{ll}
I: & T \mapsto F \\
 & F \mapsto T \\
 & N \mapsto N \\
E: & T \mapsto F \\
 & F \mapsto T \\
 & N \mapsto T
\end{array}
$$

These operations were devised to interpret two unary connectives in a three-valued propositional logic. As such, they have no role to play in the problem before us. Rather, what we require is their set-theoretic counterparts. To arrive at this, we must alter the model theory for monadic predicate logic from a classical model theory to one that I shall call apoha model theory. An apoha model for monadic predicate logic, like a standard model, comprises a nonempty universe *U* and an interpretation function that assigns to each constant an individual from the universe and that assigns to each monadic predicate a pair of disjoint subsets of the universe.

Definition 5: *Apoha Model for MPL*

Let CN and PD_1 be the constants and the monadic predicates. Then, $\langle U, i\rangle$ is a model for monadic predicate logic iff U is a nonempty set and i is an interpretation function from CN into U and from PD_1 into $\{ \langle X, Y\rangle : X, Y$ are disjoint subsets of U$\}$.

The idea is that with each predicate must be associated not only the set of things of which the predicate is true but also the set of things of which it is false. In other words, each predicate is associated with an ordered pair. Let us call the ordered pair the predicate's denotation, while we shall call the first set its positive denotation and the second its negative denotation.

There are features of natural language that seem to warrant this increase in the complexity of a general term's denotation. These are the cases of presuppositional failure and category mistakes. Consider the sentence "four is blue" by way of an illustration of a category mistake. "Blue" is a predicate that applies to physical objects that can have color. It is true of things that have color and are blue and it is false of things that have color and are not blue. Many think that it is neither true nor false of things that have no color. If one interprets natural language negation as internal negation and one assigns to the word "blue" a positive denotation of the set of colored blue things and a negative denotation of the set of colored nonblue things, then neither the sentence "four is blue" nor the sentence "four is not blue" is true.

Now, the set-theoretical counterparts of the two kinds of propositional negation given earlier can now be defined as follows:

I: $(X, Y) \mapsto (Y, X)$

E: $(X, Y) \mapsto (-X, X)$

(where - is complementation with respect to *U*). The first function simply swaps the positive and negative denotation associated with a monadic predicate, while the second function maps the positive denotation to its set-theoretic complement and maps the negative denotation to the set whose complement has become the positive denotation.

Now the question arises: how do these two kinds of negation permit universals to be dispensed with in lexical semantics in favor of just individuals? To answer this question, let us distinguish between an expression's denotation, which is dictated by our semantic theory, and an expression's extension, that of which an expression is true and can be ascertained empirically by asking a speaker of the expression's language whether or not the expression applies to some entity. In light of this distinction, we can rephrase our question as follows: how do these two kinds of negation permit universals to be dispensed with in lexical semantics of common nouns in favor of just individuals so that, given a common noun's denotation, one can arrive at its extension?

The answer of the non-*apoha-vādin* is that the denotation of a common noun is a universal and its extension is just the set of individuals in which the universal inheres. The answer of the *apoha-vādin* is that the denotation of a common noun is an individual (or what is set-theoretically equivalent, the singleton set containing that individual) and its extension is obtained by the application of the two negation operations.

As we shall now see, the answer of the *apoha-vādin* does not furnish a suitable extension for any common noun whose extension contains at least two individuals. There are exactly four possible combinations of the operations I and E: namely, II, EE, IE, and EI. Four propositions can be proved, one corresponding to each of the combinations. They are the following:

PROPOSITION 1:
$II(X, Y) = (X, Y)$,

PROPOSITION 2:
$EE(X, Y) = (X, -X)$,

PROPOSITION 3:
$IE(X, Y) = (X, -X)$,

PROPOSITION 4:
$EI(X, Y) = (-Y, Y)$

(where X ranges over the positive denotations, Y over negative denotation, and "-"is set theoretic complementation). As the first three propositions show for the combinations of II, EE, and IE, the positive denotation remains fixed under the application of the combined negations. Thus, under these forms of negation, the positive denotation and the positive extension are the same. Thus, for example, suppose the common noun *puruṣa* (man), whose extension is M and one of whose members, among others, is m, is assigned the denotation of m (that is, $X = \{m\}$), then the result of applying any of the first three combinations of negation only yields the set $\{m\}$, and not the set *M*, as its extension, which is, by hypothesis, different from *M*. In other words, given {*m*} as the positive denotation, one obtains {*m*} as the extension—not, of course, what is desired.

The only remaining possibility is the combination of *EI*. But here the fourth proposition shows us that only the negative denotation is relevant to the determination of positive and negative extension, the positive denotation is utterly irrelevant. Thus, to arrive at the extension of *M*, one must know to start with -*M*, the set of all nonmen. But this is just the problem of finding the denotation of a general term all over again. *Anya-apoha*, understood as a combination of internal and external negation, does not provide any semantic alternative to universals. Perhaps another way of spelling out *anya-apoha* will. But that remains to be seen.

Notes

1. I attributed this view to Mark Siderits (1991, 93–100), perhaps mistakenly, when I wrote my review (Gillon 1992) of his important and insightful book on the philosophy of language in classical Indian philosophy.
2. For details, see Gillon 2007.
3. MB (ed. Kielhorn) v. 1, 5–6; translated in Staal 1969, 501–502.

14

Śrughna by Dusk

Mark Siderits

When I first began thinking about the apoha theory many years ago, it struck me as an interesting test case for the interpretive approach to Buddhist philosophy that I favored. In my work on classical Indian philosophy, I have always made liberal use of the principle of charity of interpretation. The basic idea is that the classical Indian philosophers must have been pretty smart people, so there is some reason to be suspicious of interpretations of their work that leave them open to simple and powerful objections. The apoha theory claims that we can start with the cognition of a unique particular, grasp the class of things that are distinct from that particular, and then use the complement of that class as the basis for a conception of the kind to which the particular belongs. This allegedly nominalist account of concept formation is open to the obvious and devastating objection that if the particular is truly unique, then the complement of the class of things distinct from that particular will consist of just that particular itself; we get nowhere by invoking two negations. The principle of charity suggests that there must be something more going on here.

What I have written about the apoha theory over the intervening years has been motivated by the desire to find that something more. But sometimes the principle of charity can lead one astray. Sometimes the theory in question really is readily defeated by an obvious objection, so that attempts to come up with a more charitable interpretation are just flights of fancy. Might that have happened in my own efforts to understand the apoha theory? The present volume, and the conference out of which it grew, struck me as an opportunity to find out. The apoha theory has a rich and complex history in India and Tibet. The experts whose views are presented in this volume collectively have a far deeper and more complete knowledge of that history than I could ever pretend to. So if my attempts to make sense of the apoha theory led me to see things in it that are not really there, this would likely come out in comparing my understanding of the theory with what those experts have to say about it. This paper represents my attempt to do some of that comparison. I have laid out my current understanding of the theory and tried to indicate whether it seems to fit with what others have to say about the theory's history and its prospects. But I draw no final conclusions. I leave it to the reader to judge whether this episode in comparative philosophy represents a cautionary tale, or instead a vindication of the liberal use of the principle of charity.

The apoha theorist claims that the meaning of "cow" is "not noncow," or more generally that the extension of a concept consists in what is excluded from that which is other. There are two quite distinct things that might be meant by this. The first involves a point that is widely acknowledged, even by those who reject the nominalist motivation behind the Buddhist program. It is that an item may be said to have a meaning or to have representational content only by virtue of its standing in contrastive relations to some range of distinct items. Since this amounts to the claim that meaning and representational content are features that an item may have only by virtue of its place within some structure, let us call this the "structural features" understanding of apoha. The second thing that might be meant by the apoha theorist's claim is that it is possible to understand concept possession entirely in terms of facts about the unique particulars that do not fall under the concept. So the claim is that a particular falls under the concept *cow* by virtue of the fact that it is other than those particulars that are noncows. Since this is tantamount to the claim that concept possession can be explained using resources derived from a world containing only unique particulars, and it is just such an explanation that is required for what Tillemans (in this volume) calls a "bottom-up" approach to apoha to succeed, let us call this the "bottom-up" understanding of apoha.

The structural features understanding involves an insight similar to that of Saussure concerning a language's phonological structure: that from a complete physical description of a given articulate sound we cannot tell what phoneme it represents, that phonetic value within a language is always a function of a system of contrast relations.[1] What is essentially the same insight would be available to classical Indian philosophers through reflection on their own practice. When Indian philosophers proceed by first describing and then rejecting the commonly held alternatives to their own position, they are clarifying the meaning of their thesis by establishing what it is not. Now the structural features insight would have struck the Buddhist as important. The nominalist is faced with the task of explaining how kind terms such as "cow" and "yellow" could have meaning if there are no universals such as *cowness* and *yellowness* common to all their instances. It would have occurred to Buddhist nominalists that the structural features insight calls into question the realist's claim that real universals could explain how such terms can be meaningful. If something can be a property only relative to a property space within which it may be seen to stand in contrast relations with other properties, then it is unclear how "yellow" could perform its semantic function just by denoting the property of *yellowness*. For the structural features insight suggests that there is no such property apart from some set of structural relations. This makes the property of being yellow something that is not intrinsic but borrowed (in Buddhist philosophical terms, not *svabhāva* but *parabhāva*): a property something could not have were it the only existing thing. Moreover, this suggests that *yellowness* can be construed as just the absence of being other than yellow, hence as the mere absence of difference. Since absences are, for the Buddhist, mere conceptual constructions, this treatment looks like a good way of purging commitment to universals from our ultimate ontology. The nature that is common to all instances of yellow might just be exclusion from what is other than yellow.[2]

It will be apparent that this will not really solve the deepest problem confronting the nominalist. The Nyāya realist will object that since there can be no absence without an existing counterpositive, there can be no absence of difference from yellow unless yellow is itself a real entity. Of course the apoha theorist disputes the Nyāya constraint on meaningful talk of absences, but we need not enter into this debate. For it will be evident in any case that this strategy will not work on its own. The nominalist claims that only unique particulars exist. Given that no properties are shared among particulars, and that there are no resemblance relations among particulars,[3] it must be explained how there can be intersubjective agreement about which particulars to call yellow, and how the ability to see particulars as similar with respect to yellowness might prove useful.

It is of course trivially true that the class of nonyellow things contains no member of the class of things that are correctly called "yellow." But it is difficult to see how we might agree on which particulars go in which class unless there were some objective fact of the matter discernible by us that determined the boundaries of one or the other class. It is Dignāga's account of apoha that is most suggestive of a structural features approach. But it has long been thought that he smuggles in conceptual resources to which a nominalist is not entitled; if there are only unique particulars, then it is illegitimate to help oneself to classes or to real resemblances. Of course this approach may have useful applications in a nominalist semantics, for instance in explaining how compositionality might work and why certain inferences go through. But the nominalist still requires an account of how we come to possess at least the most basic protoconceptual resources, given that there are neither shared universals nor similarities in the world. This I take to be what Dharmakīrti's bottom-up approach is meant to do.

The basic strategy of Dharmakīrti's approach is well described by Dunne (in this volume), and I shall not repeat the details here. In looking at how it might apply to the explanation of concept possession, let us confine our attention to the case of an organism's coming to possess certain rudimentary protoconcepts, such as those that would be necessary for the organism to operate with something like the feature-placing conceptual scheme discussed by Ganeri (in this volume).[4] The idea is that if a bottom-up approach can be made to work here, then the apoha theorist can use the structural features approach to ascend to higher levels, such as going from a basic set of protoconcepts to a feature-placing language, or from the conceptual resources of a feature-placing language to the conceptual scheme of common sense (with its ontology of reidentifiable objects and its associated subject-predicate syntactic structure).[5] Let us consider how the bottom-up formulation of the apoha theory might explain an organism's coming to experience a variety of tastes as similar in respect of being bitter. The organism we are imagining has taste buds on its tongue, and these contain structures that might usefully be thought of as locks. Each such lock is shaped so as to accept a variety of molecular keys, any one of which triggers a neural firing. Suppose such a molecular binding takes place, leading to a certain brain event. And let us suppose that this brain event either causes or else just is a certain perception (understood as a mental image). In the story that Dharmakīrti tells, this perception will be both fully distinct or vivid and nonconceptual. To call it distinct is to say that its apprehension by the organism does not lead it to act in any way that shows the organism to have confused it with any other perception. The perception is sufficiently detailed so that its difference from all other perceptions will be evident. And this seems to be what is meant by the claim that it is nonconceptual as well. Conceptuality is here understood as basically a matter of regarding as similar (and so liable to be confused), items that are in fact unique. Now this perception causes a second perception that is indistinct or obscure, and its occurrence is understood to be the entry point into conceptuality. Its indistinctness consists in its not being evidently incompatible with certain other perceptions; these turn out to be perceptions caused when other compounds bind on this receptor of the taste bud in question.[6] In effect this enables the organism to blur the distinction between the present perception and certain other perceptions that have occurred in the past or will occur in the future. It thus comes to be able to overlook differences among a class of perceptions and treat them all as similar in some respect. And this ability in turn contributes to the organism's capacity to modify its behavior based on past encounters with other molecules. If past occurrences of those perceptions that it is now disposed to treat as similar to the present perception were associated with toxicity effects, it will now be more likely to engage in avoidance behavior. And this despite the fact that the triggers involved are all distinct (the binding molecules have different shapes), and the perceptions immediately brought about by these triggers are likewise all distinct.

The obvious question to ask at this point is how this second, conception-laden perception comes about. The short answer is that the concept-free perception triggers a trace that then brings about the concept-laden perception. This is a short answer because we are also told that the organism has such a trace due to beginningless ignorance. This seems not to be an answer at all. But here we can bring in the apoha theorist's tale of the two medicinal plants (discussed by both Dreyfus and Hattori in this volume). And we may conjoin this, if we so choose, with the theory of natural selection instead of the rebirth hypothesis. The story will then go like this.[7] The trace in question may be identified as a disposition of the organism. This disposition causes the organism to overlook the differences among certain distinct perceptions, treating them as similar with respect to what we call bitterness. These perceptions happen to be triggered by molecules, many of which have toxic effects when ingested by the organism. The presence of the disposition thus enables the organism to learn to avoid ingesting certain substances. It has such a disposition because its immediate ancestors did. Its immediate ancestors had such a disposition because one of their remote ancestors happened to acquire the trait through mutation, transcription error, or some other process of genetic recombination, and this trait conferred greater reproductive success on the organism in its typical environment: organisms possessing an innate similarity space with respect to this class of taste-bud triggerings were better able to learn to avoid ingesting toxic substances. And this despite the fact that (1) there is nothing common to the shapes of the different molecules that bind with the receptor in question; and (2) there is nothing common to the perceptions actually triggered by such bindings. Thus the apoha theorist claims to have accounted for the organism's possessing the protoconcept of bitterness without making use of universals.

This last claim will, of course, be disputed. One might for instance wonder how one can speak of a given receptor's being triggered by distinct molecules if the receptor does not retain certain features from one triggering episode to the next.[8] And this sameness of features will look like something a nominalist is not entitled to. But the apoha theorist can here make essentially the same move that they just used in explaining why different perceptions seem alike with respect to bitterness. They can say that it is only relative to our interests and cognitive limitations that a receptor at one time and a receptor at another time will appear to share certain features. One might next wonder how it is possible for the apoha theorist to employ the concept of neural stimulation, if there are no properties common to the many neurons of a single organism, or to neurons across organisms of the species in question. Likewise for the apoha theorist's appeal to the theory of natural selection, which would be difficult to formulate if there were no such things as genotypes and phenotypes. But once again the apoha theorist will reply that its seeming to us that distinct particulars are similar in all being neurons or belonging to a single genotype or phenotype is just a function of our interests and cognitive limitations. The opponent will no doubt find this move quite annoying, but they may have to concede that if the appeal to interests and cognitive limitations works in the original case, it can be used by the apoha theorist to endlessly defer invocation of genuinely shared natures.

And so the opponent will try a new tack. The bottom-up formulation of apoha requires that there be real causal relations in the world. Contact between sense faculty and object must cause a perception; this perception in conjunction with a certain trace must cause a second perception; this second perception must causally contribute to successful practice; and so on. And since causal relations are best seen as relations between universals, the nominalist project must fail in the end. The apoha theorist responds to this objection by quite forthrightly conceding that causation must therefore be a myth or mere useful fiction: since there can strictly speaking be no such kinds as those of the sense faculty and perception, it cannot be true that perceptions are caused by contact with a sense faculty.[9] But their reply cannot stop here. If we are wrong to believe there are real causal regularities in the world, then we need some explanation of why this error should uniformly lead to successful practice. And so the apoha theorist must hold that there are event pairs (e_1, e_2), (e_3, e_4), (e_5, e_6), etc., such that the occurrence of the first member brings about the production of the second member; the error lies only in supposing there to be something common among e_1, e_3, etc., and something common among e_2, e_4, etc.

The question is whether this is comprehensible. One could, of course, claim that since all explanation must end somewhere, there is no ignominy in saying that this is just how things are. But it is not clear that all appeals to universals have been successfully discharged. Buddhists appear to recognize a counterfactual element in causal claims: to say that *C* causes *E* is in part to say that had *C* not occurred then *E* would not have occurred. So long as *C* and *E* can be thought of as tokens of event types, this counterfactual can be construed as a universal generalization: events of type *E* never occur in the absence of events of type *C*. But when particular events do not share in common kinds, this way of understanding the counterfactual is closed off. If it is true that e_1 produces e_2, this requires that it be true of e_1 that, had it not occurred, then e_2 would not have occurred. And given that e_1 actually did occur, it is not clear how this could be true of it in the absence of some nature that grounds event identity across possible situations. This would suggest that there cannot be the kinds of facts that would explain the usefulness of belief in causal regularities in the absence of shared natures.

At this point the apoha theorist might usefully invoke the notion of a God's-eye point of view.[10] To say this is not to say that the apoha theory depends for its coherence on the notion of an all-perfect being, or even on the notion of an omniscient being. The notion of the God's-eye point of view is introduced here only as a heuristic device. All we are asked to do is imagine how things might seem to a being that had our interests but lacked our cognitive limitations—that was timelessly aware of all states of affairs in their full particularity. We are not asked to suppose that there actually is such a being. The point is just to see how our belief in causal regularities could be a mere concession to our cognitive limitations and yet still have objective grounding. This is, after all, a common Buddhist strategy: our erroneous belief in the existence of a chariot that is not ultimately real is explained as reflecting concessions to our interests and cognitive limitations. (About this strategy, more later.) We might thus think of the present use of the notion of an omniscient being as just a colorful way of putting the idea of there being a "how things are anyway," a nature of reality that prehends from all our interests and cognitive limitations.

An omniscient being would be both memorious (like Borges's "Funes the Memorious") and prescient. Such a being would also cognize each particular in its full individuality. And it would cognize each of the infinitely many particulars all at once. Being prescient, such a being would not have to use predictions about what is likely to come next in order to obtain what it wants and avoid what it doesn't want. Thus it would not need to look for regularities among event-type pairs. And absent the need to look for such regularities, there would be no need to overlook differences among individual events and see them as falling into kinds. Such a being would have no use for the notion of a causal law. Such a being would see the belief that there are causal laws as a superimposition that falsifies reality in important ways. Yet such a being could also be brought to understand why such a belief would prove useful for creatures with our cognitive limitations. Indeed it could see that in some cases we even get things right—that the trends we think we see are grounded in the event pairs that actually, timelessly obtain. It could see why it should be conventionally true that contact with a sense faculty causes perceptions.[11]

There is, it is true, something decidedly odd in the Buddhist's saying that causation is just a myth or useful fiction. From the outset, the Buddhist tradition has used causation to explain our erroneous belief in the reality of entities that turn out, on their analysis, to be mere conceptual fictions. The chariot is, we are told, not ultimately real because it makes no causal contribution to the world over and above that of its parts. And our belief that there are chariots turns out to be useful because when those parts are arranged in the right way, they collectively cause effects in which we take an interest, such as transporting persons. So it may be somewhat disconcerting to now be told that causation is likewise something we project onto the world. And yet the resulting view is not obviously incoherent. It might be that we usefully believe there are chariots because of facts that also make it useful for us to believe that certain causal regularities obtain. The claim about causation must be the last step in the explanatory chain. For once this step is taken, causation will no longer be available to help explain why some error should prove useful. But we can perhaps see why this should be where the Buddhist pursuit of absolute objectivity would end up. If it is true that all is impermanent, and also true that our belief in enduring things is an error made understandable by the facts about the world and about us, then it might have been foreseen that causation would turn out to be a concession to our limitations.

Another question that might be raised about the bottom-up formulation of the apoha theory concerns the role that perceptions or mental images play in it. Hale (in this volume) says that so far as he can tell, they do no useful work in explaining our linguistic behavior and so may as well be dropped. And Chakrabarti complains that since the apoha theorists never dropped talk of mental images, their theory must be based on a problematic code conception of meaning. Perhaps both concerns can be addressed by considering something Dreyfus (in this volume) tells us about the Tibetan understanding of Dharmakīrti's account. This is that they took the core claim to be that a mental image is mistakenly taken for what is common to all the instances of a kind term. And the mistake is said to be twofold: a mental image, which exists only in a mind, cannot exist where the instances of an extramental kind exist; and a mental image, being a real particular, cannot be something that is shared among many instances.[12] Now to identify the second as a mistake seems tantamount to conceding that there is nothing in the nature of the mental image itself that explains the ability of the organism to treat various stimulations as similar with respect to bitterness. To point out that the mental image is itself a particular is, for instance, to deny that it can serve to represent a variety of stimulations by virtue of being a "blurry" image, one that lacks determinacy in some dimension.[13] And this is to say that all the work is actually being done by the "taking" here: that the move from nonconceptual to conceptual perceptual cognition represents a difference not in what the organism cognizes but in how the organism behaves with respect to what it cognizes. The difference between the first ("vivid") perception and the second ("obscure") perception does not lie in their natures but in how the organism treats them: as excluding all other presented particulars, or as excluding only certain other presented particulars. So the image itself does "cancel out." What matters is just that the organism behave in ways that increase its overall success rate, by behaving as if it took the presented particular as an instance of a kind, something with shared causal capacities. The claim that this comes about through the second perception being indistinct or obscure may just be a colorful way of saying that the organism is disposed to behave in a way that shows it to possess the concept of bitterness.

If this is correct, then it may go some way toward dispelling Chakrabarti's concern that Dharmakīrti gives us no more than a code conception of language. On such a conception, words function as a kind of code that communicates to the hearer the ideas in the mind of the speaker. Such a conception is clearly problematic for any number of reasons.[14] But this need not have been the apoha theorist's conception. Let us ignore for the moment the distance that actually separates what is required for possession of the sorts of protoconcepts involved in a feature-placing conceptual scheme and what is required for full-blown linguistic competence. Let us suppose that the bottom-up approach can tell us what is meant by an utterance of "bitter." On the present understanding of this approach, the meaning of "bitter" would not be the "obscure" mental image in the speaker's mind that caused the utterance of the word. For there is no such "obscure" mental image—not in the relevant (Lockean) sense. There is just the taking of an image as if it were obscure or indistinct—as if it were not incompatible with certain other images. And this taking is something that may be manifested in behavior, and thus something about which there can be intersubjective agreement. Apoha theorists repeatedly claim that in their view there is no real entity that is the meaning of a word.[15] By this they mean that there is nothing in their ultimate ontology that may be identified as the referent of a meaningful expression—neither the particular, nor the universal, nor the particular as inhered in by the universal, etc. But this might also be construed as a way of saying that instead of looking for something that could be identified as the meaning of a word, we should look at how language use is tied to other forms of behavior.

On this way of looking at the bottom-up account, there is no such entity as the concept of bitterness. There is, though, a cognitive process that manifests the relevant concept possession. This process consists in the taking of a percept as an instance of the repeatable feature of being bitter.[16] And one may legitimately ask what it is like to engage in this process. We have already discussed how this process is said to come about, namely, through activation of traces that reflect the interests and cognitive limitations of the organism. But it may still be quite mysterious how this process yields up the result that it does. Given the resources that are available to the nominalist, how can we understand the transition from taking the percept as the particular that it is to taking the percept as an instance of a kind? A key point here is that in an ontology of pure particulars, the only thing remotely like a property is the uniqueness of the particular. It is like a property in that it effects a kind of bipartition of the universe: the particular itself on one side, everything else on the other. But being in principle unrepeatable, this is not yet a property.[17] And if it were the basis of the cognitive process, no progress would be made toward concept possession. To possess the concept is to treat as similar anything excluded by what is "other." If the particular's uniqueness were the basis for what is to count as "other," then since what is excluded from this "other" is just the particular itself, nothing could count as similar to it. And yet if we suppose that the organism is able to treat as "other" not all distinct percepts but just those percepts that do not resemble the given percept with respect to bitterness, then the detour by way of "exclusion from the other" will look superfluous. For it would seem that it can possess this ability only by possessing the concept of bitterness. And we are still left in the dark as to how the organism might have come by this concept given that there are no real resemblances in the world.[18]

I believe it is in response to this dilemma that bottom-up apoha theorists devised a strategy employing two distinct ways of excluding or rendering "other." The first involves the notion of the other derived from the uniqueness of particular *p*. In this way we arrive at the notion of a class of entities having in common their distinctness from the particular in question. Since this distinctness is just the uniqueness of the particular, no suspect entities have been smuggled in. The second involves a kind of selective overlooking. Due to activation of the relevant trace, the organism overlooks the distinctness of particulars *q* and *r* from the particular in question, *p*. But this is not expressed through any positive treatment of *q* and *r*. It rather comes out in how the organism expresses its taking particulars as other than *p*, namely, by rejecting them as incompatible with *p*; its overlooking of difference consists just in its not so rejecting *q* and *r*. This overlooking in effect creates a second class: what is excluded from the "other" with respect to *p* without being identical to *p*. This excluding lacks the sort of direct grounding in reality possessed by the first sort. But its being a mere overlooking is thought to exempt it from the need for such grounding. And the claim about our interests being equally satisfied by *p*, *q*, and *r* is meant to go some way toward giving it indirect grounding. Now when the organism acquires the disposition to fail to reject *q* and *r* as incompatible with *p*, it thereby manifests its taking what is other than *p* to include less than all the particulars distinct from *p*. But this in turn means that the class of what is excluded from the other will now include *q* and *r* as well as *p*. Its disposition to treat *q* and *r* as similar to *p* may thus be expressed as its taking all three as included in what is excluded from what is other than *p*.[19]

This is the situation that I thought might usefully be compared to what happens when we combine classical negation with three-valued negation. The classical negation of proposition *p* effects a bipartition on the world: all those states of affairs compatible with *p* and all those states of affairs incompatible with *p*, with the two sets together comprising the totality of states of affairs. Three-valued negation effects a tripartition: in addition to the aforementioned types, there are those states of affairs that are neither compatible nor incompatible with *p*. Suppose we first effect a tripartition among all possible states of affairs: those states of affairs that are such that *p*, those that are such that non-*p*, and those that are such that neither *p* nor non-*p*. Suppose we then effect a bipartition on the world, dividing it into those that are such that non-*p* and those that are not such that non-*p*. The states of affairs that are not such that non-*p* may then include states of affairs that were not among those of which *p* was true. This is so because if the tripartition yields three mutually disjoint sets, then the complement of one such set will include the other two. Now when this analogy is applied to the bottom-up apoha theory, all the heavy lifting is actually done by the way in which the third set of the tripartition is generated, namely, through an overlooking of difference between particular *p* on the one hand and particulars *q* and *r* on the other. Still the analogy does, I believe, help shed some light on how the process might work.

Some cautions are in order about this analogy. This is meant to be no more than an account of the metaphysics behind the process, an explanation of how it might pay to act as if there are patterns in a world of unique particulars. It is not meant to explain why the process gives the right results, namely, results that conform to our pretheoretic intuitions about kind membership. Nor is it meant to explain how we go about mastering kind concepts. That explanatory work is to be done by the tale of the two medicinal plants, or more generally by appeal to causal processes shaped by interest-relative factors. The analogy is also not meant to suggest that the bottom-up approach might be formulated as an "apoha semantics" expressed in model-theoretic terms. This is clearly ruled out by the fact that the "bottom" for this account consists in just pure particulars. We understand negation as an operation applied to propositions or to sets. But propositions and sets are abstract objects. It is not clear what it would mean to apply negation to a pure particular.[20] Thus it is not clear how a model-theoretic account could be given that involved set-theoretic operations on the apoha theorist's ultimate domain.[21] Finally I should say that I suspect Hugon may be right about the historical record: Śāntarakṣita probably did not intend to answer Kumārila's formulation of the first horn of the dilemma by invoking two kinds of negation. His immediate aim in bringing up the distinction was probably to answer an objection that involves an appeal to the psychological feel of linguistically mediated cognition. Nonetheless he was aware of the distinction between two kinds of negation, so-called verbally bound and nominally bound; and these do in fact behave in the manner of choice and exclusion negation respectively. I know of no smoking gun that proves the apoha theorists modeled their "exclusion of the other" on what happens when we combine two styles of negation. It does still strike me as plausible that they may have had some such idea in mind.

One intriguing bit of evidence comes from the Navya-Nyāya philosopher Mathuranātha, who maintained that the constant absence of the mutual absence of pot is *potness*. Absences are, for Nyāya, the truth makers for negative statements. The statement "there is no pot on the table" is made true by an occurrence of absence of pot in the locus of the table. A constant absence is what holds when the counterpositive (e.g., pot in the case of absence of pot) never occurs in the locus in question. "There are no horned hares" is made true by the constant absence of horn from head of hare. Mutual absences are the truth makers for statements of difference. That a pot is not a cloth is made true by the occurrence in pot of mutual absence of cloth (i.e., difference from cloth).

The proof of the equivalence of *potness* and constant absence of mutual absence of pot runs as follows: The mutual absence of pot occurs in all and only the nonpots. Hence the constant absence of the mutual absence of pot must occur in all and only pots. But it is *potness* that occurs in all and only pots. Hence the constant absence of the mutual absence of pot is *potness*. This claim is intriguing in that it involves deriving a universal from a particular substance using two distinct sorts of absence, and the two absences it uses can be seen as truth makers for two different sorts of negation. Mutual absence seems to correspond to nominally bound negation, while constant absence better fits the model of verbally bound or sentential negation. This suggests some awareness on Mathuranātha's part that combining these two types of negation does not yield what we would expect from classical double negation.

But there is more. Just as the apoha theorist was faced with the non-generalization or circularity objection,[22] so Mathuranātha was confronted with the objection that the constant absence of the mutual absence of pot is not *potness* but rather pot. So in both cases there is said to be the difficulty that a two-negations strategy will only land us back where we began. In the case of the Navya-Nyāya doctrine, the objection goes by way of the principle of double negation for constant absence, or DA. Letting "$-x$" stand for the constant absence of x, the principle is:

DA: $--x = x$

Letting "~" stand for mutual absence, and "c" for counterpositive, the proof runs as follows:

1. $--{\sim}x = {\sim}x$	by DA
2. c of $--{\sim}x$ = c of ${\sim}x$	by 1 and the identity of counterpositives of identicals
3. c of $--{\sim}x = -{\sim}x$	by the definition of "counterpositive"
4. c of ${\sim}x = x$	by the definition of "counterpositive"
5. $-{\sim}x = x$	by 2, 3, 4

This poses a difficulty for Mathuranātha, who had argued that $-{\sim}$ x is identical not with x but with x-ness.

Raghunātha diagnosed the problem as stemming from DA. For Nyāya, negations represent absences, which have a distinct place in their ontology. DA suggests that by combining absences we arrive at something that is not an absence, a positive real of some sort. So DA seems to violate the basic rules of the Nyāya categorial scheme. Rejecting DA would block the difficulty pointed out earlier, but it would also block the derivation of *potness* from constant absence of mutual absence of pot. So this would not help Mathuranātha. A Buddhist nominalist might offer a more useful suggestion. On their view absences are mere conceptual constructions, and consequently not truth makers for negative statements. It follows that DA is false, since it asserts the equivalence of the truth maker for "there is a pot" with whatever makes assertible "there is the absence of the absence of pot." Since there is no truth maker for the latter statement, there cannot be such an equivalence. This reason for rejecting DA leaves untouched the derivation of *potness* from the constant absence of mutual absence of pot.

It will not have escaped the reader, however, that Mathuranātha's derivation depends on resources that are not available to the Buddhist nominalist. For the derivation to work, the mutual absence of pot that is in play must have *potness* as delimitor (*avacchedaka*); otherwise the mutual absence would occur in other pots. Can this commitment also be avoided? The suggestion would be that the tale of the two medicinal plants and their distinct causal powers might be put to use here. In that case the overlap between Mathuranātha's derivation and the two-negations interpretation of apoha would be almost complete. Of course there remains the problem that Mathuranātha was active several centuries after Buddhism had been extirpated from India. So this can be at best no more than an intriguingly suggestive bit of evidence. Still it is interesting to note that some Indian philosophers are on record as having explored a two-negations strategy. Perhaps this was also in the minds of some apoha theorists.

Tillemans has argued, both in this volume and elsewhere, that the apoha theory should be seen as a sort of error theory. I might agree with this characterization, but it depends on what is meant by an error theory. An error theorist claims that when we take there to be Xs we commit an error, but presumably they will go on to say something more about why we commit the error. And that something more may take the error theorist in the direction of something like reductionism, or of something like eliminativism. I think the apoha theory is best thought of as something like a reductionism about kinds, but before defending this answer, I should say what I take reductionism and eliminativism to be. Most basically, these represent stances one might take toward things of some kind *K*, with "kind" understood in a quite wide-ranging and varied way, so as to cover such types as mental events, demons, enduring physical objects, mereological sums, phlogiston, chariots, diachronic identity for artifacts, moral duties, the causal powers of aggregates, etc. Toward any of these kinds one might take one or another of three possible stances: realist, reductionist, or eliminativist. To take a realist stance would be to hold that things of that sort will be found in our final ontology, that they are ingredients of the world as it is independently of the concepts we happen to employ in representing the world. To be a realist about mental events, for instance, is to hold that an inventory of the world would be incomplete if it failed to include such things as pain sensations, even if it did contain such things as brain events. A realist about mental events, in other words, takes it that the view that there are such things is more than just an artifact of our way of representing the world, that when we put "pain sensation" down on our inventory sheet we are not simply repeating an item that has already been entered elsewhere, such as under "C-fiber firings."

Reductionists and eliminativists alike reject the realist stance toward a given kind. They deny that that sort of item is in our final ontology. The reductionist and the eliminativist do, however, agree with the realist that it makes sense to ask what belongs in our final ontology. This sets all three apart from those (such as Mādhyamikas) who deny that coherent sense can be made of the idea of an ultimate ontology. Realists, reductionists, and eliminativists about items of kind *K* all share the semantic realist attitude, to the effect that the world divides up into facts that are a certain way quite apart from the conceptual scheme we happen to employ. They disagree about whether items of kind *K* belong in the final inventory, but they all reject the semantic antirealist position that questions of an ontological character lack determinate sense apart from considerations of our interests and limitations.[23]

Reductionism about the *K*s is often put as the claim that *K*s "just consist in" or "just are" things of some quite different sort. So reductionism about chariots might be put as the view that a chariot just consists in a set of parts arranged in a certain way. Likewise a reductionist about heaps might claim that a heap of sand just is all the grains of sand in close proximity to one another. And a reductionist about mental events might hold that, say, a pain sensation just is the firing of C-fibers. This way of talking suggests that the item undergoing reduction is in some sense identical with the base items to which it is being reduced. But identity cannot be the right relation between the reduced and the reduction base. First, the reduced is one while the reduction base is typically many: the one chariot consists of many parts. And it is quite impossible to see how one thing could be identical with many things. Second, identity would leave unexplained the "just" in "just is" or "just consists in." If the heap is identical with the grains of sand, which are in our final ontology, why not say that the heap is as well? The reductionist's "just" is meant to separate their view about the *K*s from the realist's. But identity between the reduced and the reducers would make the two positions indistinguishable.

The position I call Buddhist Reductionism handles this problem by distinguishing between two ways in which something might be said to exist: conventionally and ultimately.[24] Those things that are in the final ontology are said to be ultimately real, while those things that we commonly take to be real but that are not in the final ontology may instead be conventionally real. What undergoes reduction is then said to be conventionally but not ultimately real, while the base items to which it is reduced are ultimately real. Since identity cannot hold between what is merely conventionally real and what is ultimately real, the chariot is not identical with the parts.[25]

This means, though, that the chariot does not really (i.e., ultimately) exist, it is only thought to exist because of a certain set of conventional practices. Thus one will now want to know how reductionism about *K*s is distinct from eliminativism, which likewise maintains that *K*s are not ultimately real. The answer is that while both take the belief that there are *K*s to result from certain practices (typically linguistic practices), they disagree about the usefulness of these practices. The reductionist holds that, given the facts concerning those things that are ultimately real, beliefs based on such conventional practices reliably lead us to obtain our goals. This is because these beliefs' not being ultimately true stems from their representing concessions to our interests and cognitive limitations. The folk theory that commits us to the existence of chariots reflects our interest in having means of transportation, as well as our inability to keep track of all the distinct parts that make up the chariot. The eliminativist about *K*s, on the other hand, holds that the conventional practices that generate belief in *K*s do not reliably lead to successful practice, and hence should be replaced. Stock examples of objects of such eliminative treatment are disease-causing demons, phlogiston, and planetary epicycles. In all such cases it is held that the theory or practice that commits us to the items in question fails to capture aspects of what is ultimately real that might be usefully exploited by creatures with our interests and cognitive limitations.

Reductionist and eliminativist about *K*s agree that in believing there are *K*s we commit a kind of error: our belief does not reflect the ultimate truth. What they disagree about is what sort of error this is. The reductionist holds it to be a useful error for creatures like us, while the eliminativist sees it as at best useless if not positively harmful. It should now be clear why I think that the apoha theory is a kind of reductionism about universals. Indeed it is difficult to see how one might espouse eliminativism about universals. While nominalists hold that universals are not ultimately real, they must agree that our interests would not be served if we failed to treat individuals as belonging to kinds. Indeed the apoha theory turns on precisely the claim that it is our interests and cognitive limitations that explain our seeing particulars as belonging to kinds. Perhaps the point of calling apoha theory an error theory is to say that we just do see things that way, but that in doing so we commit an error. In that case, though, we are owed an explanation of why it is that so seeing things leads to success. Why suppose that the next time we put our hand in what we take to be fire, equally painful results will follow, if all that can be said about the kinds *fire* and *pain* is that our belief in them is the result of beginningless ignorance?[26]

One last point is in order on this matter. The Buddhist Reductionist requires that, for a reduction to go through, there be an explanation of the error's success. It matters at what level—conventional or ultimate—this explanation is to be couched. For if the explanation must be put in terms of the ultimate truth, then the apoha theorist's account of our conceptual resources cannot count as a reduction of universals. Since on their account only particulars are ultimately real, there can be no adequate conception of the ultimate nature of reality.[27] Hence no statement could be ultimately true. This rules out an ultimate-level explanation of our predilection for employing kind concepts. Fortunately for the present account, however, the Buddhist Reductionist practice is to give explanations of the required sort at the conventional level. This can be seen in the fact that in the Abhidharma form of Buddhist Reductionism, reductive explanations are generally formulated so as to bring in facts about "our" interests. Since the person is the primary target of Buddhist Reductionism, such explanations could not be ultimately true. Still the account is not circular, since if the reduction goes through, then all talk of "our" interests can be understood as a mere *façon de parler* for certain highly complex regularities that obtain among the ultimate reals.

We set out at dawn. The road was easy enough to follow. There was just one fork, with the left branch clearly marked in such a way as to indicate we should not take it. (It went to Janīpha.) And yet we only reached Śrughna at dusk. We traveled eagerly, since we knew of Śrughna's reputation as a pottery center. But the light was failing when we entered the village. In the crepuscular gloom we could only obscurely perceive the large objects by the side of the road, just barely making out their forms. Still we knew that they were not cows, not even the newly fashionable clay cows. We knew they were pots. And so at the end of the day we knew that we had attained our goal.[28]

Notes

1. Śāntarakṣita puts the point quite pithily: "no affirmation without distinction" (*nānvayo 'vyatirekavān*) (TS 1020).

2. As Ole Pind's contribution to this volume makes clear, Dignāga does not explicitly defend his theory of apoha in this way. He does, though, hint at this at PS V.31a, when he defends the claim that the word *siṃśapā* excludes particulars such as *palaśā* trees and does not merely denote the property of being a *siṃśapā* by giving the analogy that boastful persons, though seeming to refer to their own nature, are understood to do so by distinguishing themselves from others. My thanks to Ole Pind for bringing this passage to my attention.
3. Dignāga appears not to have recognized this constraint on the nominalist. He makes free use of the thesis that the instances of a kind term resemble one another, for instance in this account of designation: "The method of joint presence and absence is the means whereby a word designates its meaning, and this is occurrence in the similar and nonoccurrence in the dissimilar" (anvayavyatirekau hi śabdasyārthābhidhāne dvāram, tau ca tulyātulyayor vṛttyavṛttī) (PSV ad PS v. 34). This suggests that Dignāga might rightly be called an "ostrich nominalist." Dharmakīrti on the other hand is explicit in his denial of resemblances among the ultimate reals. See, for example, PV II.2a.
4. One way to help guard against unwittingly smuggling in conceptual resources to which the nominalist is not entitled is to locate the discussion at the prelinguistic or nonlinguistic level. So let us suppose that the organism in question is a human infant, or a frog. My thanks to Amita Chatterjee for pointing out the importance of this restriction.
5. I take Abhidharma to have made the last move. That is, one way of understanding Abhidharma is as a systematic attempt to work out how to reduce our folk ontology of enduring substances to an ontology of tropes or property particulars (*dharmas*, which are roughly what Nyāya means by *guṇas*). In working out this reduction they showed how the conceptual resources behind our ordinary talk of enduring objects could be acquired and used by creatures with the conceptual resources of a feature-placing language. What Ābhidharmikas seem not to have realized, though, is that they had not yet expunged universals from their ontology. The bottom-up approach of the apoha theorist might be seen as an attempt to remedy that defect.
6. Here, in the description of the second perceptual image as indistinct, is the point at which one fears Dharmakīrti may have made Locke's mistake, of conferring on the mind the mysterious ability to form abstract ideas, mental images lacking in certain of the features normal to images formed through the relevant sense modality. (For the difficulties with this view, see Berkeley's critique in the introduction to *The Principles of Human Knowledge*.) Certain Abhidharma accounts of *saṃjñā* (perceptual identifications) appear to go down this road. But I shall proceed on the assumption that Dharmakīrti did not commit Locke's error.
7. I intend this story to be broadly consistent with the account of concept acquisition discussed by Amita Chatterjee (in this volume).
8. The apoha theorist of course denies that such sense faculties as taste receptors endure longer than an instant. But then they must explain why it is useful for us to treat what are actually distinct entities as if they were one enduring particular. This explanation will start with the claim that these entities are sufficiently alike that they will seem to us to make up a single enduring thing. But now it seems as if real similarities are once again being invoked, so the explanation cannot

end here. The same explanatory burden arises, only at a different point in the story.

9. Long before the development of the apoha theory, Buddhist philosophers were already acknowledging that on their view our ordinary conception of causation can never obtain. But the part of our ordinary conception they had in mind is the part that sees the cause as performing the action of producing the effect. Given the Buddhist theory of momentariness, no existent can bring about a change, so nothing can be a cause in this sense. But this still leaves open the possibility that there is causation in the Humean sense of constant conjunction between event types. The present point is that for a radical nominalist this would seem to be a problem as well.
10. To my knowledge no apoha theorist actually did. It was a conversation about the apoha theory with Kent Machina some years ago that led me to see this as a possible move. But it was the invocation of Borges's "Funes the Memorious" by Amita Chatterjee (in this volume) that made me realize its full potential.
11. We are used to the idea that there is an element of idealization in the causal laws we discover through scientific investigation. Thus we know that when we arrive at laws of motion after measuring the speed at which a ball runs down an inclined plane, we are abstracting away from such real-world factors as friction and air resistance. Plato would have us believe that in idealizing we are coming closer to apprehending that which is most fully real. The present thought is that things are quite the other way around: that to the extent that idealization means overlooking real difference, it takes us away from our knowledge of how things really are. The move to idealization is an epistemic descent, not an ascent.
12. The first mistake is just the one that representationalists accuse naive realists of making. Representationalists typically claim that in perception we project what is actually a feature of a perceptual mental state onto the external object. For a useful discussion of projectivism, see Shoemaker, 1990, 109–131.
13. This conception of the mental image as a "blurry" picture that by virtue of its indeterminacy is able to represent a plurality of distinct objects is suggested by Kamalaśīla's description of the mental image invoked by the word "blue" as "having the nature of hovering over" (*plavamānarūpatā*) the lotus and other blue objects (TSP *ad* TS 1120–1121). But he and Śāntarakṣita follow Dharmakīrti (e.g., at PV I.109) in maintaining that qua real particular, the mental image is just as determinate as any other particular; it is how the image is treated in reflective judgment that makes it appear to accord with a plurality of objects.
14. It presupposes, for instance, that both speaker and hearer already possess a kind of language of thought with all the conceptual resources that the apoha theory was supposed to explain. This in turn leads to the problem of explaining the coordination that is necessary among speakers for establishing linguistic conventions, when no one can be assured that others share the same language of thought.
15. For example, Śāntarakṣita at TS 1090–1092.
16. This is closely related to the processes involved in thoughts about the bitter, such as hoping that there not be a bitter percept. But let us confine our attention here to a process that is tied to representation of a particular. The apoha theorists seem to have held that once we have an explanation of the ability to perceive a

particular as an instance of a kind *K*, we can use resources from this account to go on to explain the ability to have thoughts about a *K* or about what they call the object-in-general.

17. Thus, Vaiśeṣika puts its individuators into a category distinct from those of universal and quality.
18. I take this to be the heart of the misgivings expressed by Hale (in this volume) concerning the strategy of two negations. The first horn of this dilemma—the difficulty that arises from trying to make the uniqueness of the particular the basis of the exclusion—is what I have elsewhere called the circularity problem (e.g., in Siderits 1999) and that Hugon (in this volume) calls the nongeneralization problem. I take the second horn of the dilemma to pose an insuperable obstacle. And I take it this became evident to apoha theorists after Dignāga. So my working hypothesis has been that they were trying to find a way around the first horn of the dilemma.
19. Another way to put this would be to say that the transition from nonconceptual to conceptual perception involves a change in what is treated by the organism as "other" with respect to the percept. In the first stage *p* and what is taken as "other" together exhaust the domain; in the second they do not. And it is traces fueled by interest-relative factors that bring about this change in the character of the "other," through a kind of nescience or overlooking.
20. Jonardon Ganeri made this point to me in conversation in 1997. It has taken me an embarrassingly long time to appreciate its full significance. Hale's appeal to nonidentity (in this volume) now strikes me as a promising alternative, since (as Dignāga would point out) this is the apoha-theoretic equivalent of Nyāya's individuators.
21. Consequently I am not sure how relevant Gillon's assessment of a two-negations strategy (in this volume) is to my hypothesis.
22. See note 17.
23. I here follow the usage of Dummett 1993b. Somewhat confusingly, the reductionist about items of kind *K* is also called a semantic antirealist, namely, a local semantic antirealist. This is because the reductionist about *K*s maintains that with respect to items of this sort, there may be cases where there is simply no fact of the matter concerning whether or not some situation obtains. Thus it may be because the Buddha has a reductionist view of persons that he can maintain that there is no fact of the matter concerning whether or not the enlightened person either exists or does not exist after death. But the reductionist or local antirealist is typically a global semantic realist concerning statements about things other than those of kind *K*: for such statements there is always a fact of the matter that gives the statement a determinate truth-value, even if we can never know what it is. The Mādhyamika, by contrast, espouses a global semantic antirealism: no statement has verification-transcendent truth conditions.
24. Parfit (1984) introduced the convention of using "Reductionism" to mean reductionism about persons and personal identity. The view I call Buddhist Reductionism concerns more than just persons, but the primary motivation behind its formulation lay in showing that persons are reducible to thoroughly impersonal entities and events. For details see my (2003).
25. Neither is the chariot distinct from the parts, for the same reason.

26. As Tillemans (this volume; 1999) points out, however, there may be Tibetan formulations of apoha theory that are best seen as error theories in the eliminativist sense. These are theories that make little or no use of causation and would not be amenable to replacing talk of beginningless ignorance with an account relying on natural selection.
27. It might seem as if there could be a language that is fully appropriate for a domain of unique particulars, namely, one containing only proper names. But even if ephemeral particulars could be given proper names (something Russell affirmed but the apoha theorists denied), a language consisting of no more than such names would be pointless.
28. At *Apohasiddhi* (AS) 60.11–14, Ratnakīrti uses the sentence, "this road goes to Śrughna" to illustrate his claim that the apoha theory can be used to explain compositionality. (See Patil's translation at www.cup.columbia.edu/apoha-translation.) And Brendan Gillon, the resident linguist at the Crêt-Bérard conference, remarked on a bit of philosophical jargon that kept turning up in many of the discussions: "at the end of the day."

Bibliography

ABBREVIATIONS FOR INDIAN AND TIBETAN TEXTS

Sanskrit and Tibetan editions mentioned are those used by the authors of the articles in this volume. Where need be, the authors' articles specify the choice of editions used. References to translations are occasionally given for the convenience of the reader.

AJP	*Anekāntajayapaṭāka* of Haribhadra Sūri. Skt. in Kapadia 1940.
AK	*Abhidharmakośa* of Vasubandhu. Skt. in Pradhan 1967.
AP	*Apohaprakaraṇa* of Jñānaśrīmitra. Skt. in Thakur 1987 (JNĀ).
APD	*Apohaprakaraṇa* of Dharmottara. Tib. edited in Frauwallner 1937.
AS	*Apohasiddhi* of Ratnakīrti. Skt. in Thakur 1975 (RNA).
Astdh	*Aṣṭādhyāyī* of Pāṇini. See Böhtlingk 1964.
BCA	*Bodhicaryāvatāra* of Śāntideva. Skt. in Vaidya 1960; Śānti Bhikṣu Śāstrī 1983.
CAPV	*Citrādvaitaprakāśavāda* of Ratnakīrti. Skt. in Thakur 1975 (RNA).
CS	*Catuḥśataka* of Āryadeva. Skt. in Jain Bhaskar 1971, Tib. text and Skt. fragments in Suzuki 1994.
D	Derge (*sde dge*) edition of the Buddhist canon (*tripiṭaka*) in Tib.
DhPr	*Dharmottarapradīpa* of Paṇḍita Durvekamiśra. Skt. in Malvania 1971.
HB	*Hetubindu* of Dharmakīrti. Tib. and Skt. texts and German translation in Steinkellner 1967.
HBṬ	*Hetubinduṭīkā* of Arcaṭa. Skt. in Sanghavi and Jinavijayaji 1949.
HBṬĀ	*Hetubinduṭīkāloka* of Durvekamiśra. Skt. in Sanghavi and Jinavijayaji 1949.
ĪPK	*Īśvarapratyabhijñākārikā* of Abhinavagupta. See ĪPV.
ĪPV	*Īśvarapratyabhijñāvimarśinī* of Abhinavagupta. Skt. in Subramania and Pandey 1986.
JNĀ	*Jñānaśrīmitranibandhāvali*. Works of Jñānaśrīmitra. Skt. in Thakur 1987.

KSA	*Kṣaṇabhaṅgasiddhi Anvayātmikā* of Ratnakīrti. Skt. in Thakur 1975 (RNA); transl. in Woo 1999.
LPrP	*Laghuprāmāṇyaparīkṣā* of Dharmottara. Tib. text, Skt. collected material, and German translation in Krasser 1991.
LS	*Laṅkāvatārasūtra*. Skt. in Nanjio 1923.
MB	*Mahābhāṣya* of Patañjali. Skt. in Shastri and Kuddala 1938; Kielhorn 1962/85.
MMK	*Mūlamadhyamakakārikā* of Nāgārjuna. Skt. in La Vallée Poussin (1903) 1970.
N	Narthang (*snar thang*) edition of the Buddhist canon (*tripiṭaka*) in Tib.
NB	*Nyāyabindu* of Dharmakīrti. Skt. in Shastri 1985; Malvania 1971.
NBh	*Nyāyabhāṣya* of Vātsyāyana (= Pakṣilasvāmin). Skt. in Thakur 1997; Taranatha Nyaya-Tarkatirtha et al. 1982.
NBhū	*Nyāyabhūṣaṇa* of Bhāsarvajña. Skt. in Yogindrananda 1968.
NBhVP	*Nyāyabhāṣyavārttikaṭīkāvivaraṇapañjikā* of Aniruddha. Skt. in Thakur 1969.
NBṬ	*Nyāyabinduṭīkā* of Dharmottara. Skt. in Shastri 1985; Malvania 1971.
NBṬṬ	*Nyāyabinduṭīkāṭippaṇa*. Skt. in Shastri 1985.
NC	*Nayacakra* of Mallavādin. Skt. in Jambūvijayajī 1966, 1976.
NCV	*Nayacakravṛtti Nyāyāgamānusāriṇī* of Siṃhasūri in Jambūvijayajī 1966, 1976.
NK	*Nyāyakandalī* of Śrīdharabhaṭṭa. Skt. in Jhā 1977.
NM	*Nyāyamukha* of Dignāga. T. 1628. See Katsura 1977, 1978, 1979, 1981, 1982, 1984, 1987. Translation in Tucci 1930.
NMJ	*Nyāyamañjarī* of Jayantabhaṭṭa. Skt. in Shastri 1982, 1983, 1984; Shah 1972; Sukla 1936; Varadacharya 1983.
NMJG	*Nyāyamañjarīgranthibhaṅga* of Cakradhara. Skt. in Shastri 1982, 1983, 1984; Shah 1972.
NR	*Nyāyaratnākara* of Pārthasārathi Miśra (commentary on ŚV). Skt. in Shastri 1978.
NS	*Nyāyasūtra* of Gautama/Akṣapāda. Skt. in Thakur 1997; Jhā 1984; Taranatha Nyaya-Tarkatirtha et al. 1982.
NU	*Nirukta* of Yāska. Skt. in Sarup 1967.
NV	*Nyāyavārttika* of Uddyotakara. Skt. edited in Thakur 1997; Jhā 1984; Taranatha Nyaya-Tarkatirtha et al. 1982.
NVTP	*Nyāyavārttikatātparyapariśuddhi* of Udayana. Skt. in Thakur 1996b.
NVTṬ	*Nyāyavārttikatātparyaṭīkā* of Vācaspatimiśra. Skt. in Thakur 1996a; Taranatha Nyaya-Tarkatirtha et al. 1982.
P	Peking edition of the Buddhist canon (*tripiṭaka*) in Tib.
PP	*Prasannapadā* of Candrakīrti. Skt. in La Vallée Poussin (1903) 1970.
PS	*Pramāṇasamuccaya* of Dignāga. Skt. of chapter 1 in Steinkellner, Krasser, Lasic 2005. Tib. of chapter 5 (apoha) edited, with Skt. fragments, in Hattori 1982. Some Skt. retranslated from the Tib. in Jambūvijayajī 1976.
PSṬ	*Pramāṇasamuccayaṭīkā* of Jinendrabuddhi. Skt. of chapter 1 in Steinkellner, Krasser, Lasic 2005. Skt. retranslations from the Tib. in Jambūvijayajī 1976. Tib. of chapter 5 (apoha) edited, with Skt. fragments, in Hattori 1982.
PSV	*Pramāṇasamuccayavṛtti* of Dignāga. Skt. of chapter 1 in Steinkellner, Krasser, Lasic 2005. Tib. of chapter 5 (apoha) edited, with Skt. fragments, in Hattori 1982. Skt. retranslations from the Tib. in Jambūvijayajī 1976.
PV	*Pramāṇavārttika* of Dharmakīrti. Editions used by authors in this book: Skt. of PV I and PVSV in Gnoli 1960; Skt. of PV I with PVSV and PVSVṬ in

Sāṃkṛtyāyana 1943/82; Skt. of PV I–IV and PVV in Sāṃkṛtyāyana 1938–40; Skt. of PV I–IV and PVV in Shastri 1968; Skt. and Tib. of PV I–IV in Miyasaka 1971–72; Skt. text and Japanese translation of PV III in Tosaki 1979, 1985; Skt. text and English translation of PV IV, k. 1–148 in Tillemans 2000. Nota bene: the chapter order of PV I–IV is as given below and is *not* the order found in Miyasaka 1971–72 or PVV.

PV I — *Pramāṇavārttika*, Svārthānumāna.

PV II — *Pramāṇavārttika*, Pramāṇasiddhi.

PV III — *Pramāṇavārttika*, Pratyakṣa.

PV IV — *Pramāṇavārttika*, Parārthānumāna.

PVBh — *Pramāṇavārttikabhāṣya Vārttikālaṅkāra* of Prajñākaragupta. Skt. in Sāṃkṛtyāyana 1953.

PVin — *Pramāṇaviniścaya* of Dharmakīrti.

PVin I — Chapter 1 of *Pramāṇaviniścaya* of Dharmakīrti. Tib. text, Skt. fragments, and German translation in Vetter 1966.

PVin II — Chapter 2 of *Pramāṇaviniścaya* of Dharmakīrti. Tib. text and Skt. text in Steinkellner 1973, translation in Steinkellner 1979.

PVinṬ — *Pramāṇaviniścayaṭīkā* of Dharmottara. Extract of Tib. text, collected Skt. material, and German translation of the extract in Steinkellner and Krasser 1989.

PVP — *Pramāṇavārttikapañjikā* of Devendrabuddhi. P. 5717.

PVSV — *Pramāṇavārttikasvavṛtti* of Dharmakīrti. Skt. in Gnoli 1960.

PVSVṬ — *Pramāṇavārttikasvavṛttiṭīkā* of Karṇakagomin. Skt. in Sāṃkṛtyāyana 1943,1982.

PVṬ$_{\text{Śāk}}$ — *Pramāṇavārttikaṭīkā* of Śākyabuddhi. Tib. in D. 4220, P. 5718.

PVṬ$_{\text{Śaṅk}}$ — *Pramāṇavārttikaṭīkā* or **Pramāṇavārttikānusāra* of Śaṅkaranandana. Tib. in D. 4223, P. 5721.

PVV — *Pramāṇavārttikavṛtti* of Manorathanandin. Skt. in Sāṃkṛtyāyana 1938–40, Shastri 1968.

RNA — *Ratnakīrtinibandhāvalī*. Works of Ratnakīrti. Skt. in Thakur 1975.

RT — *Tshad ma rigs gter* of Sakya Pandita Kunga Gyaltsen (*sa skya paṇḍita kun dga' rgyal mtshan*), in SKB vol. V.

RTGG — *Tshad ma rigs gter gyi dgongs rgyan rigs pa'i 'khor los lugs ngan pham byed* of Shākya Chokden (*śā kya mchog ldan*), in Complete Works, vol. IX and X. See Tobgey 1975.

RTKN — *Tshad ma'i rigs gter gyi dka' gnas rnam par bshad pa sde bdun rab gsal* of Go-ram-ba Sönam Sengge (*go ram pa bsod nams seng ge*), in SKB, vol. XII.

RTRG — *Tshad ma rigs gter rang 'grel* of Sakya Pandita, in SKB vol. V; Tib. text also in Nordrang Ogyen 1989.

SD — *Sāmānyadūṣaṇa* of Paṇḍit Aśoka. Skt. in Shastri 1910.

SKB — *Sa skya pa'i bka' 'bum. The Complete Works of the Great Masters of the Sa skya Sect of Tibetan Buddhism*. See Sönam Gyatso 1969.

SP — *Sāmānyaparīkṣā* of Dignāga. For fragments, see Pind (in this volume).

ŚV — *Ślokavārttika* of Kumārilabhaṭṭa. Skt. in Kunhan Raja 1946; Shastri 1978.

ŚVṬ — *Ślokavārttikaṭīkā* (*Śarkarikā*) of Jayamiśra. Skt. in Kunhan Raja 1946.

TBh — *Tarkabhāṣā* of Mokṣākaragupta. Skt. in Rangaswami Iyengar 1952; transl. in Kajiyama 1966, 1998.

TBh_{ST}	*Tarkabhāṣā* of Mokṣākaragupta. Skt. and Tib. in Shastri 2004.
TS	*Tattvasaṃgraha* of Śāntarakṣita. Skt. in Shastri 1981–82.
TSP	*Tattvasaṃgrahapañjikā* of Kamalaśīla. Skt. in Shastri 1981–82.
VC	*Vyāpticarcā* (Jñānaśrīmitra) in JNĀ.
VNi	*Vyāptinirṇaya* of Ratnakīrti. Skt. in Thakur 1975 (RNA).
VP	*Vākyapadīya* of Bhartṛhari. Skt. in Abhyankar and Limaye 1963.
VS	*Vaiśeṣikasūtras* of Kaṇāda. Skt. in Jambūvijayajī 1961.
VV	*Vidhiviveka* of Maṇḍanamiśra. Skt. in Gosvāmī 1984.

MISCELLANEOUS ABBREVIATIONS

GOS	Gaekwad's Oriental Series.
HDBK	Hiroshima Daigaku Bungakubu Kiyō.
JIP	Journal of Indian Philosophy
k.	kārikā
Skt.	Sanskrit
T	Taishō Shinshū Daizōkyō. The Buddhist canon (*tripiṭaka*) in Chinese
Tib.	Tibetan
transl.	translated
WSTB	Wiener Studien zur Tibetologie und Buddhismuskunde
WZKM	Wiener Zeitschrift für die Kunde des Morgenlandes
WZKS	Wiener Zeitschrift für die Kunde Südasiens
WZKSO	Wiener Zeitschrift für die Kunde Süd- und Ostasiens

Abhyankar, K. P. and V. P. Limaye, eds. 1963. *Vākyapadīya* of Bhartṛhari. University of Poona Sanskrit and Prakrit Series, vol. 2. Poona: University of Poona Press.

Akamatsu, A. 1980. "Dharmakīrti no apoha- ron." *Tetsugaku Kenkyū* (*The Journal of Philosophical Studies*) 540: 87–115.

Akamatsu, A. 1981. "Karṇakagomin and Śāntarakṣita: On Thirteen *kārikās* Common to the *Pramāṇavārttikasvavṛttiṭīkā* and the *Tattvasaṃgraha*." *Indological Review* 3: 53–58.

Akamatsu, A. 1983. *L'Evolution de la Theorie de l'Apoha*. Unpublished PhD thesis. Paris: Université de la Sorbonne Nouvelle.

Armstrong, David M. 1978. *Nominalism and Realism*. Cambridge: Cambridge University Press.

Armstrong, David M. 1980. "Against 'Ostrich Nominalism': A Reply to Michael Devitt." *Pacific Philosophical Quarterly* 61: 440–449.

Arnold, Dan. 2006. "On Semantics and *Saṃketa*: Thoughts on a Neglected Problem with Buddhist *Apoha* Doctrine." JIP 34: 415–478.

Bandyopadhyay, Nandita. 1982. "The Concept of Similarity in Indian Philosophy." JIP 10: 239–275.

Bergmann, Gustav. 1958. "Frege's Hidden Nominalism." *The Philosophical Review* 67: 437–59.

Bermúdez, José. 2003. "Nonconceptual Mental Content." Entry in the *Stanford Encyclopedia of Philosophy*: http://plato.stanford.edu/entries/content-nonconceptual/.

Block, Ned. 1997. "Inverted Earth." In *The Nature of Consciousness: Philosophical Debates*, edited by Ned Block, Owen Flanagan, and Güven Güzeldere, 677–693. Cambridge, MA: MIT Press.

Block, Ned. 2003. "The Harder Problem of Consciousness." *Disputatio* 15: 5–49.

Bloom, Paul. 2004. *Descartes' Baby*. New York: Basic Books.

Böhtlingk, Otto. 1964. *Pâṇini's Grammatik (herausgegeben, übersetzt, erläutert und mit verschiedenen Indices versehen)*. Hildesheim: Georg Olms Verlags-buchhandlung.

Borges, Jorge Luis. 1964. "Funes the Memorious." In *Labyrinths, Selected Stories and Other Writings*. New York: New Directions.

Brandom, R. 1994. *Making It Explicit: Reasoning, Representing, and Discursive Commitment*. Cambridge, MA: Harvard University Press.

Bronkhorst, J. 1999. "Nagārjuna and Apoha." In Katsura 1999, 17–24.

Cardona, George. 1981. "On Reasoning from *Anvaya* and *Vyatireka* in Early Advaita." In *Studies in Indian Philosophy, A Memorial Volume in Honour of Pundit Sukhalji Sanghvi*, edited by D. Malvania and N. J. Shah, 79–104. Ahmedabad: L. D. Institute of Indology.

Chakrabarti, A. 2006. "Universal Properties in Indian Philosophical Traditions." In *Encyclopedia of Philosophy*, 2nd edition, edited by Donald Borchert, vol. 9, 580–587. Detroit: Macmillan Reference USA, Thomson Gale.

Chalmers, D. 2003. "The Content and Epistemology of Phenomenal Belief." In *Consciousness: New Perspectives*, edited by Q. Smith and A. Jokic, 220–272. New York: Oxford University Press.

Chattopadhyay, M. 2002. *Ratnakīrti on Apoha*. Kolkata: Centre of Advanced Study in Philosophy.

Clark, Austen. 1993. *Sensory Qualities*. Oxford: Clarendon Press.

Clark, Austen. 2000. *A Theory of Sentience*. Oxford: Clarendon Press.

Clark, Austen. 2004a. "Feature-placing and Proto-objects." *Philosophical Psychology* 17 (4): 443–469.

Clark, Austen. 2004b. "Sensing, Objects and Awareness: Reply to Commentators." *Philosophical Psychology* 17 (4): 553–579.

Cohen, Henri, and Claire Lefebvre, eds. 2005. *Handbook of Categorization in Cognitive Science*. Amsterdam: Elsevier.

Dalai Lama, and Paul Ekman. 2008. *Emotional Awareness: Overcoming the Obstacles to Psychological Balance and Compassion*. New York: Times Books.

Davidson, Donald. 1984. "On the Very Idea of a Conceptual Scheme." In *Inquiries into Truth and Interpretation*, 183–198. Oxford: Clarendon Press.

Dhammajoti, Bhikkhu K. L. 2007. *Abhidharma Doctrines and Controversies on Perception*. Hong Kong: Centre of Buddhist Studies, University of Hong Kong.

Dravid, Raja Ram. 1972. *The Problem of Universals in Indian Philosophy*. Delhi: Motilal Banarsidas.

Dreyfus, Georges. 1997. *Recognizing Reality*. Albany, NY: SUNY Press.

Dummett, Michael. 1993a. *Origins of Analytical Philosophy*. Cambridge, MA: Harvard University Press.

Dummett, Michael. 1993b. "Realism and Anti-Realism." In M. Dummett, *The Seas of Language*, 462–478. Oxford: Oxford University Press.

Dunne, John. 1998. "Nominalism, Buddhist Doctrine of." In *Routledge Encyclopedia of Philosophy*, edited by E. Craig, vol. 7, 23–27. London: Routledge.

Dunne, John. 2004. *The Foundations of Dharmakīrti's Philosophy*. Studies in Indian and Tibetan Buddhism. Boston: Wisdom.

Dunne, John. 2006. "Realizing the Unreal: Dharmakīrti's Theory of Yogic Perception." JIP 34: 497–519.

Eckel, David. 2008. *Bhāviveka and his Buddhist Opponents.* Harvard Oriental Series. Cambridge, MA: Harvard University Press.

Fodor, Jerry A. 1998. *Concepts: Where Cognitive Science Went Wrong.* Oxford: Oxford University Press.

Fodor, Jerry A. 2004. "Having Concepts: a Brief Refutation of the Twentieth Century." *Mind and Language* 19: 29–47.

Fodor, Jerry A. 2006. *Hume Variations.* Oxford: Oxford University Press.

Frauwallner, Erich. 1932–33. "Beiträge zur Apohalehre. I. Dharmakīrti. Übersetzung." WZKM 39: 247– 285; ibid, 40: 51– 94.

Frauwallner, Erich. 1935. "Beiträge zur Apohalehre I. Dharmakīrti. Zusammenfassung." WZKM 42: 93– 102.

Frauwallner, Erich. 1937. "Beiträge zur Apohalehre II. Dharmottara." WZKM 44: 233– 287.

Frauwallner, Erich. 1959. "Dignāga, sein Werk und seine Entwicklung." WZKSO 3: 83– 164.

Frauwallner, Erich. 1992. *Erich Frauwallner, Nachgelassene Werke* II: *Philosophische Texte des Hinduismus.* Hrsg. von G. Oberhammer und C. H. Werba. Vienna: Verlag der Österreichischen Akademie der Wissenschaften.

Funayama, Toru. 2000. "Mental cognition (*mānasa*) in Kamalaśīla's Theory of Direct Perception." *Tetsugaku Kenkyū* (*The Journal of Philosophical Studies*) 569: 105–132.

Ganeri, Jonardon. 1999a. "Self-intimation, Memory and Personal Identity." JIP 27: 469–483.

Ganeri, Jonardon. 1999b. "Dharmakīrti's Semantics for the Particle *eva*." In Katsura 1999, 101–115.

Ganeri, Jonardon, 2001. *Philosophy in Classical India: The Proper Work of Reason.* London: Routledge.

Gibson, James J. 1979. *The Ecological Approach to Visual Perception.* Boston, MA: Houghton Mifflin.

Gillon, Brendan S. 1987. "On Two Kinds of Negation in Sanskrit." *Lokaprajñā* 1: 85–99.

Gillon, Brendan S. 1992. "A Review of *Indian Philosophy of Language* by Mark Siderits." *Canadian Philosophical Reviews* 12 (5): 359–360.

Gillon, Brendan S. 1999. "Another Look at the Sanskrit Particle *eva*." In Katsura 1999, 117–130.

Gillon, Brendan S. 2007. "Pāṇini's *Aṣṭādhyāyī* and Linguistic Theory." JIP 35: 445–468.

Gillon, Brendan S., and Richard P. Hayes 1982. "The Role of the Particle *eva* in (Logical) Quantification in Sanskrit." WZKS 26: 195–203.

Gnoli, R. 1960. *The Pramāṇavārttikam of Dharmakīrti: the First Chapter with the Autocommentary.* Edited by Raniero Gnoli. Serie Orientale Roma, vol. 23. Roma: Instituto italiano per il medio ed estremo oriente.

Gopnik, Alison, and A. N. Meltzoff. 1997. *Words, Thoughts and Theories.* Cambridge, MA: MIT Press.

Gosvāmī, M. 1984. *Vidhiviveka of Maṇḍanamiśra, with Nyāyakaṇikā of Vācaspatimiśra.* Edited by Mahāprabhulāl Gosvāmī. Varanasi: Tara Printing Works.

Hahn, Michael. 1971. *Jñânashrîmitras Vrttamâlâstuti. Ein Beispielsammlung zur altindischen Metrik. Nach dem tibetischen Tanjur zusammen mit der mongolischen Version herausgegeben, übersetzt und erläutert.* Wiesbaden: Harrassowitz.

Hahn, Michael. 1989. "Sanskrit Metrics—As Studied At Buddhist Universities in the Eleventh and Twelfth Century A.D." *Adyar Library Bulletin* 28: 30–60.

Haldane, John, and Crispin Wright, eds. 1993. *Reality, Representation and Projection.* Oxford: Oxford University Press.

Hale, Bob. 1979. "Strawson, Geach and Dummett on Singular Terms and Predicates." *Synthese* 42: 275–295.

Hale, Bob. 1996. "Singular terms (1)." In *Frege: Importance and Legacy*, edited by Matthia Schirn. Berlin: Walter de Gruyter, 438–457.

Hale, Bob. 1999. "Realism and its Oppositions." In *A Companion to the Philosophy of Language*, edited by Bob Hale and Crispin Wright. Oxford: Blackwell.

Harnad, Stevan. 2005. "To Cognize is to Categorize: Cognition is Categorization." In *Handbook of Categorization in Cognitive Science*, edited by Henri Cohen and Claire Lefebvre, 20–42. Amsterdam: Elsevier.

Hattori, Masaaki. 1968. *Dignāga, On Perception*. Cambridge, MA: Harvard University Press.

Hattori, Masaaki. 1977. "The Sautrāntrika Background of the Apoha Theory." In *Buddhist Thought and Asian Civilization: Essays in Honor of Herbert V. Guenther on his Sixtieth Birthday*, edited by H. V. Guenther, L. S. Kawamura, and K. Scott, 47–58. Emeryville, CA: Dharma Publications.

Hattori, Masaaki. 1980. "Apoha and Pratibhā." *Sanskrit and Indian Studies: Essays in Honour of Daniel H. H. Ingalls*, edited by M. Nagatomi et al., 61–73. Studies of Classical India, vol. 2. Dordrecht: D. Reidel Publishing Co.

Hattori, Masaaki. 1982. *The Pramāṇasamuccayavṛtti of Dignāga with Jinendrabuddhi's Commentary, Chapter Five: Anyāpoha-parīkṣā. Tibetan Text with Sanskrit Fragments*. Memoirs of the Faculty of Letters 21. Kyoto: Kyoto University.

Hattori, Masaaki. 1996. "Discussions on *Jātimat* as the Meaning of a Word." In *Śrījñānāmṛtam: A Memorial Volume in Honour of Professor Shri Niwas Shastri*, edited by Vijaya Rani, 387–394. Delhi: Parimal Publications.

Hattori, Masaaki. 2000. "Dignāga's Theory of Meaning. An Annotated Translation of the *Pramāṇasamuccayavṛtti*, Chapter V: Anyāpoha-parīkṣā (I)." In *Wisdom, Compassion, and the Search for Understanding: The Buddhist Studies Legacy of Gadjin M. Nagao*, edited by Jonathan A. Silk, 137–146. Honolulu: University of Hawai'i Press.

Hayes, Richard P. 1980. "Dignāga's View on Reasoning (*svārthānumāna*)." JIP 8: 219–277.

Hayes, Richard P. 1986. "An Interpretation of *Anyāpoha* in Diṅnāga's General Theory of Inference." In *Buddhist Logic and Epistemology: Studies in the Buddhist Analysis of Inference and Language*, edited B. K. Matilal and R. D. Evans, 31–58. Dordrecht, Holland: D. Reidel Pub. Co.

Hayes, Richard P. 1988. *Dignāga on the Interpretation of Signs*. Studies of Classical India, vol. 9. Dordrecht, Holland: Kluwer Academic Publishers.

Hayes, Steven C, Kirk Strosahl, and Kelly G. Wilson. 1999. *Acceptance and Commitment Therapy: an Experiential Approach to Behavior Change*. New York: Guilford Press.

Hernstein, R. J., and D. H. Loveland. 1964. "Complex Visual Concepts in the Pigeon." *Science* 146: 549–550.

Hernstein, R. J., Donald H. Loveland, and Cynthia Cable. 1976. "Natural Concepts in Pigeons." *Journal of Experimental Psychology: Animal Behavioral Processes* 2 (4): 285–302.

Herzberger, Hans. 1975. "Double Negation in Buddhist Logic." *JIP* 3: 3–16.

Hoornaert, P. 2001. "Bhāviveka's Critique of *Parikalpitasvabhāva* and of Dignāga's *Anyāpoha* Theory." *Religion and Culture* 13 (Hokuriku Society for Religious and Cultural Studies): 12–47.

Horgan, T., and J. Woodward. 1985. "Folk Psychology is here to Stay." *Philosophical Review* 94: 197–226.

Hugon, Pascale. 2008. *Trésors du raisonnement. Sa skya Paṇḍita et ses prédécesseurs tibétains sur les modes de fonctionnement de la pensée et le fondement de l'inférence. Edition et traduction annotée du quatrième chapitre et d'une section du dixième chapitre du* Tshad ma rigs pa'i gter. WSTB 69 (1) and 69 (2). Vienna: Arbeitskreis für tibetische und buddhistische Studien.

Hume, David. 1960. *A Treatise of Human Nature.* Edited by L. A. Selby-Bigge. Oxford: Clarendon Press.

Inami, Masahiro. 2000. "Astu yathā tathā." In *Indo no Bunka to Ronri, Tosaki Hiromasa Hakase Koki Kinen Ronbunshū* (Culture and Logic in India, Festschrift for Dr. Hiromasa Tosaki), 359–398. Fukuoka: Kyūshū University Press.

Ishida, Hisataka. (forthcoming). "On the Classification of *anyāpoha*." In *Religion and Logic in Buddhist Philosophical Analysis: The Proceedings of the Fourth International Dharmakīrti Conference, Vienna, August 23–27, 2005*, edited by H. Krasser, H. Lasic, E. Franco and B. Kellner. Vienna: Verlag der Österreichischen Akademie der Wissenschaften.

Jackson, Frank. 1977. *Perception: A Representative Theory.* Cambridge: Cambridge University Press.

Jain Bhaskar, Bhagchandra. 1971. *Āryadeva's Catuḥśatakam, Along with the Candrakīrti vṛtti and Hindi Translation.* Edited by Bhagchandra Jain. Nagpur: Alok Prakashan. This book contains verses and commentary retranslated from Tibetan into Sanskrit.

Jambūvijayajī, Muni. 1961. *Vaiśeṣikasūtras of Kaṇāda, with Candrānanda's Vṛtti.* Edited by Muni Jambūvijayajī. GOS, vol. 136. Baroda: Oriental Institute.

Jambūvijayajī, Muni 1966–1976. *Dvādaśāraṃ Nayacakram of Ācārya Śrī Mallavādi Kṣamāśramaṇa with the commenary Nyāyāgamānusāriṇī of Śrī Siṃhasūri Gaṇi Vādi Kṣamāśramaṇa.* 2 vols. Edited with critical notes by Muni Jambūvijayajī. Bhavnagar, India: Sri Jain Atmanand Sabha.

Jhā, Durgādhara. 1977. *Padārthadharmasaṅgraha with Nyāyakandalī of Śrīdhara Bhaṭṭa.* Edited with Hindi translation by Durgādhara Jhā. 2nd ed. Varanasi: Sampurnananda Sanskrit University.

Jhā, Gāṅgānāṭha 1984. *The Nyāya-sūtras of Gautama. With The Bhāṣya of Vātsyāyana and the Vārtika of Uddyotakara.* Edited and translated by Gāṅgānāṭha Jhā. Delhi: Motilal Banarsidass.

Jhā, Gāṅgānāṭha. 1986. *The Tattvasaṃgraha of Śāntarakṣita with the commentary of Kamalaśīla.* Edited and translated by Gāṅgānāṭha Jhā. Delhi: Motilal Banarsidass.

Kajiyama, Yūichi. 1966. *Introduction to Buddhist Philosophy.* Kyoto: Kyoto University. Reprinted in Kajiyama 1989.

Kajiyama, Yūichi. 1973. "Three Kinds of Affirmation and Two Kinds of Negation in Buddhist Philosophy." WZKS 17: 161–175. Reprinted in Kajiyama 1989.

Kajiyama, Yūichi. 1989. *Studies in Buddhist Philosophy.* Kyoto: Rinsen Book Co., Ltd.

Kajiyama, Yūichi. 1998. *An Introduction to Buddhist Philosophy: An Annotated Translation of the Tarkabhāṣā of Mokṣākaragupta: Reprint with Corrections in the Author's Hand.* Vienna: Arbeitskreis für Tibetische und Buddhistische Studien Universität Wien.

Kapadia, H. R. 1940. *Anekāntajayapatakā of Haribhadra Sūri.* Edited with *Vṛtti* and Municandra Sūri's *Vivaraṇa* by H. R. Kapadia. 2 volumes. GOS, vol. 88. Baroda: Oriental Institute.

Katsura, Shōryū. 1977, 1978, 1979, 1981, 1982, 1984, 1987. *Inmyōshōrimonron Kenkyū [A Study of the Nyāyamukha].* (1) in HDBK 37, 1977; (2) in 38, 1978; (3) in 39, 1979; (4) in 41, 1981; (5) in 42, 1982; (6) in 44, 1984; (7) in 46, 1987.

Katsura, Shōryū. 1979. "The *Apoha* Theory of Dignāga." IBK 28: 16–20.

Katsura, Shōryū. 1986. "Jñānaśrīmitra on Apoha." In *Buddhist Logic and Epistemology: Studies in the Buddhist Analysis of Inference and Language*, edited b y B. K. Matilal and R. D. Evans, 171–184. Dordrecht, Holland: D. Reidel Publishing Co.

Katsura, Shōryū. 1989. "Gainen—apoha-ron o chūshin ni" (Concepts: an Essay on *Apoha* Theory). Iwanami-kōza Tōyō-shisō, vol. 10. *Indo-Bukkyō* 3, 135–159

Katsura, Shōryū. 1991. "Dignāga and Dharmakīrti on *Apoha*." In *Studies in the Buddhist Epistemological Tradition: Proceedings of the Second International Dharmakīrti Conference (Vienna, June 11–16, 1989)*, edited by Ernst Steinkellner, 129–44. Vienna: Verlag der Österreichischen Akademie der Wissenschaften.

Katsura, Shōryū. 1993. "On Perceptual Judgment." In *Studies in Buddhism in Honour of A. K. Warder*, edited by N. K. Wagle and F. Watanabe, 66–75. Toronto: Centre for South Asian Studies at the University of Toronto.

Katsura, Shōryū 1999. *Dharmakīrti's Thought and its Impact on Indian and Tibetan Philosophy: Proceedings of the Third International Dharmakīrti Conference*, edited by Shōryū Katsura. Beiträge zur Kultur- und Geistesgeschichte Asiens. Vienna: Verlag der Österreichischen Akademie der Wissenschaften:

Kellner, Birgit. 2004. "Why Infer and Not Just Look?" In *The Role of the Example (dṛṣṭānta) in Classical Indian Logic*, edited S. Katsura and E. Steinkellner, 1–51. WSTB 58. Vienna: Arbeitskreis für tibetische und buddhistische Studien, Universität Wien.

Kenny, Anthony. 1963. *Action, Emotion and Will*. London: Routledge and Kegan Paul.

Keyt, C. M. 1980. *Dharmakīrti's Concept of the Svalakṣaṇa*. Unpublished PhD dissertation, University of Washington.

Kielhorn, Franz. 1985. *The Vyākaraṇa-Mahābhāṣya of Patañjali*. 3 vols. Poona: Bhandarkar Oriental Research Institute. This is the 4th edition. The 3rd edition, revised and annotated by K.V. Abhyankar, was published in 1962.

Kim, Jaegwon. 1988. "What is 'Naturalized Epistemology'?" *Philosophical Perspectives* 2: 381–405. The issue was entitled "Epistemology."

Kitagawa, Hidenori. 1973. *Indo-Koten Ronrigaku no kenkyū: Jinna no taikei*. Rev. ed. Tokyo: Suzuki Gakujutsu Zaidan.

Krasser, Helmut. 1991. *Dharmottara's kurze untersuchung der Gültigkeit einer Erkenntnis (Laghuprāmāṇyaparīkṣā)*. Materialen zur Definition gültiger Erkenntnis in der Tradition Dharmakīrtis, vol. 2. Vienna: Verlag der Österreichischen Akademie der Wissenschaften.

Krasser, Helmut. 1995. "Dharmottara's Theory of Knowledge in his *Laghuprāmāṇyaparīkṣā*." JIP 23: 247–271.

Krasser, Helmut. 2002. *Śaṅkaranandanas Īśvarāpākaraṇasaṅkṣepa*. Vienna: Verlag der Österreichischen Akademie der Wissenschaften.

Kumar, Mahendra., ed. 1939, 1941. *Laghīyastraya of Akalaṅka, with Prabhācandra's Nyāyakumudacandra*. Manikacandra Digambara Jain Granthamālā, vol. 38, 1939; 39, 1941. 2 volumes. Reprinted Sri Garib Das Oriental Series, vols. 121 and 122, Delhi 1991

Kunhan Raja, C. 1946. *Ślokavārttika of Kumārila Bhaṭṭa*. Edited, with *Śarkarikā* of Bhaṭṭaputra Jayamiśra, by C. Kunhan Raja. Madras University Sanskrit Series, vol. 17. Madras: University of Madras.

La Vallée Poussin, Louis de. (1903) 1970. *Mūlamadhyamakakārikās (Mādhyamikasūtras) de Nāgārjuna, avec la Prasannapadā commentaire de Candrakīrti*. St. Petersburg: Bibliotheca Buddhica IV. Reprint, Osnabrück: Biblio Verlag.

Laine, J. 1998. "Vācaspatimiśra." In *Routledge Encyclopedia of Philosophy*, edited by Edward Craig. New York: Routledge.

Lasic, Horst. 2000a. *Jñānaśrīmitra's Vyāpticarcā: Sanskrit text, Übersetzung, Analyse*. WSTB 48. Vienna: Arbeitskreis für Tibetische und Buddhistische Studien.

Lasic, Horst. 2000b. *Ratnakīrti's Vyāptinirṇaya: Sanskrit Text, Übersetzung, Analyse*. WSTB 49. Vienna: Arbeitskreis für Tibetische und Buddhistische Studien Universität Wien.

Lawrence, S., and E. Margolis. 1999. *Concepts: Core Readings*. Cambridge, MA: MIT Press.

Levine, Joseph. 2004. "Thoughts on Sensory Representation: A Commentary on Austen Clark's *A Theory of Sentience*." *Philosophical Psychology* 17 (4): 541–551.

Lewis, David. 1983. "New Work for a Theory of Universals." *Australasian Journal of Philosophy* 61: 343–377.

Longuenesse, Beatrice. 1998. *Kant and the Capacity to Judge*. Translated by Charles T. Wolfe. Princeton: Princeton University Press.

Malvania, D. 1971. *Dharmottara-pradīpa. Paṇḍita Durveka Miśra's Dharmottara-pradīpa. Being a sub-commentary on Dharmottara's Nyāya-bindu-ṭīkā, a commentary on Dharmakīrti's Nyāya-bindu*. Edited by D. Malvania. Patna: Kashi Prasad Jayaswal.

Matilal, B. K. 1971. *Epistemology, Logic, and Grammar in Indian Philosophical Analysis*. Series Minor, vol. 111. The Hague: Mouton and Co.

Matilal, B. K.1986. *Perception: An Essay on Classical Indian Theories of Knowledge*. Oxford: Clarendon Press.

Matilal, B. K. 1998. *The Character of Logic in India*. Albany: SUNY Press.

Matilal, B. K. 2002. *Mind, Language and World*. The Collected Essays of Bimal Krishna Matilal, vol. 2. Edited by Jonardon Ganeri. New Delhi: Oxford University Press.

Matthen, Mohan. 2004. "Features, Places and Things: Reflections on Austen Clark's Theory of Sentience." *Philosophical Psychology* 17 (4): 497–518.

Maturana, Humberto. 1980. "Biology of Cognition." In *Autopoiesis and Cognition: The Realization of the Living*, edited by Humberto Maturana and Francisco Varela, 5–62. Boston: Reidel.

McCrea, L., and Parimal Patil. 2006. "Traditionalism and Innovation." JIP 34: 303–366.

McCrea, L., and Parimal Patil. 2010. *Buddhist Philosophy of Language in India: Jñānaśrīmitra's Monograph on Exclusion*. New York: Columbia University Press.

Metzinger, Thomas. 2003. *Being No One*. Cambridge, MA: MIT Press.

Millar, Alan. 1991. *Reason and Experience*. Oxford: Clarendon Press.

Millikan, Ruth. 2000. *On Clear and Confused Ideas*. Cambridge: Cambridge University Press.

Mimaki, K. 1976. *La Réfutation bouddhique de la permanence des choses [Sthirasiddhidūṣaṇa] et la preuve de la momentaneité des choses [Kṣaṇabhaṅgasiddhi]*. Publications de l'Institut de civilisation Indienne, vol. 41. Paris: A. Maisonneuve.

Miyasaka, Y. 1971–72. *Pramāṇa-vārttika-kārikā* (Sanskrit and Tibetan). Edited by Y. Miyasaka. *Acta Indologica* 2: 1–206.

Mookerjee, S. (1935) 1993. *The Buddhist Philosophy of Universal Flux: an Exposition of the Philosophy of Critical Realism as Expounded by the School of Dignāga*. Calcutta: University of Calcutta. Reprint, Delhi: Motilal Banarsidass.

Mookerjee, S., and H. Nagasaki. 1964. *The Pramāṇavārttikam of Dharmakīrti: An English Translation of the First Chapter with the Autocommentary and with Elaborate Comments* [k. 1–51]. Nava Nālandā Research Publication, vol. 4. Patna: Nava Nālandā Mahāvihāra.

Much, Michael Torsten. 1994. "Uddyotakaras Kritik der *apoha*-Lehre (*Nyāyavārttika* ad NS II, 66)." WZKS 38: 351–366.

Mukhopadhyaya, P. K. 1984. *Indian Realism*. Calcutta: K. P. Bagchi.

Murphy, Gregory L. 2004. *The Big Book of Concepts*. Cambridge, MA: MIT Press.

Nagatomi, M. 1968. "*Mānasa-pratyakṣa*: A Conundrum in the Buddhist Pramāṇa System." In *Sanskrit and Indian Studies: Essays in Honor of Daniel H. H. Ingalls*, edited by M. Nagatomi et al., 243–260. Dordrecht, Holland: D. Reidel.

Nanjio, B. 1923. *Laṅkāvatārasūtra.* Edited by Bunyu Nanjio. Kyoto: Otani University Press.

Nordrang Ogyen (*nor brang o rgyan*). 1989. *Tshad ma rigs pa'i gter gyi rang gi 'grel pa* of Sa skya Paṇḍita Kun dga' rgyal mtshan. Edited by Nor brang o rgyan. Lhasa: Bod ljongs mi dmangs dpe skrun khang.

Oberhammer, Gerhard et al. 1991–96. *Terminologie der frühen philosophischen Scholastik in Indien. Ein Begriffswörterbuch zur altindischen Dialektik, Erkenntnislehre und Methodologie.* Vols. 1 and 2. Vienna: Verlag der Österreichischen Akademie der Wissenschaften.

Oetke, Claus. 1993. *Bemerkungen zur Buddhistischen Doktrin der Momentanheit des Seienden: Dharmakīrtis Sattvānumāna.* WSTB 29. Vienna: Arbeitskreis für Tibetische und Buddhistische Studien, Universität Wien.

Parfit, Derek. 1984. *Reasons and Persons.* Oxford: Oxford University Press.

Patil, Parimal G. 2003. "On What It Is that Buddhists Think About: *Apoha* in the *Ratnakīrtinibandhāvalī.*" JIP 31: 229–256.

Patil, Parimal G. 2007. "Dharmakīrti's White-Lie." In *Pramāṇakīrtiḥ. Papers Dedicated to Ernst Steinkellner on the Occasion of his 70th birthday*, part 2, edited by B. Kellner, H. Krasser, and M. T. Much, and H. Tauscher, 597–619. WSTB, vol. 70.2. Vienna: Arbeitskreis für Tibetische und Buddhistische Studien Universität Wien.

Patil, Parimal G. 2009. *Against a Hindu God: Buddhist Philosophy of Religion in India.* New York: Columbia University Press.

Patil, Parimal G. (forthcoming). "History, Philology, and the Philosophical Study of Sanskrit Texts." JIP.

Peacocke, Christopher. 1992a. "Scenarios, concepts, and perception." In *The Contents of Experience: Essays on Perception*, edited by Tim Crane. Cambridge: Cambridge University Press.

Peacocke, Christopher. 1992b. *A Study of Concepts.* Cambridge MA: MIT Press.

Peacocke, Christopher. 1994. "Nonconceptual Content: Kinds, Rationales and Relations." *Mind and Language* 9: 419–429.

Peacocke, Christopher. 2001. "Does Perception Have a Nonconceptual Content?" *Journal of Philosophy* 98: 239–264.

Pind, O. 1991. "Dignāga on *Śabdasāmānya* and *Śabdaviśeṣa.*" In *Studies in the Buddhist Epistemological Tradition*, edited by Ernst Steinkellner, 268–280. Vienna: Österreichische Akademie der Wissenschaften.

Pind, O. 1999. "Dharmakīrti's Interpretation of Pramāṇasamuccayavṛtti V 36: *śabdo 'rthāntaranivṛttiviśiṣṭān eva bhāvān āha.*" In Katsura 1999, 317–332.

Potter, Karl H.. 1963. *Presuppositions of India's Philosophies.* Englewood Cliffs, N. J.: Prentice Hall.

Potter, Karl H., ed. 1977. *Encyclopedia of Indian Philosophy.* Vol. 2. Delhi: Motilal Banarsidass.

Potter, Karl H., Robert E. Buswell, Padmanabh S. Jaini, and Noble Ross Reat, ed. 1998. *Abhidharma Buddhism to 150 A.D.* Encyclopedia of Indian Philosophies, vol. 7. Delhi: Motilal Banarsidass.

Pradhan, P. 1967. *Abhidharmakośa and Abhidharmakośabhāṣya of Vasubandhu.* Edited by Prahlad Pradhan. Patna: K. P. Jayaswal Research Institute.

Prinz, Jesse. 2002: *Furnishing the Mind.* Cambridge, MA: MIT Press.

Putnam, Hillary. 1987. *The Many Faces of Realism.* LaSalle, IL: Open Court.

Quine, W. V. O. 1953. *From a Logical Point of View.* Cambridge, MA: Harvard University Press.

Quine, W. V. O. 1966. "Quantifiers and Propositional Attitudes." In Quine's *The Ways of Paradox and Other Essays*, 183–94. New York: Random House.

Raja, K. K. 1986. "*Apoha* Theory and Pre-Diṅnāga Views on Sentence Meaning." In *Buddhist Logic and Epistemology: Studies in the Buddhist Analysis of Inference and Language*, edited by B. K. Matilal and R. D. Evans, 185–92. Dordrecht, Holland: D. Reidel Pub. Co.

Rangaswami Iyengar, H. R., ed. 1952. *Tarkabhāṣā of Mokṣākaragupta*. Mysore

Rospatt, A. V. 1995. *The Buddhist Doctrine of Momentariness: A Survey of the Origins and Early Phase of this Doctrine up to Vasubandhu*. Stuttgart: Steiner.

Russell, Bertrand. 1921.*The Analysis of Mind*. London: George Allen & Unwin.

Saito, A. 2004. "Bhāviveka's Theory of Meaning." *Journal of Indian and Buddhist Studies* 52 (2): 24–31.

Sakurai, Yoshihiko. 2000. "Dharmakīrti, Śākyabuddhi, and Śāntarakṣita on Apoha." *Ryūkoku Daigaku Daigakuin Kiyō, Bungakubu Kenkyūkai* 22.

Sāṃkṛtyāyana, R. 1953. *Pramāṇavārttikabhāṣya of Prajñākaragupta*. Patna: Kashi Prasad Jayaswal Research Institute.

Sāṃkṛtyāyana, R. (1943) 1982, ed.. *Pramāṇavārttikasvavṛttiṭīkā of Karṇakagomin*. In *Karṇakagomin's Commentary on the Pramāṇavārttikavṛtti of Dharmakīrti*. Allahabad: Kitab Mahal. Reprint, Kyoto: Rinsen Books.

Sanghavi, S., and Muni Shri Jinavijayaji. 1949. *Hetubinduṭīkā of Bhaṭṭa Arcaṭa: with the Sub-commentary Entitled Āloka of Durveka Miśra*. Baroda: Oriental Institute.

Śānti Bhikṣu Śāstrī, ed. and trans. 1983. *Bodhicaryāvatāra*. Edited with Hindi translation. Lucknow: Buddhavihara.

Sarup, L, ed.. 1967. *Nirukta of Yāska*. Reprint. Delhi: Motilal Banarsidass.

Scharf, P. M. 1996. *The Denotation of Generic Terms in Ancient Indian Philosophy: Grammar, Nyāya and Mīmāṃsā*. Philadelphia: American Philosophical Society.

Schmithausen, L. 1965. *Maṇḍanamiśra's Vibhramavivekaḥ. Mit einer Studie Zur Entwicklung der indischen Irrtumslehre*. OAWV 2. Vienna: Verlag der Österreichische Akademie der Wissenschaften.

Schmithausen, L. 1987. *Ālayavijñāna*. Tokyo: International Institute for Buddhist Studies.

Shah, N. J., ed. 1972. *Nyāyamañjarīgranthibhaṅga of Cakradhara*. Ahmedabad: L. D. Institute of Indology.

Sharma, D. 1969. *The Differentiation Theory of Meaning*. The Hague: Mouton.

Shastri, D., ed. 1978. *Ślokavārttika of Śrī Kumārila Bhaṭṭa. With the Commentary Nyāyaratnākara of Pārthasārathi Miśra*. Edited by Dvārikādāsa Śāstrī. Prāchyabhārati Series, vol. 10. Varanasi: Tara Publications.

Shastri, D., ed. 1981–82. *Tattvasaṅgraha of Śāntaraksita with the Tattvasaṅgrahapañjikā of Kamalaśīla*. Reprint. Varanasi: Bauddha Bharati.

Shastri, D., ed. 1985. *Nyāyabindu of Ācārya Dharmakīrti with the Commentaries by Ārya Vinitadeva and Dharmottara and Dharmottara-Ṭīkā-Ṭippaṇa*. Dharmakīrtinibandhāvali 3. Bauddha Bhāratī vol. 18. Varanasi: Bauddha Bharati.

Shastri, D. N. 1964. *Critique of Indian Realism*. Agra: Agra University.

Shastri, G., ed. 1982–1984. *Nyāyamañjarī of Jayanta Bhaṭṭa, with Nyāyamañjarīgranthibhaṅga of Cakradhara*. Edited by Gaurīnatha Śāstrī. 3 vols. Varanasi: Sampurnananda Sanskrit University.

Shastri, H., ed. 1910. *Six Buddhist Nyāya Tracts in Sanskrit*. Edited by Haraprasād Śāstrī. Bibliotheca India, vol. 185. Calcutta: The Asiatic Society.

Shastri, Lobsang Norbu, ed. 2004. *Tarkabhāṣā of Mokṣākaragupta, Sanskrit and Tibetan Texts*. Varanasi: Central Institute of Higher Tibetan Studies.

Shastri, R., and S. D. Kuddala., eds. 1938. *Mahābhāṣya of Patañjali, with the Pradīpa of Kaiyaṭa and Uddyota of Nāgeśa.* Edited by Raghunātha K. Śāstrī and Śivadatta D. Kuddāla. Vol. 1. Bombay: Nirnaya Sagar Press.

Shoemaker, Sydney. 1990. "Qualities and Qualia: What's in the Mind?" *Philosophy and Phenomenological Research* 50: 109–131.

Siderits, Mark. 1982. "More Things in Heaven and Earth." JIP 10: 187–208.

Siderits, Mark. 1986. "Word Meaning, Sentence Meaning, and Apoha." JIP 13: 133–151.

Siderits, Mark. 1991. *Indian Philosophy of Language: Studies in Selected Issues.* Studies in Linguistics and Philosophy, vol. 46. Dordrecht: Kluwer Academic Publishers.

Siderits, Mark. 1999. "Apohavāda, Nominalism and Resemblance Theories." In Katsura 1999, 341–348.

Siderits, Mark. 2003. *Personal Identity and Buddhist Philosophy: Empty Persons.* Aldershot, UK: Ashgate.

Siderits, Mark. 2005. "Buddhist Nominalism and Desert Ornithology." In *Universals, Concepts and Qualities,* edited by Arindam Chakrabarti and Peter Strawson, 91–103. Abingdon: Ashgate.

Siderits, Mark. 2006. "Apohavāda." In *Philosophical Concepts Relevant to Sciences in Indian Tradition,* PHISPC III.5, edited by Prabal Kumar Sen, 727–736. Delhi: Munshiram Manoharlal.

Siderits, Mark. 2007. *Buddhism as Philosophy: an Introduction.* Indianapolis: Hackett Publishing, 2007.

Siegel, R. K., and W. K. Honig. 1970. "Pigeon Concept Formation: Successive and Simultaneous Acquisition." *Journal of the Experimental Analysis of Behavior* 13 (3): 385–390.

Siegel, S. 2005. "The Contents of Perception." *The Stanford Encyclopedia of Philosophy,* edited by Edward N. Zalta: http://plato.stanford.edu/archives/win2008/entries/perception-contents.

Sönam Gyatso (*bsod nams rgya mtsho*), ed. 1969. *Sa skya pa'i bka' 'bum.* The Complete Works of the Great Masters of the Sa skya Sect of Tibetan Buddhism. Tokyo: The Toyo Bunko.

Staal, J. F. 1969. "Sanskrit Philosophy of Language." In *Linguistics in South Asia,* edited by Thomas A. Sebeok, 499–531. Current Trends in Linguistics, vol. 5. The Hague: Mouton.

Stcherbatsky, T. 1984. *Buddhist Logic.* New Delhi, Munshiram Manoharlal.

Steinkellner, E. 1966. "Bemerkungen zu Īśvarasenas Lehre vom Grund." WZKSO 10: 73–85.

Steinkellner, Ernst. 1967. *Dharmakīrti's Hetubinduḥ. Teil I, Tibetischer Text und rekonstruierter Sanskrit-Text. Teil II, Übersetzung und Anmerkungen* Vienna: Verlag der Österreichischen Akademie der Wissenschaften.

Steinkellner, Ernst. 1969. "Die Entwicklung des Kṣaṇikatvānumānam bei Dharmakīrti." WZKSO 13: 361–377.

Steinkellner, Ernst. 1973. *Dharmakīrti's Pramāṇaviniścaya. Zweites Kapitel: Svārthānumāna. Teil 1: Tibetischer Text und Sanskrittexte.* Vienna: Verlag der Österreichischen Akademie der Wissenschaften.

Steinkellner, Ernst. 1976. "Der Einleitungsvers von Dharmottara's Apohaprakaraṇam." *Wiener Zeitschrift für die Kunde Südasiens und Archiv für indische Philosophie* 20: 123–124.

Steinkellner, E. 1977 "Jñānaśrīmitra's Sarvajñasiddhiḥ." In *Prajñāpāramitā and Related Systems,* edited by Lewis Lancaster and Luis Gomez, 383–393. Berkeley, CA: University of California Press.

Steinkellner, E. 1979. *Dharmakīrti's Pramāṇaviniścaya. Zweites Kapitel: Svārthānumāna. Teil 2: Übersetzung und Anmerkungen*. Vienna: Verlag der Österreichischen Akademie der Wissenschaften.

Steinkellner, E., and H. Krasser. 1989. *Dharmottaras Exkurs zur Definition gültiger Erkenntnis im Pramāṇaviniścaya: (Materialien zur Definition gültiger Erkenntnis in der Tradition Dharmakīrtis 1): Tibetischer Text, Sanskritmaterialien und Übersetzung*. Vienna: Verlag der Österreichischen Akademie der Wissenschaften.

Steinkellner, E., H. Krasser, and H. Lasic. 2005. *Jinendrabuddhi's Viśālāmalavatī Pramāṇasamuccayaṭīkā: Chapter 1*. Part 1: Critical Edition. Part II: Diplomatic Edition. Beijing-Vienna: China Tibetology Publishing House and Austrian Academy of Sciences Press.

Steinkellner, Ernst, and M. T. Much. 1995. *Texte der erkenntnistheoretischen Schule des Buddhismus*. Göttingen: Vandenhoeck & Ruprecht.

Strawson, Galen. 1989. "Red and 'Red.'" *Synthese* 78: 193–232.

Strawson, Peter F. 1963. *Individuals*. New York: Anchor Books.

Strawson, Peter F. 1966. *The Bounds of Sense*. London: Methuen.

Strawson, Peter F. 1974. *Subject and Predicate in Logic and Grammar*. London: Methuen.

Subramania Iyer, K. A. and K. C. Pandey. 1986. *Īśvara-pratyabhijñā-vimarśinī of Abhinavagupta: Doctrine of divine recognition*. Sanskrit text with the commentary Bhāskarī, edited by K. A. Subramania Iyer and K. C. Pandey. Delhi: Motilal Banarsidass.

Sukla, N. S. 1936. *Nyāyamañjarī of Jayantabhaṭṭa*. Edited by Surya Narayana Sukla. Benares: Kashi Sanskrit Series, Nyaya Section 15.

Suzuki, K., ed. 1994. *Sanskrit Fragments and Tibetan Translation of Candrakīrti's Bodhisattvayogācāracatuḥśataka*. Tokyo: Sankibo Press.

Taber, John. 2005. *A Hindu Critique of Buddhist Epistemology*. London: Routledge.

Tani, T. 1997. "Problems of Interpretation in Dharmottara's Kṣaṇabhaṅgasiddhi (1), (2) and (3)." *Bulletin of Kochi National College of Technology* 41: 19–77.

Tani, T. 1999. "Reinstatement of the Theory of External Determination of Pervasion (*Bahirvyāptivāda*): Jñānaśrīmitra's Proof of Momentary Existence." In Katsura 1999, 363–386.

Taranatha Nyaya-Tarkatirtha et al. 1982. *Nyāyadarśanam: With Vātsyāyana's Bhāṣya, Uddyotakara's Vārttika, Vācaspati Miśra's Tātparyaṭīkā and Viśvanātha's Vṛtti*, edited by Taranatha Nyaya-Tarkatirtha, Amarendramohan Tarkatirtha, Hemantakumar Tarkatirtha. 2 vols. Calcutta (Calcutta Sanskrit Series 18 and 29). Reprint. Kyoto: Rinsen Book Co.

Tarski, Alfred. 1944. "The Semantic Conception of Truth and the Foundations of Semantics." *Philosophy and Phenomenological Research* 4: 341–376.

Tarski, Alfred. 1983. "The Concept of Truth in Formalized Languages." In *Logic, Semantics, Metamathematics*, edited by John Corcoran, 2nd ed., 152–278. Indianapolis, IN: Hackett Publishing Company.

Thakur, A., ed. 1969. *Nyāyabhāṣyavārttikaṭīkāvivaraṇapañjikā of Aniruddha*. Darbhanga: Mithila Research Institute.

Thakur, A. 1975. *Ratnakīrtinibandhāvali*. Patna: Kashiprasad Jayaswal Research Institute.

Thakur, A. 1987. *Jñānaśrīmitranibandhāvali*. Patna: Kashiprasad Jayaswal Research Institute.

Thakur, A., ed. 1996a. *Nyāyavārttikatātparyaṭīkā of Vācaspatimiśra*. Nyāyacaturgraṅthikā 3. New Delhi: Indian Council of Philosophical Research.

Thakur, A., ed. 1996b. *Nyāyavārttikatātparyapariśuddhi of Udayana*. Nyāyacaturgraṅthikā 4. New Delhi: Indian Council of Philosophical Research.

Thakur, A., ed. 1997. *Nyāyasūtras with Nyāyabhāṣya of Vātsyāyana and Nyāyavārttika of Uddyotakara*. Nyāyacaturgraṅthikā 1–2.. New Delhi: Indian Council of Philosophical Research.

Tillemans, Tom J. F. 1986. "Identity and Referential Opacity in Tibetan Buddhist *apoha* Theory." In *Buddhist Logic and Epistemology*, edited by Bimal K. Matilal and R. D. Evans, 207–27. Studies of Classical India, vol. 7. Dordrecht: D. Reidel.

Tillemans, Tom J. F. 1990. *Materials for the Study of Āryadeva, Dharmapāla and Candrakīrti*. WSTB 24.1–2. Vienna: Arbeitskreis für Tibetische und Buddhistische Studien Universität Wien.

Tillemans, Tom J. F. 1995. "On the So-Called Difficult Point of the *Apoha* Theory." *Asiatische Studien/Etudes Asiatiques* 59: 854–889.

Tillemans, Tom J. F. 1999. *Scripture, Logic, Language: Essays on Dharmakīrti and His Tibetan Successors*. Studies in Indian and Tibetan Buddhism. Boston: Wisdom Publications.

Tillemans, Tom J. F. 2000. *Dharmakīrti's Pramāṇavārttika: An Annotated Translation of the Fourth Chapter (parārthānumāna)*. Vol. 1 (k. 1–148). Vienna: Verlag der Österreichischen Akademie der Wissenschaften.

Tillemans, Tom J. F. (forthcoming). "The Theory of Apoha: What does Bhāviveka Have to Do with It?" In *Religion and Logic in Buddhist Philosophical Analysis: The Proceedings of the Fourth International Dharmakīrti Conference, Vienna, August 23–27, 2005*, edited by H. Krasser, H. Lasic, E. Franco, and B. Kellner. Vienna: Verlag der Österreichischen Akademie der Wissenschaften 2011.

Tobgey, Kunzang., ed. 1975. *The Complete Works (gsung 'bum) of gSer-mdog Paṇ-chen Śākya-mchog-ldan*. Thimphu, Bhutan.

Tosaki, Hiromasa. 1979–1985. *Bukkyō Ninshikiron no Kenkyū.—Hosshō-cho Pramāṇavārttika no Genryo-ron* (A Study of Buddhist Epistemology—The Theory of Perception in the *Pramāṇavārttika* of Dharmakīrti). 2 vols. Tokyo: Daitō Shuppansha. This is an edition and Japanese translation of PV III.

Treisman, A., and Gelade, G. 1980. "A Feature-integration Theory of Attention." *Cognitive Psychology* 12: 97–136.

Tucci, Giuseppe. (1930) 1976. *The Nyāyamukha of Dignāga: The Oldest Buddhist Text on Logic, After Chinese and Tibetan Materials*. Heidelberg: Materialen zur Kunde des Buddhismus. Reprint, San Francisco: Chinese Materials Center.

Vaidya, P. L. 1960. *Śāntideva, Bodhicaryāvatāra, with the Commentary Pañjikā of Prajñākaramati*. Buddhist Sanskrit Texts, vol. 12. Darbhanga: Mithila Institute.

Van Cleve, James. 1999. *Problems from Kant*. Oxford: Oxford University Press.

Vetter, T. 1966. *Dharmakīrti's Pramāṇaviniścaya, 1. Kapitel: Pratyakṣam. Einleitung, Text der Tibetischen Übersetzung, Sanskritfragmente, deutsche Übersetzung*. Vienna: H. Böhlaus Nachf., Kommissionsverlag der Österreichischen Akademie der Wissenschaften.

Von Rospatt, A. 1995. *The Buddhist Doctrine of Momentariness*. Stuttgart: Steiner.

Waldron, William S. 2003. *The Buddhist Unconscious: The Ālaya-vijñāna in the Context of Indian Buddhist Thought*. Curzon Critical Studies in Buddhism. London: Routledge.

Watanabe, Satoshi. 1969. *Knowing and Guessing*. New York: Wiley.

Watson, Alex. 2006. *The Self's Awareness of Itself: Bhaṭṭa Rāmakaṇṭha's Arguments Against the Buddhist Doctrine of No-Self*. Vienna: The De Nobili Research Library.

Woo, Jeson. 1999. *The Kṣaṇabhaṅgasiddhi-Anvayātmika: An Eleventh-Century Buddhist Work on Existence and Causal Theory*. Unpublished PhD Dissertation, University of Pennsylvania.

Woo, Jeson. 2003. "Dharmakīrti and his Commentators on *Yogipratyakṣa*." JIP 31: 439–448.

Varadacharya, V. K. S., ed. 1983. *Nyāyamañjarī of Jayantabhaṭṭa*. Part 2. Mysore: Oriental Research Institute, University of Mysore.

Yogindrananda 1968. *Nyāyabhūṣaṇa of Bhāsarvajña*: In *Nyāyasāra* of Bhāsarvajña. Edited, with autocommentary *Nyāyabhūṣaṇa* and editor's commentary, by Yogindrananda. Varanasi.

Yoshimizu, C. 1999. "The Development of Sattvānumāna from the Refutation of a Permanent Existent in the Sautrāntika Tradition." WZKS 43: 231–254.

Yoshimizu, C. (forthcoming). "Buddhist Inquiries into the Nature of an Object's Determinate Existence in terms of Space, Time, and Defining Essence." unpublished book manuscript.

Contributors

ARINDAM CHAKRABARTI is professor of philosophy at University of Hawaii at Manoa, where he teaches analytic and comparative philosophy. After receiving his doctorate from Oxford University in 1982, he was trained in the traditional Sanskrit method in Indian logic, epistemology, and philosophy of language, and continues to do research in Nyaya, Buddhism, Kashmir Shaivism, and Vedanta. In addition to his book *Denying Existence* (1997) on the problem of singular negative existentials, he coedited, with the late P. F. Strawson, *Universals, Concepts and Qualities* (2006), and, with the late B. K. Matilal, *Knowing from Words* (1994). Currently, he is working on a book on the moral psychology of the emotions, having published two books of philosophical essays in Bengali, and a book on contemporary Western epistemology in Sanskrit.

AMITA CHATTERJEE recently became vice chancellor of Presidency University, Kolkata, India. Before that she was professor of philosophy and the coordinator of the Centre for Cognitive Science, at Jadavpur University, Kolkata. Among the books she has authored and edited are *Understanding Vagueness* (1994); *Bharatiya Dharmaniti* (ed.,1998); *Perspectives on Consciousness* (ed., 2003); *Physicalism and Its Alternatives* (Bengali, jointly ed. with M. N. Mitra and P. Sarkar, 2003); and *Mental Reasoning, Experiments and Theories* (jointly authored with Smita Sirker, 2009). She is consulting editor of *Indian Philosophical Quarterly*, *Philosophy East and West*, and the *International Journal of Artificial Intelligence and Soft Computing*. She is currently engaged in research on inconsistency-tolerance logics, consciousness studies, human reasoning ability and perception and cognition.

GEORGES DREYFUS was the first Westerner to receive the title of Geshe after spending fifteen years studying in Tibetan Buddhist monasteries. He received his Ph.D. in the History of Religions Program at the University of Virginia. He is currently a professor in the Department of Religion at Williams College. His publications include *Recognizing Reality: Dharmakīrti and his Tibetan Interpreters* (1997), *The Svātantrika-Prāsaṅgika Distinction* (coedited with Sara McClintock, 2003), and *The Sound of Two Hands Clapping: The*

Education of a Tibetan Buddhist Monk (2003), as well as many articles on various aspects of Buddhist philosophy and Tibetan culture.

JOHN D. DUNNE is associate professor in the Department of Religion at Emory University, where he is codirector of the Encyclopedia of Contemplative Practices and the Emory Collaborative for Contemplative Studies. His work focuses on Buddhist philosophy and contemplative practice. His publications include *Foundations of Dharmakīrti's Philosophy* (2004). His current research includes an inquiry into the notion of "mindfulness" in both classical Buddhist and contemporary contexts, and he is also engaged in a study of Candrakirti's *Prasannapadā*, a major Buddhist philosophical work on the metaphysics of "Emptiness." He currently serves on the board of directors of the Mind and Life Institute, and is currently a consultant or coinvestigator on various scientific studies of contemplative practices.

JONARDON GANERI is professor of philosophy at the University of Sussex. His work is at the border between Indian and analytical philosophy. He has published four books: *The Last Age of Reason: Philosophy in Early Modern India* (2011); *The Concealed Art of the Soul: Theories of Self and Practices of Truth in Indian Ethics and Epistemology* (2007); *Philosophy in Classical India: The Proper Work of Reason* (2001); and *Semantic Powers: Meaning and the Means of Knowing in Classical Indian Philosophy* (1999). He is currently working on the philosophy of mind in classical Indian thought.

BRENDAN S. GILLON teaches semantics in McGill University's department of linguistics. He is also an affiliate of McGill University's department of philosophy. His research interests include Indian logic and epistemology. He is the editor of *Logic in Early Classical India* (2008) as well as the author of a number of articles, including two translations, in collaboration with Richard P. Hayes, of the beginning of Dharmakīrti's *Svārthānumāna* chapter of the *Pramāṇavārttika*.

BOB HALE has been a member of the philosophy department in the University of Sheffield since 2006. Before that he taught in the universities of Lancaster, St. Andrews, and Glasgow. Most of his work has been in the philosophy of logic and mathematics. His published work includes *Abstract Objects* (1987); *The Blackwell Companion to the Philosophy of Language* (coedited with Crispin Wright, 1997); and *The Reason's Proper Study: Essays Towards a Neo-Fregean Philosophy of Mathematics* (coauthored with Crispin Wright, 2001), together with numerous articles in journals and edited collections. His current research is divided between his two main areas of interest—mathematics and modality—and connections between them.

MASAAKI HATTORI is professor emeritus of Kyoto University, where he taught Indian philosophy for many years. His major publications include *Dignāga, on Perception* (1968), *Ninshiki to Chōetsu: Yuishiki* (Cognition and Transcendence: Consciousness-Only, 1970), *Kodai Indo no Sinpi shiso* (Mysticism in Ancient India, 1979), and *The Pramāṇasamuccayavṛtti of Dignāga with Jinendrabuddhi's Commentary, Chapter Five: Anyāpoha-parīkṣā, Tibetan Text with Sanskrit Fragments* (1982). His major research interests include the history of Indian philosophy, Indian logic and epistemology, and the Upaniṣads.

PASCALE HUGON is a research fellow at the Institute for the Cultural and Intellectual History of Asia of the Austrian Academy of Sciences in Vienna. Her ongoing research addresses questions linked with the transmission of the Indian Buddhist religious and philosophical corpus to Tibet, with focus on the epistemological school of Buddhism. Her latest publications include a study of Saskya Paṇḍita's epistemological masterwork, the *Treasure of Reasoning*, and several articles based on newly available works by his famous predecessor Phya pa Chos kyi seng ge.

SHŌRYŪ KATSURA is now professor of Indian philosophy at Ryukoku University, Kyoto, having taught at Hiroshima University for many years. His major publications include *The Role of the Example (Dṛṣṭānta) in Classical Indian Logic* (coedited with E. Steinkellner, 2004); *Indojin no Ronrigaku* (Indian Logic, 1998); and *Inmyōshōrimonron Kenkyū* (A Study of the *Nyāyamukha*, 1977–1987). His major research interests include Buddhist logic and epistemology, Abhidharma philosophy, Madhyamaka and Yogācāra philosophy, and Mahāyana sūtras.

PARIMAL G. PATIL is the John L. Loeb Associate Professor of the Humanities in the Committee on the Study of Religion and the Department of Sanskrit and Indian Studies at Harvard University. His first two books, *Against a Hindu God* (2009) and *Buddhist Philosophy of Language in India* (forthcoming), focus on the final phase of Buddhist philosophy in India. He is currently writing a book on Indian philosophy of religion in late premodernity and early modernity.

OLE PIND was educated at University of Arhus, Denmark. His chief areas of expertise are early Buddhism, the indigenous traditions of the Pali and Sanskrit grammarians, and Indian philosophy of language. He was a coeditor of *A Critical Pali Dictionary*, and has written numerous articles on the Grammarians. He is currently preparing a study of the fifth chapter of Dignāga's *Pramāṇasamuccaya*, based on a newly available Sanskrit manuscript of an important commentary.

PRABAL KUMAR SEN received his education at Presidency College, Calcutta; Government Sanskrit College, Calcutta; and University of Calcutta. At present, he is professor of philosophy at University of Calcutta. He has been visiting fellow at Wolfson College, Oxford, and Department of Philosophy, University of Poona. He has edited Nyāya texts like *Ākhyatavādavyākhyā*, *Nyāyarahasya*, and *Mokṣavāda* by Rāmabhādra Sārvabhauma and *Anvikṣikītattvavivaraṇa* of Jānakīnātha Cūḍāmaṇi and has contributed several articles to journals, festschrifts, commemoration volumes, and anthologies.

MARK SIDERITS is currently in the philosophy department of Seoul National University. Before that he taught philosophy for many years at Illinois State University. He is the author of *Indian Philosophy of Language* (1991); *Personal Identity and Buddhist Philosophy: Empty Persons* (2003); and *Buddhism as Philosophy* (2007), as well as numerous articles. His principal area of interest is analytic metaphysics as it plays out in the intersection between contemporary analytic philosophy and classical Indian and Buddhist philosophy.

TOM TILLEMANS is an expatriate Dutch Canadian who since 1992 has occupied the chair of Buddhist Studies at the University of Lausanne in Switzerland. His initial training was in analytic philosophy, with a second training in Sanskrit, Tibetan, and Chinese. Published work has been in Buddhist Madhyamaka and Buddhist epistemology, with an increasing emphasis on issues of comparative philosophy. There have been occasional forays into Tibetan grammar and poetry. Books include: *Materials for the Study of Āryadeva, Dharmapāla, and Candrakīrti* (1990); *Dharmakīrti's Pramāṇavārttika: An Annotated Translation of the Fourth Chapter* (2000); *Persons of Authority* (1993); *Agents and Actions in Classical Tibetan* (with D. Herforth, 1989); *Scripture, Logic, Language: Essays on Dharmakīrti and his Tibetan Successors* (1999); *Pointing at the Moon: Buddhism, Logic, Analytic Philosophy* (coedited with Jay Garfield and Mario D'Amato, 2009).

Index

Bold highlighting shows importance.

abhāva. *See* absence
abheda (nondifference), 72, 97
Abhidharmakośa (AK) of Vasubandhu, 133n6
abhidheya (what is signified/expressed), 52, 78, 111
absence, 9, 30, 35, 43, 51, 74, 75, 76, 112, 121n4, 167n31, 186, 187, 197, 237, 285, 289, 295, 296; according to Kumārila (abhāva), 137, 141, 147n9, 184; of perception (upalambhābhāvamātra), 72; coabsence (vyatireka), 128; constant, 295, 296; joint, 70, 72, 73, 74 (*see* vyatireka); mutual, 295, 296, 297; synonymous with apoha, vyāvṛtti, etc., 167n31; the property of being an (abhāvatva), 198
action (karman, kriyā, pravṛtti), 3, 80, 90, 100, 102, 114, 127, 130, 131, 173, 174, 178, 179, 222, 255, 274
actionable objects, 153, 154, 156, 158
action word (kriyāśabda), 172, 173, 179
adarśanamātra (mere nonobservation), 71, 72, 82n17, 129
ādhāra (substratum), 139, 187
adhyavasāya (determination), 55, 154, 189
agovyāvṛtti (difference from noncows), 172, 191
ākāra (aspect, image, form). *See* aspect
ākṛti (configuration, characteristic of a species, universal), 2, **3**, 85, 146n1, 173, 180, 254
anekāntavāda (Jaina nonabsolutism), 12–13
antirealism, 11, 24, 253, 303n23
anugatavyavahāra (application of the same term to different things), 201
anumāna. *See* inference
anumānānumeyasambandha (inference-inferendum connection), 79
anvaya: concordance, joint presence, copresence, 70–74, 79, 128, 129; distribution, repeatability, 93, 94
anyāpoha (exclusion of what is other): and the doctrine of momentariness, 193, 194; as "apohist double negation," 54; as involving two kinds of negation, 45, 46, 47, 54; three types of, 34, 125–28. *See also* agovyāvṛtti; anyavyāvṛtti; arthātmakasvalakṣaṇānyāpoha; atadvyāvṛtti; vyavaccheda; vyāvṛtti

anyavyavacchedamātra (mere exclusion of others), 125
anyavyāvṛtti (exclusion of what is other), 136, 190, 191, 195
anyayogavyavaccheda (elimination of the link with other things), 123n19
anyonyasaṃśraya. *See* interdependence
Apohaprakaraṇa (AP) of Jñānaśrīmitra, 82n17, 161, 165n9, 166nn17, 19, 23, 167nn25, 30, 168nn41, 45
Apohaprakaraṇa (APD) of Dharmottara, 141, 142, 143
Apohasiddhi (AS) of Ratnakīrti, 36, 47, 304n28
Aristotle, 1, 14, 15, 16, 243, 245n8
Armstrong, David, 11, 25, 60
arthakriyā (efficacy, telic function), 91, 157
arthakriyākārin (causally potent), 248
arthakriyāśakti (causal powers), 126
arthakriyāsāmarthya (causal efficacy), 171
arthakriyāyogya (potential for efficacy), 145
arthāntaravyavaccheda (exclusion of other signified objects), 69
arthaśabdaviśeṣa (individual object and individual word), 67, 68
arthasāmānya (the signified object's general type, object universal), 67, 217
arthātmakasvalakṣaṇānyāpoha (objective elimination), 216
arthaviśeṣa (the individual object), 67
asādhāraṇaviśeṣaṇa (uncommon property), 178
aspect (ākāra), 55–56, 188, 67, 71, 141, 151, 187, 188, 190, 194, 196, 237–39, 242–43
Aṣṭādhyāyī (Astdh) of Pāṇini, 2, 72, 73, 123n28
asvātantrya (lack of independence), 175, 176
atadvyāvṛtti (difference from what is not that), 191, 252
avacchedaka. *See* limitor
avadhāraṇa (restriction), 73, 74
avidyā (ignorance), 26, 28, **33**, 37, 44, 102, 196, 200, 287, 300, 304n26. *See also* beginningless
ayaṃśabda ([the demonstrative pronoun] "this"), 78
beginningless (anādi): ignorance (avidyā), 28, **33**, 37, 44, 196, 200, 287, 300, 304n26; imprints, tendencies (vāsanā), 33, 62n10, 100, 101, **108n32**, 144, 220; Vedic words, 3
Berkeley, George, 17, 212, 301n6
Bhartṛhari, 4, 78, 79, 81n6, 140, 173
Bhāviveka, 51, 52, 163n4
Bodhicaryāvatāra (BCA) of Śāntideva, 201n2
bottom-up approach to apoha, 27–28, 31, 44, 46–47, 53, **54–56**, 59–61, 221, 228, 237, 243, 284, 286, 288–89, 291–94

Candrakīrti, 170
Cardona, George, 128
Catuḥśataka (CS) of Āryadeva, 51, 201n2
"causally inert" argument against universals, 4–5, 29
Chaba (phya pa chos kyi seng ge), 223
'chad pa'i tshe/'jug pa'i tshe (theoretical perspective/practical perspective; critical explanation/practical application), 62n7, 223, 227n20
charity, principle of, 283
choice negation, 295
Churchland, Patricia and Paul, 56
circularity, 45, 63, **109–11**, 117–18, 123n28, 269, 271n2, 303n18; according to Dharmakīrti, 112, 117–18; according to Kumārila, 111–112; according to Udyottakara, 110
Citrādvaitaprakāśavāda (CAPV) of Ratnakīrti, 166n20
Clark, Austen, 41–42, 229–36, 240–43
coabsence. *See* absence; vyatireka
code conception of language, 46
compositionality, 21, 26, 45, 49n8, 260, **262–63**, 271
concept. *See* vikalpa
concept formation, 12, 40
concepts, theories of: ability, 23–24; classical definition, 15–16; family resemblance, 17–18; function, 19–20; idea, 17; Kantian, 18–19; paradigmatism, 36; resemblance, 5, 12–13, 212–14; similarity-based (prototype, proxytype, exemplar), 20–21, 39–40; theory, 21–22

conceptualism, 15, 17, 208, 209, 211
convention, 35, 99, 151, 166n19, 244, 246n11, 268; conventional practices, 152, 232, 299; conventional truth or reality (*see also* saṃvṛtisat), 32, 37, 44, 45, 48n3, 88, 131, 171, 257, 290, 298, 299, 300; social, 151, 152, 155, 156, 157, 256, 267; verbal or linguistic (*see also* saṅketa; vyavahāra), 109, **111**, 113, **114–19**, 126, 131, 132, 154, 155, 156, 272n2, 302n14
copresence. *See* anvaya

Davidson, Donald, 23, 61n1
delimitor (*avacchedaka*). *See* limitor
desires, 269–70
Devendrabuddhi, 55, 57, 84, 85
Dharmakīrti: on apoha's circularity, 58, 62n14, **113–21**, 166n19, 271n2, 303n18; on causal laws, 47; causal approach to link language to world, 55, 57, 60; on concepts, 32, 40, 49n7, 87, 88, 103–5, 143–45; versus Dignāga on apoha, 27, 28, 31, 32, 55, 58, 59; on imprints, 99–102; and naturalistic accounts, 57, 58, 59, 86, 105n2, 157, 161, **207–25**, 254; and noncomputationalism, 43; objections to universals, 11, 106n14; ontology, 51, 88–90, 195, 211, 285, 286, 292, 293, 298; on sameness of effect, 94–99 (*see also* ekakārya); Tibetan interpreters, 40, 41, 57, 62n7, 120, 209, **217**, **222–25**, 226n3, 257, 291
Dharmapāla, 51
Dharmottara, 34, 36, 40, 87, 127, **141–46**, 149, 162, 163n7, 172, 190–91, 194, 197, 238
Dharmottarapradīpa (DhPr) of Paṇḍita Durvekamiśra, 164n7
difference. *See* vyatireka
difficulties, for apoha theory. *See* circularity; nongeneralization
Dignāga: on anyāpoha possessing the essential characteristics of a universal, 130, 131; apoha theory of, general formulation, 54, 66–68, 135–36, 138, 139; arguments for apoha, 173–178; on invariable connections between words and objects, 68–73, 129; on mere nonobservation (adarśanamātra), 71, 72, 82n7, 129; ontology, 171; on sentence meaning, 140, 141, 197, 253; on tadvatpakṣa, 61n3; top-down approach to apoha, 27, 28, 53, 54; and Vaiśeṣika taxonomy of universals, 30, 80, 130, 131; works on apoha, 65, 66; writing of *Pramāṇasamuccaya*, 65
distribution, repeatability. *See* anvaya
don rang mtshan gyi gzhan sel. *See* arthātmakasvalakṣaṇānyāpoha
don spyi (object universal), 217. *See also* arthasāmānya
Dummett, Michael, 2, **23–24**, 46, 239–40, 303n23
Dvādaśaśatikā, of Dignāga, 66, 81n4
'dzin stangs kyi yul (the apprehended object of application), 223

ekadharman (single property), 76, 77
ekakārya (same effect), 57. *See also* Dharmakīrti, on sameness of effects
ekapratyavamarśa: (translated as "same judgment"), 56, 57, 59, 60, **62n11**, 98, **137**, 148n30; (translated as "judgment of sameness"), 34, 35, 95–97, 113, 116–18, 120, 124n30, 213, 237, 239, 251
eliminativism, 56, **297–99**
error, error theories, 12, **13**, 41, **56**, 57, 62nn7– 8, 97, 144, 145, 168n44, 193, 200, 208, 237, **240–241**, 289, 291, **297–300**, 304n26
exclusion negation, 295

family resemblance theory (Wittgenstein). *See* concepts, theories of
feature-placing, 229, **232–37**, 240, 244, 286. *See also* Clark, Austen
Fodor, Jerry, 22–24, 26, 250
fourfold classifications of words, 172, 173
Frege, Gottlob, 2, 10, 19, 20 24, 25, 26, 240, 243

gcig tu rtogs (conception of oneness), 226n7. *See* ekapratyavamarśa
Geluk (dge lugs), 62n9, 223, 226n3

Gibson, James J. 43, 246n11, 251, 254, 255, 256
God's-eye point of view, 289–90
Goodman, Nelson, 14
Gopnik, Alison, 22
Goramba (go ram pa bsod nams seng ge), 217
Grammarians (Indian), 2–3, 88n6, 128, 172, 273
guṇaśabda (quality word), 172, 173, 179

hastasaṃjñā (ostention), 78
Herzberger, Hans, 54, 58
hetu (logical reason) 67, 72, 73, 77
Hetubindu (HB) of Dharmakīrti, 49n7, 82n16, 126
Hetubinduṭīkā (HBṬ) of Arcaṭa, 49n7
Hetubinduṭīkāloka (HBṬĀ) of Durvekamiśra, 49n7
Hetumukha of Dignāga, 66, 80, 82n12, 203n20
horizontal universal. *See* tiryaksāmānya
Hume, David, 17, 22–24, 26, 28, 212

ignorance. *See* avidyā; beginningless
image (pratibhāsa), 125, 165n16, 188, 190, 250. *See also* aspect
imprint, trace, tendency. *See* beginningless, imprints, tendencies; vāsanā
individuator (viśeṣa), 5, 9, 13, 65
inference (anumāna), 67, 69, 71, 76–79, 164n8, 171
inherence (samavāya), 7–**8**, 9, 14, 48n6, 179, 186, 198, 212
interdependence (anyonyasaṃśraya), 33, 62n14, 110, 112, 114–15, 118, **123n28**, 271
internal realism, 44, 256
Īśvarapratyabhijñākārikā (ĪPK) of Abhinavagupta, 238
Īśvarapratyabhijñāvimarśinī (ĪPV) of Abhinavagupta, 238
itaretarāśraya. *See* interdependence

Jaina, 4, 8, 12–13, 14, 48n6, 178
Janus-faced nature of concepts, 32, 34, 41
jāti (universal, genus, caste), 2, 3, 8–9, 68, 76, 85, 127, 128, 130, 134, 135, 138, 145, 146n1, 172, 173, 174, 179, 180, 183, 184, 197, 198, 202n12, 252, 254
jātiśabda (universal word), 172–74, 178–80, 183, 184, 202n12
jātivat (possessing a universal), 52, 81n6. *See* sāmānyavat; tadvatpakṣa
jātiviśiṣṭavyakti (an individual characterized by a universal), 134, 135, 184, 254
Jayanta, 35–36, 39, **134–46**, 178, 191, 197, 198
Jñānaśrīmitra, 85, 125, 150–54, 158–59, 161–62, 162n1, 172, 192, 197; on mental content, 37, 38, 158–59, 161–62
Jñānaśrīmitranibandhāvali (JNĀ) (works of Jñānaśrīmitra), 165n13, 166n16, 166nn21, 23, 167n25, 167n30, 168nn41, 45, 169n46, 206n33
joint presence. *See* anvaya
judgment of sameness/same judgment. *See* ekapratyavamarśa

Kamalaśīla, 85, 122n12, 146, 163n4, 192, 302n13
Kaṇāda, 5
Kant, Immanuel, 18–19, 26, 200
kāraka (participant), 274
karman. *See* action
Kātyāyana, 201n7
kriyā. *See* action
Kṣaṇabhaṅgasiddhi Anvayātmikā (KSA) of Ratnakīrti, 164n8, 167n31 & 32
Kumārilabhaṭṭa, 13, 35–36, 39, 45, 54, 62n14, 64, 75, 76, 77, 82n17, 111–12, 118, 120, 137–41, 143, 145–46, 178, 183–85, 295

lakṣyārtha (secondary meaning), 176
Laṅkāvatārasūtra (LS), 170, 201n1
Lewis, David, 60, 61
limitor (*avacchedaka*), 6, 297
liṅga (inferential indicator, inferential mark), 67, 81n7, 125–126, 135, 224. *See also* hetu
Locke, John, 21, 23, 26

Maddhva Vedānta, 5, 14, 48n6
Madhyamaka/Mādhyamika, 45
Madhyamakahṛdayakārikā of Bhāviveka, 51, 52

Madhyamakaśāstra/Mūlamadhyamakakārikā (MMK) of Nāgārjuna, 170, 171
Mahābhāṣya (MB) of Patañjali, 2, 172, 173
Mallavādin, 64, 65, 178, 184
many properties problem, 232–33, 236
Mathuranātha, 198, 295–97
Matilal, Bimal K. 53, 171, 273
mere nonobservation. *See* adarśanamātra
Mīmāṃsā, 3, 4, 39, 40, 130, 134, 198, 219; Bhāṭṭa Mīmāṃsā, 4, 8, 226n8; Prābhākara Mīmāṃsā, 48n6, 199, 226n8; Pūrva-Mīmāṃsakas, 180
model theory, 274–79
moderate realism, 223, 226n3
Mokṣākaragupta, 85, 165n12, 219, 238
mukhyārtha (primary meaning) 176

Nāgārjuna, 170, 171
natural selection/selectionism, 37, 41, 44, 157, 158, 246n11, 251, 287–88
naturalized epistemology, 57, 105n2
Nayacakra (NC) of Mallavādin, 65, 83n20, 184, 203n18
Nayacakravṛtti Nyāyāgamānusāriṇī (NCV) of Siṃhasūri, 65
negation: choice and exclusion, 295; external, 46, 274, 279, 282; internal, 46, 274, 279, 280, 282; nominally vs. verbally bound, 259, 261–62; terms vs. sentential, 261–62. *See also* paryudāsa; prasajyapratiṣedha; pratiṣedha
nirākaraṇa (elimination, exclusion), 114
Nirukta (NU) of Yāska, 172
nirvikalpa (nonconceptual, nondetermined), 10, 13, 41, 49n7, 87, 103, 105, 116, 150–52, 158, 165nn15–16, 168n39, 213, 218, 228, 236, 238, 245n1, 255, 287, 291, 303n19
nominalism: adequate (D. Lewis), 61; apoha account of concept formation, 12; apoha, "in a nutshell," 12, 90–94, 171–73; flatus vocis, 58; happy, 28, 32, 58; ostrich, 25–26, 28, 60, 249, 301n3; resemblance theory of, 5, 12–14, 35, 40, 132, 212–214; resourceful (H. Herzberger), 54. *See also* anyāpoha; universals
nominalistic semantics, 263–66
nominally bound negation. *See* paryudāsa
nonconceptual, nondetermined. *See* nirvikalpa
nongeneralization, 45, 47, **111**, 115, 296
Nyāya, 4, **5–10**, 11, 14, 25, 29, 30, 39, 40, 47, 52, 61n3, 65, 68, 70, 74, 81n6, 85, 134, 141, 174, 180, 186, 194, 209, 226n3, 285, 295, 296, 303n20
Nyāyabhāṣya (NBh) of Vātsyāyana (= Pakṣilasvāmin), 123n28, 172
Nyāyabhāṣyavārttikaṭīkāvivaraṇapañjikā (NBhVP) of Aniruddha, 205nn26–27
Nyāyabindu (NB) of Dharmakīrti, 106n6, 163n7, 164, 207n14, 238, 257n4
Nyāyabinduṭīkā (NBṬ) of Dharmottara. 106n7, 163n7, 227n14
Nyāyabinduṭīkāṭippaṇa (NBṬṬ) (author unknown), 164n7
Nyāyakandalī (NK) of Śrīdharabhaṭṭa, 190
Nyāyakaṇikā of Vācaspatimiśra, 188, 205n27
Nyāyamañjarī (NMJ) of Jayantabhaṭṭa, **134–46**, 146n1, 191, 197
Nyāyamañjarīgranthibhaṅga (NMJG) of Cakradhara, 147n24, 201n8, 205n27
Nyāyamukha (NM) of Dignāga, 65, 82n15
Nyāyaratnākara (NR) of Pārthasārathimiśra (commentary on ŚV), 122n8,n14, 183, 184, 202n16, 203nn16, 17, 20
Nyāyasūtra (NS) of Gautama/Akṣapāda, 121n3, 134, 146n1, 173
Nyāyavārttika (NV) of Uddyotakara, 109, 110, 121n1, 121n3, 202n13
Nyāyavārttikatātparyapariśuddhi (NVTP) of Udayana, 206n34
Nyāyavārttikatātparyaṭīkā (NVṬṬ) of Vācaspatimiśra, 188, 205n27

one over many problem, 1, 4, 28, 32, 61, 63n15
ostrich nominalism. *See under* nominalism

Paṇḍit Aśoka, 11
Pāṇini, 2–3, 72, 73–74, 123n28, 274
paradigm imagery, 266–68
paratantrasvabhāva (dependent nature), 51
parikalpitasvabhāva (imagined nature), 51
Pārthasārathimīśra, 122n8, 178, 183, 184, 202n16, 203n17

particular. *See* svalakṣaṇa; vyakti
paryudāsa (nominally bound negation), 125, 203n22, 273
Patañjali, 2–3, 123n28, 172, 173, 179, 275
Peacocke, Christopher, 19, 23, 26, 229, 238, 243, 245n7
perception (pratyakṣa), 66, 70, 81n7, 82n15, 89, 105, 126, 171, 189, 215
perceptual judgment (sāṃvṛta, smārta), 38–39, 44, 126, 128
pervaded. *See* vyāpya
pervasion. *See* vyāpti
"pet-fish" problem, 21, 26, 45. *See also* compositionality
Plato, 1, 14, 18, 62, 264, 302n11
Post, Emil, 54
Potter, Karl, 53
Prabhācandra, 13
Prajñākaragupta, 149, 152, 206n31
pramāṇa (reliable means of cognition, instrument of knowledge, means of knowledge), 29, 85, 134
Pramāṇasamuccaya (PS) of Dignāga. 29, 52, **54–55**, 58, 64, 121n3, 135, 140, 173, **183–85**, 201nn5, 8, 202n11, 259
Pramāṇasamuccayaṭīkā (PSṬ) of Jinendrabuddhi. 201n8, 202n10
Pramāṇasamuccayavṛtti (PSV) of Dignāga. 83n19, 202nn9–10, 301n3
pramāṇavāda (pramāṇa theory, Buddhist epistemology), 64, 84
Pramāṇavārttika (PV) of Dharmakīrti. **11**, **55**, 74, **91**, **95**, **103–5**, **113–14**, 110, 117, 119–20, 136, **144–45**, **186**, **204n24**, 210, 212, 215, 238
Pramāṇavārttikabhāṣya Vārttikālaṅkāra (PVBh) of Prajñākaragupta, 82n17, 206n31
Pramāṇavārttikapañjikā (PVP) of Devendrabuddhi, **55**, 61n6
Pramāṇavārttikasvavṛttiṭīkā (PVSVṬ) of Karṇakagomin, 62n11, 81nn3–4, 82n17, 121nn1, 3, 123n25
Pramāṇavārttikasvṛtti (PVSV) of Dharmakīrti, 62n11, 65, 66, **91**, **95**, 106, 108, 113
Pramāṇavārttikaṭīkā (PVṬ$_{Śāk}$) of Śākyabuddhi, 121n1
Pramāṇavārttikaṭīkā or **Pramāṇavārttikānusāra* (PVṬ$_{Śaṅk}$) of Śaṅkaranandana, 121n1
Pramāṇaviniścaya (PVin) of Dharmakīrti, 227n14
prameya (object of knowledge), 134
prasajyapratiṣedha (verbally bound negation), 125, 203n22, 273
Prasannapadā (PP) of Candrakīrti, 170, 201n2
pratibhā (flash of intuition), 141
pratibhāsa (representation, form, image), 125, 188, 190, 250
pratiṣedha (negation), 67, 70. *See also* paryudāsa; prasajyapratiṣedha
pratiṣedhamātra (mere negation), 125
pravṛtti (application), 68, 202n12, 216. *See also* action
pravṛttinimitta (ground of application), 68, 172, 202n12
Prinz, Jesse, 21
private language argument, 23, 36
protoconcept, 44, 219, 228, 235–37, 239, 240, 244, 252, 286, 288, 292
Putnam, Hilary, 23, 44, 256

Qualified Monist Vedānta, 48n6
qualifier-qualified relation (viśeṣaṇaviśeṣyabhāva), 35, 138–40
quality particular, trope (guṇa), 14, 80, 130, 138, 174, 183, 301n5
quality word (guṇaśabda), 173, 179
Quine, W. V. O., 13, 16, 23, 25, 57, 105n2, 269

Raghunātha, 296
Rāmānuja, 48n6
Ratnakīrti, 36, 37, 47, 85, 125, 149–50, 152–54, 158–59, 162n1, 165n9, 172, 192, 197
Ratnakīrtinibandhāvalī (RNA), Works of Ratnakīrti, 154, 167nn31–32, 206
realism, 1, 4, 5, 7, 8, 10–14, 20, 22, 25, 28, 32–36, 38, 39, 48n4, 56, 60, 61, 85, 90, 109, 113–16, 119, 120, 128, 130, 210, 211, 221, 223, 244, 247, 248, 253, 271n2, 285, 297, 298, 303n23. *See also* internal realism, moderate realism

reasoning through anvaya and vyatireka, 128, 301n3
recognitional capacities, 23–24. *See also* Dummett, Michael
reduction/reductionism, 42, 130, **297–300**, 301n5, 303nn23–24
representation. *See* aspect; image
resemblance theory. *See under* concepts, theories of
rlom tshod kyi chos can (assumed subject), 224
Rosch, Eleanor, 16, 20
Russell, Bertrand, 11, 13, 243, 272n5, 304n27
Ryle, Gilbert, 23

śabdasāmānya (word type), 67–68
śabdaviśeṣa (individual words), 67
sādṛṣya (similarity), 212, 226n8
sajātīya (homologous object, objects of the same class), 76, 126, 133n4, 165n12
Sakya (sa skya), 209, 217, 223, 224
Sakya Pandita (sa skya paṇḍita), 62n7, 120, 124n30, 217, 223, 227n20
sapakṣa (similar instances), 128
Shākya Chokden (śā kya mchog ldan), 224
śakya (reference), 6
Śākyabuddhi, 34, 59, 84, 85, 125, 127, 242
sāmānādhikaraṇya (colocusness, same locus of reference), 138, 139, 175
sāmānya (universal), 2, 5, 67, 68, 76, 80, 85, 92, 127, 128, 130, 135, 139, 145, 190, 198, 273
Sāmānyadūṣaṇa (SD) of Paṇḍit Aśoka, 11–12
sāmānyalakṣaṇa (universal), 51, 66, 67, 94, 126, 257, 171
Sāmānyaparīkṣā (SP) of Dignāga, 61n4, 65, 75
sāmānyavat (possessing a universal), 52
sambandhapradarśana (showing the connection), 78
samavāya. *See* inherence
saṃjñāvyutpatti (scope of the name), 78
sāṃvṛta. *See* perceptual judgment
saṃvṛtisat (conventionally real), 88, 130, 170, 171
saṅketa/saṃketa (verbal convention), 112
Śāntarakṣita, 45, 85, 117, 120, 125, 145, 172, 185, 192, 215, 219, 295
Saussure, Ferdinand de, 284
scheme-content dichotomy, 27, 29, 44, 50, 51, 53, 61n1, 249, 256–57
sel ngo'i rdzas (the substance considered from the point of view of exclusion), 224
semantic theory, 131, 160, 260, 264, 270, **274–80**; of Dharmakīrti, 214, 217; of Dignāga, 31, 66, 217; of Frege, 19
semantics, lexical, **278–82**
sense-reference distinction, 6, 19, 52–53
Ślokavārttika (ŚV) of Kumārilabhaṭṭa, 61n3, 62n14, 76, 109, 137, **183–85**, 201n8, 202n16, 203n17
Ślokavārttikaṭīkā (*Śarkarikā*) (ŚVṬ) of Jayamiśra, 201n8, 203n18
smārta. *See* perceptual judgment
speech intention (vivakṣā), 56, 77–78
song tshod kyi chos can (actual subject), 224
song tshod kyi rtags (actual reason), 224
Śrīdhara, 40, 190
superimposition/superimposed (samāropa, samāropita/āropita), 4, 12, 13, 36, 51, 116, 130, 132, **141–43**, 189, 217, 221, 240, 251, 252, 290
svalakṣaṇa (particular), 51, 66, 67, 125, 138, 150, 163n7, 171, 188, 189, 216, 238, 244, 248, 257

tadvatpakṣa (thesis of [words expressing the particulars] possessing the [universal]), 61n3, 81n6. *See* jātivat; sāmānyavat
Tarkabhāṣā (TBh) of Mokṣākaragupta, 133n4, 162n2, 165n12, 238
Tarski, Alfred, 274–75
Tattvasaṃgraha/*Tattvasaṅgraha* (TS) of Śāntarakṣita, 63n14, 146, 185
Tattvasaṃgrahapañjikā (TSP) of Kamalaśīla, 146
theory theory. *See* concepts, theories of
tiryaksāmānya (horizontal universal), 4, 154, 194

top-down approach to apoha, 27–28, 29, 31, **53–54**, 58–59, 63n15, 221, 243–44
Triṃśikā of Vasubandhu, 51
trisvabhāva (three natures), 51
Trisvabhāvanirdeśa of Vasubandhu, 51
Tshad ma rigs gter (RT) of Sakya Pandita Kunga Gyaltsen (sa skya paṇḍita kun dga' rgyal mtshan), 120, 123n19
Tshad ma rigs gter gyi dgongs rgyan rigs pa'i 'khor los lugs ngan pham byed (RTGG) of Shākya Chokden (śā kya mchog ldan), 227n21
Tshad ma'i rigs gter gyi dka' gnas rnam par bshad pa sde bdun rab gsal (RTKN) of Go-ram-ba Sönam Senge (go ram pa bsod nams seng ge), 226n13
Tshad ma rigs gter rang 'grel (RTRG) of Sakya Pandita, 123n19,n25, 124n30
tu quoque argument, 118–20

Uddyotakara, 35, 39, 54, 64, 65, 77, 85, 90, 109, **110**, 112, 140, 141, 178, **180–83**
Ugly Duckling theorem, of Satoshi Watanabe, 35, 36, 132
universal. *See* jāti; sāmānya; sāmānyalakṣaṇa
universal blocker, 8–9. *See also* upādhibādhaka
universals, Indian theories of: Buddhist theories (*see* apoha; nominalism); contrasted with Western theories, 14; Grammarians on universals, 3; how universals are known, 10–11; Jainas on universals, 4, 5, **12–14**; Mīmāṃsaka realism, 3,4, 8, 13; Nyāya-Vaiśeṣika realism, 4, **5–8**, 39, 40; Nyāya-Vaiśeṣika taxonomy of universals, 30; real universals, 8, 25–27, 29, 31, 39–40, 51, 53, 59, 60, 68, 70, 74, 89–90, 103, 115, 131, 160, 171, 208–9, 216, 285 (*see also* realism); tests for being a genuine (*see* upādhibādhaka)
upādhibādhaka (universal blocker), 8–9, 48n4
urdhvatāsāmānya (vertical universal), 4, 12, 154, 194
Vācaspatimiśra, 39, 110, 122n6, 178, 188, 203n23, 205n27
Vaiśeṣika, 4, 5, 6, 8, 9, 11, 13, 14, 16, 29, 48n6, 52, 65, 68, 70, 74, 80, 81n6, 174, 194, 303n17
Vaiśeṣikasūtras (VS) of Kaṇāda, 5
Vājapyāyana, 3, 173, 201n7
Vākyapadīya (VP) of Bhartṛhari, 173
vākyārtha (sentence meaning), 140
vāsanā (imprint, trace, tendency), **33**, 41, 44, 55, 56, 90–94, 98, **99–102**, 220, 237, 239; mistaken, 222, 225; placed (āhita), 100; beginningless (anādi), **100–103**, **108n32** (*see also* beginningless, imprints, tendencies)
Vasubandhu, 51, 130
verbally bound negation. *See* prasajyapratiṣedha
vertical universal. *See* urdhvatāsāmānya
Vidhiviveka (VV) of Maṇḍanamiśra, 88
vijātīya (objects of a different class), 126, 133n4, 165n12, 167n29, 167n31
vikalpa (conceptual cognition, concept, conceptualization, construction), 15, 49n7, 55, 85, 126, 142, 143, 144, 171, 187, 188–91, 195, 196
viśeṣa (śabda). *See* individuator
viśeṣaṇaviśeṣyabhāva. *See* qualifier-qualified relation
vivakṣā. *See* speech intention
Vyāḍi, 173, 197, 201n7
vyakti (individual, particular), 3, 139, 145, 146n1, 173, 180, 254
Vyaktiśaktivādin, 197
vyāpti (pervasion), 73–74
Vyāptinirṇaya (VNi) of Ratnakīrti, 167n31
vyāpya (pervaded), 76, 186
vyatireka (coabsence, joint absence, difference), 70–72, 73, 74, 77, 128–29
vyavaccheda (exclusion, elimination), 67, 69, 74, 114, 123n19, 126, 143
vyavacchedānumāna (inference based on exclusion), 71, 76–77, 129
vyavacchedyavivakṣā (intention of expressing [objects] to be excluded), 76
vyavahāra (verbal convention, practical activity), 125, 126, 131

vyāvṛtta (excluded, differentiated), **92**, 125, 126, 133n4, 135, 165n12, 167n31
vyāvṛttasvalakṣaṇa (the excluded particular), 125
vyāvṛtti (exclusion, differentiation), 56, 59, 67, 80, 92, **94**, 101, 103, 126, 136, 138, 143, 144, **167n31**, 172, 190, 191, 195, 252
vyutpatti (word-meaning connection), 78

Watanabe, Satoshi, 35, 132
Wittgenstein, Ludwig, 16–18, 20, 23, 25, 278

yadṛcchāśabda (arbitrary word), 172, 173, 179, 202n12
Yogācāra, 12, 42, 99, 100, 108n33, 244, 248; and apoha, 50, **51–52**
Yogācāra-Sautrāntika, 11, 27, 28, 194
yogyatā (fitness), 219

GPSR Authorized Representative: Easy Access System Europe, Mustamäe tee 50, 10621 Tallinn, Estonia, gpsr.requests@easproject.com

www.ingramcontent.com/pod-product-compliance
Lightning Source LLC
Chambersburg PA
CBHW072120290426
44111CB00012B/1729
* 9 7 8 0 2 3 1 1 5 3 6 1 4 *